VOLUME FIVE HUNDRED AND NINETEEN

# METHODS IN ENZYMOLOGY

## Fluorescence Fluctuation Spectroscopy (FFS), Part B

# METHODS IN ENZYMOLOGY


*Editors-in-Chief*

**JOHN N. ABELSON and MELVIN I. SIMON**
*Division of Biology*
*California Institute of Technology*
*Pasadena, California*

*Founding Editors*

**SIDNEY P. COLOWICK and NATHAN O. KAPLAN**

VOLUME FIVE HUNDRED AND NINETEEN

# METHODS IN ENZYMOLOGY

## Fluorescence Fluctuation Spectroscopy (FFS), Part B

Edited by

**SERGEY Y. TETIN**

*Abbott Diagnostics Division*
*Abbott Laboratories,*
*Abbott Park, IL, USA*



ELSEVIER

AMSTERDAM • BOSTON • HEIDELBERG • LONDON
NEW YORK • OXFORD • PARIS • SAN DIEGO
SAN FRANCISCO • SINGAPORE • SYDNEY • TOKYO
Academic Press is an imprint of Elsevier

Academic Press is an imprint of Elsevier
525 B Street, Suite 1900, San Diego, CA 92101-4495, USA
225 Wyman Street, Waltham, MA 02451, USA
The Boulevard, Langford Lane, Kidlington, Oxford, OX51GB, UK
32, Jamestown Road, London NW1 7BY, UK
Radarweg 29, PO Box 211, 1000 AE Amsterdam, The Netherlands

First edition 2013



ISBN: 978-0-12-405539-1
ISSN: 0076-6879

Printed and bound in United States of America
13 14 15 16 11 10 9 8 7 6 5 4 3 2 1

# CONTENTS

# CONTRIBUTORS

**Joseph P. Albanesi**
Department of Pharmacology, University of Texas Southwestern Medical Center, Dallas, Texas, USA

**Alexia I. Bachir**
Department of Cell Biology, University of Virginia, Charlottesville, Virginia, USA

**Nicolas Bertaux**
Institut Fresnel, Centre National de la Recherche Scientifique (CNRS), UMR7249, and École Centrale Marseille, Technopôle de Château-Gombert, Marseille, France

**Sebastien Besson**
Department of Cell Biology, Harvard Medical School, Boston, Massachusetts, USA

**Cyrille Billaudeau**
Centre d'Immunologie de Marseille-Luminy (CIML), Aix-Marseille University, UM2; Institut National de la Santé et de la Recherche Médicale (Inserm), U1104, and Centre National de la Recherche Scientifique (CNRS), UMR7280, Marseille, France

**Gaudenz Danuser**
Department of Cell Biology, Harvard Medical School, Boston, Massachusetts, USA

**Nathalie Declerck**
Centre de Biochimie Structurale, INSERM U1054, CNRS UMR5048, Université Montpellier 1 and 2, Montpellier, and Département de Microbiologie, Institut National pour la Recherche Agronomique, Paris, France

**Christian Eggeling**
Department of Nanobiophotonics, Max Planck Institute for Biophysical Chemistry, Göttingen, Germany, and Weatherall Institute of Molecular Medicine, University of Oxford, Headley Way, Oxford, United Kingdom

**Hunter Elliott**
Department of Cell Biology, Harvard Medical School, Boston, Massachusetts, USA

**Suren Felekyan**
Institut für Physikalische Chemie, Lehrstuhl für Molekulare Physikalische Chemie, Heinrich-Heine-Universität, Düsseldorf, Germany

**Klaus Hahn**
Department of Pharmacology and Lineberger Cancer Center, University of North Carolina at Chapel Hill, Chapel Hill, North Carolina, USA

**Nadia Halidi**
Department of Cell Biology, Harvard Medical School, Boston, Massachusetts, USA

**Yannick Hamon**
Centre d'Immunologie de Marseille-Luminy (CIML), Aix-Marseille University, UM2; Institut National de la Santé et de la Recherche Médicale (Inserm), U1104, and Centre National de la Recherche Scientifique (CNRS), UMR7280, Marseille, France

**Hai-Tao He**
Centre d'Immunologie de Marseille-Luminy (CIML), Aix-Marseille University, UM2; Institut National de la Santé et de la Recherche Médicale (Inserm), U1104, and Centre National de la Recherche Scientifique (CNRS), UMR7280, Marseille, France

**Alf Honigmann**
Department of Nanobiophotonics, Max Planck Institute for Biophysical Chemistry, Göttingen, Germany

**Alan R. Horwitz**
Department of Cell Biology, University of Virginia, Charlottesville, Virginia, USA

**Nicholas G. James**
Department of Cell and Molecular Biology, John A. Burns School of Medicine, University of Hawaii, Honolulu, Hawaii, USA

**David M. Jameson**
Department of Cell and Molecular Biology, John A. Burns School of Medicine, University of Hawaii, Honolulu, Hawaii, USA

**Stanislav Kalinin**
Institut für Physikalische Chemie, Lehrstuhl für Molekulare Physikalische Chemie, Heinrich-Heine-Universität, Düsseldorf, Germany

**Ralf Kühnemuth**
Institut für Physikalische Chemie, Lehrstuhl für Molekulare Physikalische Chemie, Heinrich-Heine-Universität, Düsseldorf, Germany

**Kristopher E. Kubow**
Department of Cell Biology, University of Virginia, Charlottesville, Virginia, USA

**Pauline Maffre**
Institute of Applied Physics and Center for Functional Nanostructures (CFN), Karlsruhe Institute of Technology (KIT), Wolfgang-Gaede-Straße 1, Karlsruhe, Germany

**Sébastien Mailfert**
Centre d'Immunologie de Marseille-Luminy (CIML), Aix-Marseille University, UM2; Institut National de la Santé et de la Recherche Médicale (Inserm), U1104, and Centre National de la Recherche Scientifique (CNRS), UMR7280, Marseille, France

**Didier Marguet**
Centre d'Immunologie de Marseille-Luminy (CIML), Aix-Marseille University, UM2; Institut National de la Santé et de la Recherche Médicale (Inserm), U1104, Marseille, and Centre National de la Recherche Scientifique (CNRS), UMR7280, Marseille, France

**Rebecca Medda**
Department of Nanobiophotonics, Max Planck Institute for Biophysical Chemistry, Göttingen, Germany

**Veronika Mueller**
Department of Nanobiophotonics, Max Planck Institute for Biophysical Chemistry, Göttingen, Germany

**G. Ulrich Nienhaus**
Institute of Applied Physics and Center for Functional Nanostructures (CFN), Karlsruhe Institute of Technology (KIT), Wolfgang-Gaede-Straße 1, Karlsruhe, Germany, and Department of Physics, University of Illinois at Urbana-Champaign, Urbana, Illinois, USA

**Karin Nienhaus**
Institute of Applied Physics and Center for Functional Nanostructures (CFN), Karlsruhe Institute of Technology (KIT), Wolfgang-Gaede-Straße 1, Karlsruhe, Germany

**Christian Ringemann**
Department of Nanobiophotonics, Max Planck Institute for Biophysical Chemistry, Göttingen, Germany

**Vincent Rouger**
Centre d'Immunologie de Marseille-Luminy (CIML), Aix-Marseille University, UM2; Institut National de la Santé et de la Recherche Médicale (Inserm), U1104, and Centre National de la Recherche Scientifique (CNRS), UMR7280, Marseille, France

**Catherine A. Royer**
Centre de Biochimie Structurale, INSERM U1054, CNRS UMR5048, Université Montpellier 1 and 2, Montpellier, France

**Qiaoqiao Ruan**
Diagnostics Research, Abbott Diagnostics Division, Abbott Park, Illinois, USA

**Hugo Sanabria**
Institut für Physikalische Chemie, Lehrstuhl für Molekulare Physikalische Chemie, Heinrich-Heine-Universität, Düsseldorf, Germany

**Günter Schwarzmann**
Life and Medical Sciences Center (LIMES) Membrane Biology and Lipid Biochemistry Unit, University of Bonn, Bonn, Germany

**Claus A.M. Seidel**
Institut für Physikalische Chemie, Lehrstuhl für Molekulare Physikalische Chemie, Heinrich-Heine-Universität, Düsseldorf, Germany

**Joseph P. Skinner**
Diagnostics Research, Abbott Diagnostics Division, Abbott Park, Illinois, USA

**Sergey Y. Tetin**
Diagnostics Research, Abbott Diagnostics Division, Abbott Park, Illinois, USA

**Tomasz Trombik**
Centre d'Immunologie de Marseille-Luminy (CIML), Aix-Marseille University, UM2; Institut National de la Santé et de la Recherche Médicale (Inserm), U1104, and Centre National de la Recherche Scientifique (CNRS), UMR7280, Marseille, France

**Jessica Tytell**
Department of Cell Biology, Harvard Medical School, Boston, Massachusetts, USA

**Marco Vilela**
Department of Cell Biology, Harvard Medical School, Boston, Massachusetts, USA

**Jerker Widengren**
Experimental Biomolecular Physics, Department of Applied Physics, Royal Institute of Technology (KTH), Albanova University Center, Stockholm, Sweden

# PREFACE

Fluorescence Fluctuation Spectroscopy (FFS) is a growing family of methods for studying molecular motions and interactions. Introduced as Fluorescence Correlation Spectroscopy (FCS) 40 years ago, this elegant technique has evolved into a large group of powerful tools of modern fluorescence spectroscopy. FFS has found numerous applications in biophysical and biochemical studies of proteins, nucleic acids, and membranes. In combination with fluorescence imaging microscopy, FFS provides unique opportunities for characterizing dynamic processes in living cells and has been increasingly used by cell and systems biologists. Significant progress in the technology was achieved in the past decade, and the number of new FFS methods is still growing. The attraction of FFS relates to fact that it is based on statistical analysis of the fluctuations occurring at the molecular level. Therefore, these methods can resolve molecular populations and provide additional information about biological systems.

The chapters in this two-volume publication are written by the authors and first-hand users of the methods under discussion. Part A is dedicated predominately to FFS methods. It contains well-established and also recently introduced experimental techniques and data analysis algorithms. Part B continues with FFS methods and also includes various FFS applications *in vivo* and *in vitro*. Here the reader will find specific examples and detailed experimental protocols developed for several biological systems. The purpose of these chapters is to show the power and potentials of the technology in biological research.

I am very thankful to all the authors who participated in this project and wrote excellent chapters that made my editorial work easy. I am expressing my special gratitude to Elliot Elson, whose seminal work on FCS began in the early 1970s, for writing a historical note and current systemization of the FFS methods in addition to his chapter on FCS conceptual basis and theory. I am also very thankful to Shaun Gamble from Elsevier for his great assistance in organizing these volumes. On a personal note, I want to express my immense gratefulness to my wife Luda for her devotion, serenity, and constant support to all my work.

FFS offers a variety of powerful techniques for studying biological molecules *in vitro* and *in vivo*. I hope that these MIE volumes will be used by many new and experienced researchers alike.

Sergey Y. Tetin

# METHODS IN ENZYMOLOGY

VOLUME I. Preparation and Assay of Enzymes
*Edited by* SIDNEY P. COLOWICK AND NATHAN O. KAPLAN

VOLUME II. Preparation and Assay of Enzymes
*Edited by* SIDNEY P. COLOWICK AND NATHAN O. KAPLAN

VOLUME III. Preparation and Assay of Substrates
*Edited by* SIDNEY P. COLOWICK AND NATHAN O. KAPLAN

VOLUME IV. Special Techniques for the Enzymologist
*Edited by* SIDNEY P. COLOWICK AND NATHAN O. KAPLAN

VOLUME V. Preparation and Assay of Enzymes
*Edited by* SIDNEY P. COLOWICK AND NATHAN O. KAPLAN

VOLUME VI. Preparation and Assay of Enzymes (*Continued*)
Preparation and Assay of Substrates
Special Techniques
*Edited by* SIDNEY P. COLOWICK AND NATHAN O. KAPLAN

VOLUME VII. Cumulative Subject Index
*Edited by* SIDNEY P. COLOWICK AND NATHAN O. KAPLAN

VOLUME VIII. Complex Carbohydrates
*Edited by* ELIZABETH F. NEUFELD AND VICTOR GINSBURG

VOLUME IX. Carbohydrate Metabolism
*Edited by* WILLIS A. WOOD

VOLUME X. Oxidation and Phosphorylation
*Edited by* RONALD W. ESTABROOK AND MAYNARD E. PULLMAN

VOLUME XI. Enzyme Structure
*Edited by* C. H. W. HIRS

VOLUME XII. Nucleic Acids (Parts A and B)
*Edited by* LAWRENCE GROSSMAN AND KIVIE MOLDAVE

VOLUME XIII. Citric Acid Cycle
*Edited by* J. M. LOWENSTEIN

VOLUME XIV. Lipids
*Edited by* J. M. LOWENSTEIN

VOLUME XV. Steroids and Terpenoids
*Edited by* RAYMOND B. CLAYTON

VOLUME XVI. Fast Reactions
*Edited by* KENNETH KUSTIN

VOLUME XVII. Metabolism of Amino Acids and Amines (Parts A and B)
*Edited by* HERBERT TABOR AND CELIA WHITE TABOR

VOLUME XVIII. Vitamins and Coenzymes (Parts A, B, and C)
*Edited by* DONALD B. MCCORMICK AND LEMUEL D. WRIGHT

VOLUME XIX. Proteolytic Enzymes
*Edited by* GERTRUDE E. PERLMANN AND LASZLO LORAND

VOLUME XX. Nucleic Acids and Protein Synthesis (Part C)
*Edited by* KIVIE MOLDAVE AND LAWRENCE GROSSMAN

VOLUME XXI. Nucleic Acids (Part D)
*Edited by* LAWRENCE GROSSMAN AND KIVIE MOLDAVE

VOLUME XXII. Enzyme Purification and Related Techniques
*Edited by* WILLIAM B. JAKOBY

VOLUME XXIII. Photosynthesis (Part A)
*Edited by* ANTHONY SAN PIETRO

VOLUME XXIV. Photosynthesis and Nitrogen Fixation (Part B)
*Edited by* ANTHONY SAN PIETRO

VOLUME XXV. Enzyme Structure (Part B)
*Edited by* C. H. W. HIRS AND SERGE N. TIMASHEFF

VOLUME XXVI. Enzyme Structure (Part C)
*Edited by* C. H. W. HIRS AND SERGE N. TIMASHEFF

VOLUME XXVII. Enzyme Structure (Part D)
*Edited by* C. H. W. HIRS AND SERGE N. TIMASHEFF

VOLUME XXVIII. Complex Carbohydrates (Part B)
*Edited by* VICTOR GINSBURG

VOLUME XXIX. Nucleic Acids and Protein Synthesis (Part E)
*Edited by* LAWRENCE GROSSMAN AND KIVIE MOLDAVE

VOLUME XXX. Nucleic Acids and Protein Synthesis (Part F)
*Edited by* KIVIE MOLDAVE AND LAWRENCE GROSSMAN

VOLUME XXXI. Biomembranes (Part A)
*Edited by* SIDNEY FLEISCHER AND LESTER PACKER

VOLUME 64. Enzyme Kinetics and Mechanism
(Part B: Isotopic Probes and Complex Enzyme Systems)
*Edited by* DANIEL L. PURICH

VOLUME 65. Nucleic Acids (Part I)
*Edited by* LAWRENCE GROSSMAN AND KIVIE MOLDAVE

VOLUME 66. Vitamins and Coenzymes (Part E)
*Edited by* DONALD B. MCCORMICK AND LEMUEL D. WRIGHT

VOLUME 67. Vitamins and Coenzymes (Part F)
*Edited by* DONALD B. MCCORMICK AND LEMUEL D. WRIGHT

VOLUME 68. Recombinant DNA
*Edited by* RAY WU

VOLUME 69. Photosynthesis and Nitrogen Fixation (Part C)
*Edited by* ANTHONY SAN PIETRO

VOLUME 70. Immunochemical Techniques (Part A)
*Edited by* HELEN VAN VUNAKIS AND JOHN J. LANGONE

VOLUME 71. Lipids (Part C)
*Edited by* JOHN M. LOWENSTEIN

VOLUME 72. Lipids (Part D)
*Edited by* JOHN M. LOWENSTEIN

VOLUME 73. Immunochemical Techniques (Part B)
*Edited by* JOHN J. LANGONE AND HELEN VAN VUNAKIS

VOLUME 74. Immunochemical Techniques (Part C)
*Edited by* JOHN J. LANGONE AND HELEN VAN VUNAKIS

VOLUME 75. Cumulative Subject Index Volumes XXXI, XXXII, XXXIV–LX
*Edited by* EDWARD A. DENNIS AND MARTHA G. DENNIS

VOLUME 76. Hemoglobins
*Edited by* ERALDO ANTONINI, LUIGI ROSSI-BERNARDI, AND EMILIA CHIANCONE

VOLUME 77. Detoxication and Drug Metabolism
*Edited by* WILLIAM B. JAKOBY

VOLUME 78. Interferons (Part A)
*Edited by* SIDNEY PESTKA

VOLUME 79. Interferons (Part B)
*Edited by* SIDNEY PESTKA

VOLUME 80. Proteolytic Enzymes (Part C)
*Edited by* LASZLO LORAND

VOLUME 81. Biomembranes (Part H: Visual Pigments and Purple Membranes, I)
*Edited by* LESTER PACKER

VOLUME 82. Structural and Contractile Proteins (Part A: Extracellular Matrix)
*Edited by* LEON W. CUNNINGHAM AND DIXIE W. FREDERIKSEN

VOLUME 83. Complex Carbohydrates (Part D)
*Edited by* VICTOR GINSBURG

VOLUME 84. Immunochemical Techniques (Part D: Selected Immunoassays)
*Edited by* JOHN J. LANGONE AND HELEN VAN VUNAKIS

VOLUME 85. Structural and Contractile Proteins (Part B: The Contractile Apparatus and the Cytoskeleton)
*Edited by* DIXIE W. FREDERIKSEN AND LEON W. CUNNINGHAM

VOLUME 86. Prostaglandins and Arachidonate Metabolites
*Edited by* WILLIAM E. M. LANDS AND WILLIAM L. SMITH

VOLUME 87. Enzyme Kinetics and Mechanism (Part C: Intermediates, Stereo-chemistry, and Rate Studies)
*Edited by* DANIEL L. PURICH

VOLUME 88. Biomembranes (Part I: Visual Pigments and Purple Membranes, II)
*Edited by* LESTER PACKER

VOLUME 89. Carbohydrate Metabolism (Part D)
*Edited by* WILLIS A. WOOD

VOLUME 90. Carbohydrate Metabolism (Part E)
*Edited by* WILLIS A. WOOD

VOLUME 91. Enzyme Structure (Part I)
*Edited by* C. H. W. HIRS AND SERGE N. TIMASHEFF

VOLUME 92. Immunochemical Techniques (Part E: Monoclonal Antibodies and General Immunoassay Methods)
*Edited by* JOHN J. LANGONE AND HELEN VAN VUNAKIS

Volume 93. Immunochemical Techniques (Part F: Conventional Antibodies, Fc Receptors, and Cytotoxicity)
*Edited by* John J. Langone and Helen Van Vunakis

Volume 94. Polyamines
*Edited by* Herbert Tabor and Celia White Tabor

Volume 95. Cumulative Subject Index Volumes 61–74, 76–80
*Edited by* Edward A. Dennis and Martha G. Dennis

Volume 96. Biomembranes [Part J: Membrane Biogenesis: Assembly and Targeting (General Methods; Eukaryotes)]
*Edited by* Sidney Fleischer and Becca Fleischer

Volume 97. Biomembranes [Part K: Membrane Biogenesis: Assembly and Targeting (Prokaryotes, Mitochondria, and Chloroplasts)]
*Edited by* Sidney Fleischer and Becca Fleischer

Volume 98. Biomembranes (Part L: Membrane Biogenesis: Processing and Recycling)
*Edited by* Sidney Fleischer and Becca Fleischer

Volume 99. Hormone Action (Part F: Protein Kinases)
*Edited by* Jackie D. Corbin and Joel G. Hardman

Volume 100. Recombinant DNA (Part B)
*Edited by* Ray Wu, Lawrence Grossman, and Kivie Moldave

Volume 101. Recombinant DNA (Part C)
*Edited by* Ray Wu, Lawrence Grossman, and Kivie Moldave

Volume 102. Hormone Action (Part G: Calmodulin and Calcium-Binding Proteins)
*Edited by* Anthony R. Means and Bert W. O'Malley

Volume 103. Hormone Action (Part H: Neuroendocrine Peptides)
*Edited by* P. Michael Conn

Volume 104. Enzyme Purification and Related Techniques (Part C)
*Edited by* William B. Jakoby

Volume 105. Oxygen Radicals in Biological Systems
*Edited by* Lester Packer

Volume 106. Posttranslational Modifications (Part A)
*Edited by* Finn Wold and Kivie Moldave

VOLUME 107. Posttranslational Modifications (Part B)
*Edited by* FINN WOLD AND KIVIE MOLDAVE

VOLUME 108. Immunochemical Techniques (Part G: Separation and Characterization of Lymphoid Cells)
*Edited by* GIOVANNI DI SABATO, JOHN J. LANGONE, AND HELEN VAN VUNAKIS

VOLUME 109. Hormone Action (Part I: Peptide Hormones)
*Edited by* LUTZ BIRNBAUMER AND BERT W. O'MALLEY

VOLUME 110. Steroids and Isoprenoids (Part A)
*Edited by* JOHN H. LAW AND HANS C. RILLING

VOLUME 111. Steroids and Isoprenoids (Part B)
*Edited by* JOHN H. LAW AND HANS C. RILLING

VOLUME 112. Drug and Enzyme Targeting (Part A)
*Edited by* KENNETH J. WIDDER AND RALPH GREEN

VOLUME 113. Glutamate, Glutamine, Glutathione, and Related Compounds
*Edited by* ALTON MEISTER

VOLUME 114. Diffraction Methods for Biological Macromolecules (Part A)
*Edited by* HAROLD W. WYCKOFF, C. H. W. HIRS, AND SERGE N. TIMASHEFF

VOLUME 115. Diffraction Methods for Biological Macromolecules (Part B)
*Edited by* HAROLD W. WYCKOFF, C. H. W. HIRS, AND SERGE N. TIMASHEFF

VOLUME 116. Immunochemical Techniques
(Part H: Effectors and Mediators of Lymphoid Cell Functions)
*Edited by* GIOVANNI DI SABATO, JOHN J. LANGONE, AND HELEN VAN VUNAKIS

VOLUME 117. Enzyme Structure (Part J)
*Edited by* C. H. W. HIRS AND SERGE N. TIMASHEFF

VOLUME 118. Plant Molecular Biology
*Edited by* ARTHUR WEISSBACH AND HERBERT WEISSBACH

VOLUME 119. Interferons (Part C)
*Edited by* SIDNEY PESTKA

VOLUME 120. Cumulative Subject Index Volumes 81–94, 96–101

VOLUME 121. Immunochemical Techniques (Part I: Hybridoma Technology and Monoclonal Antibodies)
*Edited by* JOHN J. LANGONE AND HELEN VAN VUNAKIS

Volume 122. Vitamins and Coenzymes (Part G)
*Edited by* Frank Chytil and Donald B. McCormick

Volume 123. Vitamins and Coenzymes (Part H)
*Edited by* Frank Chytil and Donald B. McCormick

Volume 124. Hormone Action (Part J: Neuroendocrine Peptides)
*Edited by* P. Michael Conn

Volume 125. Biomembranes (Part M: Transport in Bacteria, Mitochondria, and Chloroplasts: General Approaches and Transport Systems)
*Edited by* Sidney Fleischer and Becca Fleischer

Volume 126. Biomembranes (Part N: Transport in Bacteria, Mitochondria, and Chloroplasts: Protonmotive Force)
*Edited by* Sidney Fleischer and Becca Fleischer

Volume 127. Biomembranes (Part O: Protons and Water: Structure and Translocation)
*Edited by* Lester Packer

Volume 128. Plasma Lipoproteins (Part A: Preparation, Structure, and Molecular Biology)
*Edited by* Jere P. Segrest and John J. Albers

Volume 129. Plasma Lipoproteins (Part B: Characterization, Cell Biology, and Metabolism)
*Edited by* John J. Albers and Jere P. Segrest

Volume 130. Enzyme Structure (Part K)
*Edited by* C. H. W. Hirs and Serge N. Timasheff

Volume 131. Enzyme Structure (Part L)
*Edited by* C. H. W. Hirs and Serge N. Timasheff

Volume 132. Immunochemical Techniques (Part J: Phagocytosis and Cell-Mediated Cytotoxicity)
*Edited by* Giovanni Di Sabato and Johannes Everse

Volume 133. Bioluminescence and Chemiluminescence (Part B)
*Edited by* Marlene DeLuca and William D. McElroy

Volume 134. Structural and Contractile Proteins (Part C: The Contractile Apparatus and the Cytoskeleton)
*Edited by* Richard B. Vallee

Volume 135. Immobilized Enzymes and Cells (Part B)
*Edited by* Klaus Mosbach

VOLUME 136. Immobilized Enzymes and Cells (Part C)
*Edited by* KLAUS MOSBACH

VOLUME 137. Immobilized Enzymes and Cells (Part D)
*Edited by* KLAUS MOSBACH

VOLUME 138. Complex Carbohydrates (Part E)
*Edited by* VICTOR GINSBURG

VOLUME 139. Cellular Regulators (Part A: Calcium- and Calmodulin-Binding Proteins)
*Edited by* ANTHONY R. MEANS AND P. MICHAEL CONN

VOLUME 140. Cumulative Subject Index Volumes 102–119, 121–134

VOLUME 141. Cellular Regulators (Part B: Calcium and Lipids)
*Edited by* P. MICHAEL CONN AND ANTHONY R. MEANS

VOLUME 142. Metabolism of Aromatic Amino Acids and Amines
*Edited by* SEYMOUR KAUFMAN

VOLUME 143. Sulfur and Sulfur Amino Acids
*Edited by* WILLIAM B. JAKOBY AND OWEN GRIFFITH

VOLUME 144. Structural and Contractile Proteins (Part D: Extracellular Matrix)
*Edited by* LEON W. CUNNINGHAM

VOLUME 145. Structural and Contractile Proteins (Part E: Extracellular Matrix)
*Edited by* LEON W. CUNNINGHAM

VOLUME 146. Peptide Growth Factors (Part A)
*Edited by* DAVID BARNES AND DAVID A. SIRBASKU

VOLUME 147. Peptide Growth Factors (Part B)
*Edited by* DAVID BARNES AND DAVID A. SIRBASKU

VOLUME 148. Plant Cell Membranes
*Edited by* LESTER PACKER AND ROLAND DOUCE

VOLUME 149. Drug and Enzyme Targeting (Part B)
*Edited by* RALPH GREEN AND KENNETH J. WIDDER

VOLUME 150. Immunochemical Techniques (Part K: *In Vitro* Models of B and T Cell Functions and Lymphoid Cell Receptors)
*Edited by* GIOVANNI DI SABATO

VOLUME 151. Molecular Genetics of Mammalian Cells
*Edited by* MICHAEL M. GOTTESMAN

VOLUME 152. Guide to Molecular Cloning Techniques
*Edited by* SHELBY L. BERGER AND ALAN R. KIMMEL

VOLUME 153. Recombinant DNA (Part D)
*Edited by* RAY WU AND LAWRENCE GROSSMAN

VOLUME 154. Recombinant DNA (Part E)
*Edited by* RAY WU AND LAWRENCE GROSSMAN

VOLUME 155. Recombinant DNA (Part F)
*Edited by* RAY WU

VOLUME 156. Biomembranes (Part P: ATP-Driven Pumps and Related Transport: The Na, K-Pump)
*Edited by* SIDNEY FLEISCHER AND BECCA FLEISCHER

VOLUME 157. Biomembranes (Part Q: ATP-Driven Pumps and Related Transport: Calcium, Proton, and Potassium Pumps)
*Edited by* SIDNEY FLEISCHER AND BECCA FLEISCHER

VOLUME 158. Metalloproteins (Part A)
*Edited by* JAMES F. RIORDAN AND BERT L. VALLEE

VOLUME 159. Initiation and Termination of Cyclic Nucleotide Action
*Edited by* JACKIE D. CORBIN AND ROGER A. JOHNSON

VOLUME 160. Biomass (Part A: Cellulose and Hemicellulose)
*Edited by* WILLIS A. WOOD AND SCOTT T. KELLOGG

VOLUME 161. Biomass (Part B: Lignin, Pectin, and Chitin)
*Edited by* WILLIS A. WOOD AND SCOTT T. KELLOGG

VOLUME 162. Immunochemical Techniques (Part L: Chemotaxis and Inflammation)
*Edited by* GIOVANNI DI SABATO

VOLUME 163. Immunochemical Techniques (Part M: Chemotaxis and Inflammation)
*Edited by* GIOVANNI DI SABATO

VOLUME 164. Ribosomes
*Edited by* HARRY F. NOLLER, JR., AND KIVIE MOLDAVE

VOLUME 165. Microbial Toxins: Tools for Enzymology
*Edited by* SIDNEY HARSHMAN

VOLUME 180. RNA Processing (Part A: General Methods)
*Edited by* JAMES E. DAHLBERG AND JOHN N. ABELSON

VOLUME 181. RNA Processing (Part B: Specific Methods)
*Edited by* JAMES E. DAHLBERG AND JOHN N. ABELSON

VOLUME 182. Guide to Protein Purification
*Edited by* MURRAY P. DEUTSCHER

VOLUME 183. Molecular Evolution: Computer Analysis of Protein and Nucleic Acid Sequences
*Edited by* RUSSELL F. DOOLITTLE

VOLUME 184. Avidin-Biotin Technology
*Edited by* MEIR WILCHEK AND EDWARD A. BAYER

VOLUME 185. Gene Expression Technology
*Edited by* DAVID V. GOEDDEL

VOLUME 186. Oxygen Radicals in Biological Systems (Part B: Oxygen Radicals and Antioxidants)
*Edited by* LESTER PACKER AND ALEXANDER N. GLAZER

VOLUME 187. Arachidonate Related Lipid Mediators
*Edited by* ROBERT C. MURPHY AND FRANK A. FITZPATRICK

VOLUME 188. Hydrocarbons and Methylotrophy
*Edited by* MARY E. LIDSTROM

VOLUME 189. Retinoids (Part A: Molecular and Metabolic Aspects)
*Edited by* LESTER PACKER

VOLUME 190. Retinoids (Part B: Cell Differentiation and Clinical Applications)
*Edited by* LESTER PACKER

VOLUME 191. Biomembranes (Part V: Cellular and Subcellular Transport: Epithelial Cells)
*Edited by* SIDNEY FLEISCHER AND BECCA FLEISCHER

VOLUME 192. Biomembranes (Part W: Cellular and Subcellular Transport: Epithelial Cells)
*Edited by* SIDNEY FLEISCHER AND BECCA FLEISCHER

VOLUME 193. Mass Spectrometry
*Edited by* JAMES A. MCCLOSKEY

Volume 194. Guide to Yeast Genetics and Molecular Biology
*Edited by* Christine Guthrie and Gerald R. Fink

Volume 195. Adenylyl Cyclase, G Proteins, and Guanylyl Cyclase
*Edited by* Roger A. Johnson and Jackie D. Corbin

Volume 196. Molecular Motors and the Cytoskeleton
*Edited by* Richard B. Vallee

Volume 197. Phospholipases
*Edited by* Edward A. Dennis

Volume 198. Peptide Growth Factors (Part C)
*Edited by* David Barnes, J. P. Mather, and Gordon H. Sato

Volume 199. Cumulative Subject Index Volumes 168–174, 176–194

Volume 200. Protein Phosphorylation (Part A: Protein Kinases: Assays, Purification, Antibodies, Functional Analysis, Cloning, and Expression)
*Edited by* Tony Hunter and Bartholomew M. Sefton

Volume 201. Protein Phosphorylation (Part B: Analysis of Protein Phosphorylation, Protein Kinase Inhibitors, and Protein Phosphatases)
*Edited by* Tony Hunter and Bartholomew M. Sefton

Volume 202. Molecular Design and Modeling: Concepts and Applications (Part A: Proteins, Peptides, and Enzymes)
*Edited by* John J. Langone

Volume 203. Molecular Design and Modeling: Concepts and Applications (Part B: Antibodies and Antigens, Nucleic Acids, Polysaccharides, and Drugs)
*Edited by* John J. Langone

Volume 204. Bacterial Genetic Systems
*Edited by* Jeffrey H. Miller

Volume 205. Metallobiochemistry (Part B: Metallothionein and Related Molecules)
*Edited by* James F. Riordan and Bert L. Vallee

Volume 206. Cytochrome P450
*Edited by* Michael R. Waterman and Eric F. Johnson

Volume 207. Ion Channels
*Edited by* Bernardo Rudy and Linda E. Iverson

Volume 208. Protein–DNA Interactions
*Edited by* Robert T. Sauer

Volume 209. Phospholipid Biosynthesis
*Edited by* Edward A. Dennis and Dennis E. Vance

Volume 210. Numerical Computer Methods
*Edited by* Ludwig Brand and Michael L. Johnson

Volume 211. DNA Structures (Part A: Synthesis and Physical Analysis of DNA)
*Edited by* David M. J. Lilley and James E. Dahlberg

Volume 212. DNA Structures (Part B: Chemical and Electrophoretic Analysis of DNA)
*Edited by* David M. J. Lilley and James E. Dahlberg

Volume 213. Carotenoids (Part A: Chemistry, Separation, Quantitation, and Antioxidation)
*Edited by* Lester Packer

Volume 214. Carotenoids (Part B: Metabolism, Genetics, and Biosynthesis)
*Edited by* Lester Packer

Volume 215. Platelets: Receptors, Adhesion, Secretion (Part B)
*Edited by* Jacek J. Hawiger

Volume 216. Recombinant DNA (Part G)
*Edited by* Ray Wu

Volume 217. Recombinant DNA (Part H)
*Edited by* Ray Wu

Volume 218. Recombinant DNA (Part I)
*Edited by* Ray Wu

Volume 219. Reconstitution of Intracellular Transport
*Edited by* James E. Rothman

Volume 220. Membrane Fusion Techniques (Part A)
*Edited by* Nejat Düzgüneş

Volume 221. Membrane Fusion Techniques (Part B)
*Edited by* Nejat Düzgüneş

Volume 222. Proteolytic Enzymes in Coagulation, Fibrinolysis, and Complement Activation (Part A: Mammalian Blood Coagulation

Factors and Inhibitors)
*Edited by* Laszlo Lorand and Kenneth G. Mann

Volume 223. Proteolytic Enzymes in Coagulation, Fibrinolysis, and Complement Activation (Part B: Complement Activation, Fibrinolysis, and Nonmammalian Blood Coagulation Factors)
*Edited by* Laszlo Lorand and Kenneth G. Mann

Volume 224. Molecular Evolution: Producing the Biochemical Data
*Edited by* Elizabeth Anne Zimmer, Thomas J. White, Rebecca L. Cann, and Allan C. Wilson

Volume 225. Guide to Techniques in Mouse Development
*Edited by* Paul M. Wassarman and Melvin L. DePamphilis

Volume 226. Metallobiochemistry (Part C: Spectroscopic and Physical Methods for Probing Metal Ion Environments in Metalloenzymes and Metalloproteins)
*Edited by* James F. Riordan and Bert L. Vallee

Volume 227. Metallobiochemistry (Part D: Physical and Spectroscopic Methods for Probing Metal Ion Environments in Metalloproteins)
*Edited by* James F. Riordan and Bert L. Vallee

Volume 228. Aqueous Two-Phase Systems
*Edited by* Harry Walter and Göte Johansson

Volume 229. Cumulative Subject Index Volumes 195–198, 200–227

Volume 230. Guide to Techniques in Glycobiology
*Edited by* William J. Lennarz and Gerald W. Hart

Volume 231. Hemoglobins (Part B: Biochemical and Analytical Methods)
*Edited by* Johannes Everse, Kim D. Vandegriff, and Robert M. Winslow

Volume 232. Hemoglobins (Part C: Biophysical Methods)
*Edited by* Johannes Everse, Kim D. Vandegriff, and Robert M. Winslow

Volume 233. Oxygen Radicals in Biological Systems (Part C)
*Edited by* Lester Packer

Volume 234. Oxygen Radicals in Biological Systems (Part D)
*Edited by* Lester Packer

Volume 235. Bacterial Pathogenesis (Part A: Identification and Regulation of Virulence Factors)
*Edited by* Virginia L. Clark and Patrik M. Bavoil

VOLUME 236. Bacterial Pathogenesis (Part B: Integration of Pathogenic Bacteria with Host Cells)
*Edited by* VIRGINIA L. CLARK AND PATRIK M. BAVOIL

VOLUME 237. Heterotrimeric G Proteins
*Edited by* RAVI IYENGAR

VOLUME 238. Heterotrimeric G-Protein Effectors
*Edited by* RAVI IYENGAR

VOLUME 239. Nuclear Magnetic Resonance (Part C)
*Edited by* THOMAS L. JAMES AND NORMAN J. OPPENHEIMER

VOLUME 240. Numerical Computer Methods (Part B)
*Edited by* MICHAEL L. JOHNSON AND LUDWIG BRAND

VOLUME 241. Retroviral Proteases
*Edited by* LAWRENCE C. KUO AND JULES A. SHAFER

VOLUME 242. Neoglycoconjugates (Part A)
*Edited by* Y. C. LEE AND REIKO T. LEE

VOLUME 243. Inorganic Microbial Sulfur Metabolism
*Edited by* HARRY D. PECK, JR., AND JEAN LEGALL

VOLUME 244. Proteolytic Enzymes: Serine and Cysteine Peptidases
*Edited by* ALAN J. BARRETT

VOLUME 245. Extracellular Matrix Components
*Edited by* E. RUOSLAHTI AND E. ENGVALL

VOLUME 246. Biochemical Spectroscopy
*Edited by* KENNETH SAUER

VOLUME 247. Neoglycoconjugates (Part B: Biomedical Applications)
*Edited by* Y. C. LEE AND REIKO T. LEE

VOLUME 248. Proteolytic Enzymes: Aspartic and Metallo Peptidases
*Edited by* ALAN J. BARRETT

VOLUME 249. Enzyme Kinetics and Mechanism (Part D: Developments in Enzyme Dynamics)
*Edited by* DANIEL L. PURICH

VOLUME 250. Lipid Modifications of Proteins
*Edited by* PATRICK J. CASEY AND JANICE E. BUSS

Volume 251. Biothiols (Part A: Monothiols and Dithiols, Protein Thiols, and Thiyl Radicals)
*Edited by* Lester Packer

Volume 252. Biothiols (Part B: Glutathione and Thioredoxin; Thiols in Signal Transduction and Gene Regulation)
*Edited by* Lester Packer

Volume 253. Adhesion of Microbial Pathogens
*Edited by* Ron J. Doyle and Itzhak Ofek

Volume 254. Oncogene Techniques
*Edited by* Peter K. Vogt and Inder M. Verma

Volume 255. Small GTPases and Their Regulators (Part A: Ras Family)
*Edited by* W. E. Balch, Channing J. Der, and Alan Hall

Volume 256. Small GTPases and Their Regulators (Part B: Rho Family)
*Edited by* W. E. Balch, Channing J. Der, and Alan Hall

Volume 257. Small GTPases and Their Regulators (Part C: Proteins Involved in Transport)
*Edited by* W. E. Balch, Channing J. Der, and Alan Hall

Volume 258. Redox-Active Amino Acids in Biology
*Edited by* Judith P. Klinman

Volume 259. Energetics of Biological Macromolecules
*Edited by* Michael L. Johnson and Gary K. Ackers

Volume 260. Mitochondrial Biogenesis and Genetics (Part A)
*Edited by* Giuseppe M. Attardi and Anne Chomyn

Volume 261. Nuclear Magnetic Resonance and Nucleic Acids
*Edited by* Thomas L. James

Volume 262. DNA Replication
*Edited by* Judith L. Campbell

Volume 263. Plasma Lipoproteins (Part C: Quantitation)
*Edited by* William A. Bradley, Sandra H. Gianturco, and Jere P. Segrest

Volume 264. Mitochondrial Biogenesis and Genetics (Part B)
*Edited by* Giuseppe M. Attardi and Anne Chomyn

Volume 265. Cumulative Subject Index Volumes 228, 230–262

VOLUME 266. Computer Methods for Macromolecular Sequence Analysis
*Edited by* RUSSELL F. DOOLITTLE

VOLUME 267. Combinatorial Chemistry
*Edited by* JOHN N. ABELSON

VOLUME 268. Nitric Oxide (Part A: Sources and Detection of NO; NO Synthase)
*Edited by* LESTER PACKER

VOLUME 269. Nitric Oxide (Part B: Physiological and Pathological Processes)
*Edited by* LESTER PACKER

VOLUME 270. High Resolution Separation and Analysis of Biological Macromolecules (Part A: Fundamentals)
*Edited by* BARRY L. KARGER AND WILLIAM S. HANCOCK

VOLUME 271. High Resolution Separation and Analysis of Biological Macromolecules (Part B: Applications)
*Edited by* BARRY L. KARGER AND WILLIAM S. HANCOCK

VOLUME 272. Cytochrome P450 (Part B)
*Edited by* ERIC F. JOHNSON AND MICHAEL R. WATERMAN

VOLUME 273. RNA Polymerase and Associated Factors (Part A)
*Edited by* SANKAR ADHYA

VOLUME 274. RNA Polymerase and Associated Factors (Part B)
*Edited by* SANKAR ADHYA

VOLUME 275. Viral Polymerases and Related Proteins
*Edited by* LAWRENCE C. KUO, DAVID B. OLSEN, AND STEVEN S. CARROLL

VOLUME 276. Macromolecular Crystallography (Part A)
*Edited by* CHARLES W. CARTER, JR., AND ROBERT M. SWEET

VOLUME 277. Macromolecular Crystallography (Part B)
*Edited by* CHARLES W. CARTER, JR., AND ROBERT M. SWEET

VOLUME 278. Fluorescence Spectroscopy
*Edited by* LUDWIG BRAND AND MICHAEL L. JOHNSON

VOLUME 279. Vitamins and Coenzymes (Part I)
*Edited by* DONALD B. MCCORMICK, JOHN W. SUTTIE, AND CONRAD WAGNER

VOLUME 280. Vitamins and Coenzymes (Part J)
*Edited by* DONALD B. MCCORMICK, JOHN W. SUTTIE, AND CONRAD WAGNER

VOLUME 297. Photosynthesis: Molecular Biology of Energy Capture
*Edited by* LEE MCINTOSH

VOLUME 298. Molecular Motors and the Cytoskeleton (Part B)
*Edited by* RICHARD B. VALLEE

VOLUME 299. Oxidants and Antioxidants (Part A)
*Edited by* LESTER PACKER

VOLUME 300. Oxidants and Antioxidants (Part B)
*Edited by* LESTER PACKER

VOLUME 301. Nitric Oxide: Biological and Antioxidant Activities (Part C)
*Edited by* LESTER PACKER

VOLUME 302. Green Fluorescent Protein
*Edited by* P. MICHAEL CONN

VOLUME 303. cDNA Preparation and Display
*Edited by* SHERMAN M. WEISSMAN

VOLUME 304. Chromatin
*Edited by* PAUL M. WASSARMAN AND ALAN P. WOLFFE

VOLUME 305. Bioluminescence and Chemiluminescence (Part C)
*Edited by* THOMAS O. BALDWIN AND MIRIAM M. ZIEGLER

VOLUME 306. Expression of Recombinant Genes in Eukaryotic Systems
*Edited by* JOSEPH C. GLORIOSO AND MARTIN C. SCHMIDT

VOLUME 307. Confocal Microscopy
*Edited by* P. MICHAEL CONN

VOLUME 308. Enzyme Kinetics and Mechanism (Part E: Energetics of Enzyme Catalysis)
*Edited by* DANIEL L. PURICH AND VERN L. SCHRAMM

VOLUME 309. Amyloid, Prions, and Other Protein Aggregates
*Edited by* RONALD WETZEL

VOLUME 310. Biofilms
*Edited by* RON J. DOYLE

VOLUME 311. Sphingolipid Metabolism and Cell Signaling (Part A)
*Edited by* ALFRED H. MERRILL, JR., AND YUSUF A. HANNUN

VOLUME 312. Sphingolipid Metabolism and Cell Signaling (Part B)
*Edited by* ALFRED H. MERRILL, JR., AND YUSUF A. HANNUN

VOLUME 313. Antisense Technology
(Part A: General Methods, Methods of Delivery, and RNA Studies)
*Edited by* M. IAN PHILLIPS

VOLUME 314. Antisense Technology (Part B: Applications)
*Edited by* M. IAN PHILLIPS

VOLUME 315. Vertebrate Phototransduction and the Visual Cycle
(Part A)
*Edited by* KRZYSZTOF PALCZEWSKI

VOLUME 316. Vertebrate Phototransduction and the Visual Cycle (Part B)
*Edited by* KRZYSZTOF PALCZEWSKI

VOLUME 317. RNA–Ligand Interactions (Part A: Structural Biology Methods)
*Edited by* DANIEL W. CELANDER AND JOHN N. ABELSON

VOLUME 318. RNA–Ligand Interactions (Part B: Molecular Biology Methods)
*Edited by* DANIEL W. CELANDER AND JOHN N. ABELSON

VOLUME 319. Singlet Oxygen, UV-A, and Ozone
*Edited by* LESTER PACKER AND HELMUT SIES

VOLUME 320. Cumulative Subject Index Volumes 290–319

VOLUME 321. Numerical Computer Methods (Part C)
*Edited by* MICHAEL L. JOHNSON AND LUDWIG BRAND

VOLUME 322. Apoptosis
*Edited by* JOHN C. REED

VOLUME 323. Energetics of Biological Macromolecules (Part C)
*Edited by* MICHAEL L. JOHNSON AND GARY K. ACKERS

VOLUME 324. Branched-Chain Amino Acids (Part B)
*Edited by* ROBERT A. HARRIS AND JOHN R. SOKATCH

VOLUME 325. Regulators and Effectors of Small GTPases
(Part D: Rho Family)
*Edited by* W. E. BALCH, CHANNING J. DER, AND ALAN HALL

VOLUME 326. Applications of Chimeric Genes and Hybrid Proteins
(Part A: Gene Expression and Protein Purification)
*Edited by* JEREMY THORNER, SCOTT D. EMR, AND JOHN N. ABELSON

Volume 327. Applications of Chimeric Genes and Hybrid Proteins (Part B: Cell Biology and Physiology)
*Edited by* Jeremy Thorner, Scott D. Emr, and John N. Abelson

Volume 328. Applications of Chimeric Genes and Hybrid Proteins (Part C: Protein–Protein Interactions and Genomics)
*Edited by* Jeremy Thorner, Scott D. Emr, and John N. Abelson

Volume 329. Regulators and Effectors of Small GTPases (Part E: GTPases Involved in Vesicular Traffic)
*Edited by* W. E. Balch, Channing J. Der, and Alan Hall

Volume 330. Hyperthermophilic Enzymes (Part A)
*Edited by* Michael W. W. Adams and Robert M. Kelly

Volume 331. Hyperthermophilic Enzymes (Part B)
*Edited by* Michael W. W. Adams and Robert M. Kelly

Volume 332. Regulators and Effectors of Small GTPases (Part F: Ras Family I)
*Edited by* W. E. Balch, Channing J. Der, and Alan Hall

Volume 333. Regulators and Effectors of Small GTPases (Part G: Ras Family II)
*Edited by* W. E. Balch, Channing J. Der, and Alan Hall

Volume 334. Hyperthermophilic Enzymes (Part C)
*Edited by* Michael W. W. Adams and Robert M. Kelly

Volume 335. Flavonoids and Other Polyphenols
*Edited by* Lester Packer

Volume 336. Microbial Growth in Biofilms (Part A: Developmental and Molecular Biological Aspects)
*Edited by* Ron J. Doyle

Volume 337. Microbial Growth in Biofilms (Part B: Special Environments and Physicochemical Aspects)
*Edited by* Ron J. Doyle

Volume 338. Nuclear Magnetic Resonance of Biological Macromolecules (Part A)
*Edited by* Thomas L. James, Volker Dötsch, and Uli Schmitz

Volume 339. Nuclear Magnetic Resonance of Biological Macromolecules (Part B)
*Edited by* Thomas L. James, Volker Dötsch, and Uli Schmitz

VOLUME 340. Drug–Nucleic Acid Interactions
*Edited by* JONATHAN B. CHAIRES AND MICHAEL J. WARING

VOLUME 341. Ribonucleases (Part A)
*Edited by* ALLEN W. NICHOLSON

VOLUME 342. Ribonucleases (Part B)
*Edited by* ALLEN W. NICHOLSON

VOLUME 343. G Protein Pathways (Part A: Receptors)
*Edited by* RAVI IYENGAR AND JOHN D. HILDEBRANDT

VOLUME 344. G Protein Pathways (Part B: G Proteins and Their Regulators)
*Edited by* RAVI IYENGAR AND JOHN D. HILDEBRANDT

VOLUME 345. G Protein Pathways (Part C: Effector Mechanisms)
*Edited by* RAVI IYENGAR AND JOHN D. HILDEBRANDT

VOLUME 346. Gene Therapy Methods
*Edited by* M. IAN PHILLIPS

VOLUME 347. Protein Sensors and Reactive Oxygen Species (Part A: Selenoproteins and Thioredoxin)
*Edited by* HELMUT SIES AND LESTER PACKER

VOLUME 348. Protein Sensors and Reactive Oxygen Species (Part B: Thiol Enzymes and Proteins)
*Edited by* HELMUT SIES AND LESTER PACKER

VOLUME 349. Superoxide Dismutase
*Edited by* LESTER PACKER

VOLUME 350. Guide to Yeast Genetics and Molecular and Cell Biology (Part B)
*Edited by* CHRISTINE GUTHRIE AND GERALD R. FINK

VOLUME 351. Guide to Yeast Genetics and Molecular and Cell Biology (Part C)
*Edited by* CHRISTINE GUTHRIE AND GERALD R. FINK

VOLUME 352. Redox Cell Biology and Genetics (Part A)
*Edited by* CHANDAN K. SEN AND LESTER PACKER

VOLUME 353. Redox Cell Biology and Genetics (Part B)
*Edited by* CHANDAN K. SEN AND LESTER PACKER

VOLUME 354. Enzyme Kinetics and Mechanisms (Part F: Detection and Characterization of Enzyme Reaction Intermediates)
*Edited by* DANIEL L. PURICH

VOLUME 355. Cumulative Subject Index Volumes 321–354

VOLUME 356. Laser Capture Microscopy and Microdissection
*Edited by* P. MICHAEL CONN

VOLUME 357. Cytochrome P450, Part C
*Edited by* ERIC F. JOHNSON AND MICHAEL R. WATERMAN

VOLUME 358. Bacterial Pathogenesis (Part C: Identification, Regulation, and Function of Virulence Factors)
*Edited by* VIRGINIA L. CLARK AND PATRIK M. BAVOIL

VOLUME 359. Nitric Oxide (Part D)
*Edited by* ENRIQUE CADENAS AND LESTER PACKER

VOLUME 360. Biophotonics (Part A)
*Edited by* GERARD MARRIOTT AND IAN PARKER

VOLUME 361. Biophotonics (Part B)
*Edited by* GERARD MARRIOTT AND IAN PARKER

VOLUME 362. Recognition of Carbohydrates in Biological Systems (Part A)
*Edited by* YUAN C. LEE AND REIKO T. LEE

VOLUME 363. Recognition of Carbohydrates in Biological Systems (Part B)
*Edited by* YUAN C. LEE AND REIKO T. LEE

VOLUME 364. Nuclear Receptors
*Edited by* DAVID W. RUSSELL AND DAVID J. MANGELSDORF

VOLUME 365. Differentiation of Embryonic Stem Cells
*Edited by* PAUL M. WASSAUMAN AND GORDON M. KELLER

VOLUME 366. Protein Phosphatases
*Edited by* SUSANNE KLUMPP AND JOSEF KRIEGLSTEIN

VOLUME 367. Liposomes (Part A)
*Edited by* NEJAT DÜZGÜNEŞ

VOLUME 368. Macromolecular Crystallography (Part C)
*Edited by* CHARLES W. CARTER, JR., AND ROBERT M. SWEET

VOLUME 369. Combinational Chemistry (Part B)
*Edited by* GUILLERMO A. MORALES AND BARRY A. BUNIN

VOLUME 370. RNA Polymerases and Associated Factors (Part C)
*Edited by* SANKAR L. ADHYA AND SUSAN GARGES

VOLUME 371. RNA Polymerases and Associated Factors (Part D)
*Edited by* SANKAR L. ADHYA AND SUSAN GARGES

VOLUME 372. Liposomes (Part B)
*Edited by* NEJAT DÜZGÜNEŞ

VOLUME 373. Liposomes (Part C)
*Edited by* NEJAT DÜZGÜNEŞ

VOLUME 374. Macromolecular Crystallography (Part D)
*Edited by* CHARLES W. CARTER, JR., AND ROBERT W. SWEET

VOLUME 375. Chromatin and Chromatin Remodeling Enzymes (Part A)
*Edited by* C. DAVID ALLIS AND CARL WU

VOLUME 376. Chromatin and Chromatin Remodeling Enzymes (Part B)
*Edited by* C. DAVID ALLIS AND CARL WU

VOLUME 377. Chromatin and Chromatin Remodeling Enzymes (Part C)
*Edited by* C. DAVID ALLIS AND CARL WU

VOLUME 378. Quinones and Quinone Enzymes (Part A)
*Edited by* HELMUT SIES AND LESTER PACKER

VOLUME 379. Energetics of Biological Macromolecules (Part D)
*Edited by* JO M. HOLT, MICHAEL L. JOHNSON, AND GARY K. ACKERS

VOLUME 380. Energetics of Biological Macromolecules (Part E)
*Edited by* JO M. HOLT, MICHAEL L. JOHNSON, AND GARY K. ACKERS

VOLUME 381. Oxygen Sensing
*Edited by* CHANDAN K. SEN AND GREGG L. SEMENZA

VOLUME 382. Quinones and Quinone Enzymes (Part B)
*Edited by* HELMUT SIES AND LESTER PACKER

VOLUME 383. Numerical Computer Methods (Part D)
*Edited by* LUDWIG BRAND AND MICHAEL L. JOHNSON

VOLUME 384. Numerical Computer Methods (Part E)
*Edited by* LUDWIG BRAND AND MICHAEL L. JOHNSON

VOLUME 385. Imaging in Biological Research (Part A)
*Edited by* P. MICHAEL CONN

VOLUME 386. Imaging in Biological Research (Part B)
*Edited by* P. MICHAEL CONN

Volume 387. Liposomes (Part D)
*Edited by* Nejat Düzgüneş

Volume 388. Protein Engineering
*Edited by* Dan E. Robertson and Joseph P. Noel

Volume 389. Regulators of G-Protein Signaling (Part A)
*Edited by* David P. Siderovski

Volume 390. Regulators of G-Protein Signaling (Part B)
*Edited by* David P. Siderovski

Volume 391. Liposomes (Part E)
*Edited by* Nejat Düzgüneş

Volume 392. RNA Interference
*Edited by* Engelke Rossi

Volume 393. Circadian Rhythms
*Edited by* Michael W. Young

Volume 394. Nuclear Magnetic Resonance of Biological Macromolecules (Part C)
*Edited by* Thomas L. James

Volume 395. Producing the Biochemical Data (Part B)
*Edited by* Elizabeth A. Zimmer and Eric H. Roalson

Volume 396. Nitric Oxide (Part E)
*Edited by* Lester Packer and Enrique Cadenas

Volume 397. Environmental Microbiology
*Edited by* Jared R. Leadbetter

Volume 398. Ubiquitin and Protein Degradation (Part A)
*Edited by* Raymond J. Deshaies

Volume 399. Ubiquitin and Protein Degradation (Part B)
*Edited by* Raymond J. Deshaies

Volume 400. Phase II Conjugation Enzymes and Transport Systems
*Edited by* Helmut Sies and Lester Packer

Volume 401. Glutathione Transferases and Gamma Glutamyl Transpeptidases
*Edited by* Helmut Sies and Lester Packer

Volume 402. Biological Mass Spectrometry
*Edited by* A. L. Burlingame

Volume 403. GTPases Regulating Membrane Targeting and Fusion
*Edited by* William E. Balch, Channing J. Der, and Alan Hall

Volume 404. GTPases Regulating Membrane Dynamics
*Edited by* William E. Balch, Channing J. Der, and Alan Hall

Volume 405. Mass Spectrometry: Modified Proteins and Glycoconjugates
*Edited by* A. L. Burlingame

Volume 406. Regulators and Effectors of Small GTPases: Rho Family
*Edited by* William E. Balch, Channing J. Der, and Alan Hall

Volume 407. Regulators and Effectors of Small GTPases: Ras Family
*Edited by* William E. Balch, Channing J. Der, and Alan Hall

Volume 408. DNA Repair (Part A)
*Edited by* Judith L. Campbell and Paul Modrich

Volume 409. DNA Repair (Part B)
*Edited by* Judith L. Campbell and Paul Modrich

Volume 410. DNA Microarrays (Part A: Array Platforms and Web-Bench Protocols)
*Edited by* Alan Kimmel and Brian Oliver

Volume 411. DNA Microarrays (Part B: Databases and Statistics)
*Edited by* Alan Kimmel and Brian Oliver

Volume 412. Amyloid, Prions, and Other Protein Aggregates (Part B)
*Edited by* Indu Kheterpal and Ronald Wetzel

Volume 413. Amyloid, Prions, and Other Protein Aggregates (Part C)
*Edited by* Indu Kheterpal and Ronald Wetzel

Volume 414. Measuring Biological Responses with Automated Microscopy
*Edited by* James Inglese

Volume 415. Glycobiology
*Edited by* Minoru Fukuda

Volume 416. Glycomics
*Edited by* Minoru Fukuda

Volume 417. Functional Glycomics
*Edited by* Minoru Fukuda

Volume 418. Embryonic Stem Cells
*Edited by* Irina Klimanskaya and Robert Lanza

VOLUME 419. Adult Stem Cells
*Edited by* IRINA KLIMANSKAYA AND ROBERT LANZA

VOLUME 420. Stem Cell Tools and Other Experimental Protocols
*Edited by* IRINA KLIMANSKAYA AND ROBERT LANZA

VOLUME 421. Advanced Bacterial Genetics: Use of Transposons and Phage for Genomic Engineering
*Edited by* KELLY T. HUGHES

VOLUME 422. Two-Component Signaling Systems, Part A
*Edited by* MELVIN I. SIMON, BRIAN R. CRANE, AND ALEXANDRINE CRANE

VOLUME 423. Two-Component Signaling Systems, Part B
*Edited by* MELVIN I. SIMON, BRIAN R. CRANE, AND ALEXANDRINE CRANE

VOLUME 424. RNA Editing
*Edited by* JONATHA M. GOTT

VOLUME 425. RNA Modification
*Edited by* JONATHA M. GOTT

VOLUME 426. Integrins
*Edited by* DAVID CHERESH

VOLUME 427. MicroRNA Methods
*Edited by* JOHN J. ROSSI

VOLUME 428. Osmosensing and Osmosignaling
*Edited by* HELMUT SIES AND DIETER HAUSSINGER

VOLUME 429. Translation Initiation: Extract Systems and Molecular Genetics
*Edited by* JON LORSCH

VOLUME 430. Translation Initiation: Reconstituted Systems and Biophysical Methods
*Edited by* JON LORSCH

VOLUME 431. Translation Initiation: Cell Biology, High-Throughput and Chemical-Based Approaches
*Edited by* JON LORSCH

VOLUME 432. Lipidomics and Bioactive Lipids: Mass-Spectrometry–Based Lipid Analysis
*Edited by* H. ALEX BROWN

Volume 433. Lipidomics and Bioactive Lipids: Specialized Analytical Methods and Lipids in Disease
*Edited by* H. Alex Brown

Volume 434. Lipidomics and Bioactive Lipids: Lipids and Cell Signaling
*Edited by* H. Alex Brown

Volume 435. Oxygen Biology and Hypoxia
*Edited by* Helmut Sies and Bernhard Brüne

Volume 436. Globins and Other Nitric Oxide-Reactive Protiens (Part A)
*Edited by* Robert K. Poole

Volume 437. Globins and Other Nitric Oxide-Reactive Protiens (Part B)
*Edited by* Robert K. Poole

Volume 438. Small GTPases in Disease (Part A)
*Edited by* William E. Balch, Channing J. Der, and Alan Hall

Volume 439. Small GTPases in Disease (Part B)
*Edited by* William E. Balch, Channing J. Der, and Alan Hall

Volume 440. Nitric Oxide, Part F Oxidative and Nitrosative Stress in Redox Regulation of Cell Signaling
*Edited by* Enrique Cadenas and Lester Packer

Volume 441. Nitric Oxide, Part G Oxidative and Nitrosative Stress in Redox Regulation of Cell Signaling
*Edited by* Enrique Cadenas and Lester Packer

Volume 442. Programmed Cell Death, General Principles for Studying Cell Death (Part A)
*Edited by* Roya Khosravi-Far, Zahra Zakeri, Richard A. Lockshin, and Mauro Piacentini

Volume 443. Angiogenesis: *In Vitro* Systems
*Edited by* David A. Cheresh

Volume 444. Angiogenesis: *In Vivo* Systems (Part A)
*Edited by* David A. Cheresh

Volume 445. Angiogenesis: *In Vivo* Systems (Part B)
*Edited by* David A. Cheresh

Volume 446. Programmed Cell Death, The Biology and Therapeutic Implications of Cell Death (Part B)
*Edited by* Roya Khosravi-Far, Zahra Zakeri, Richard A. Lockshin, and Mauro Piacentini

VOLUME 447. RNA Turnover in Bacteria, Archaea and Organelles
*Edited by* LYNNE E. MAQUAT AND CECILIA M. ARRAIANO

VOLUME 448. RNA Turnover in Eukaryotes: Nucleases, Pathways and Analysis of mRNA Decay
*Edited by* LYNNE E. MAQUAT AND MEGERDITCH KILEDJIAN

VOLUME 449. RNA Turnover in Eukaryotes: Analysis of Specialized and Quality Control RNA Decay Pathways
*Edited by* LYNNE E. MAQUAT AND MEGERDITCH KILEDJIAN

VOLUME 450. Fluorescence Spectroscopy
*Edited by* LUDWIG BRAND AND MICHAEL L. JOHNSON

VOLUME 451. Autophagy: Lower Eukaryotes and Non-Mammalian Systems (Part A)
*Edited by* DANIEL J. KLIONSKY

VOLUME 452. Autophagy in Mammalian Systems (Part B)
*Edited by* DANIEL J. KLIONSKY

VOLUME 453. Autophagy in Disease and Clinical Applications (Part C)
*Edited by* DANIEL J. KLIONSKY

VOLUME 454. Computer Methods (Part A)
*Edited by* MICHAEL L. JOHNSON AND LUDWIG BRAND

VOLUME 455. Biothermodynamics (Part A)
*Edited by* MICHAEL L. JOHNSON, JO M. HOLT, AND GARY K. ACKERS (RETIRED)

VOLUME 456. Mitochondrial Function, Part A: Mitochondrial Electron Transport Complexes and Reactive Oxygen Species
*Edited by* WILLIAM S. ALLISON AND IMMO E. SCHEFFLER

VOLUME 457. Mitochondrial Function, Part B: Mitochondrial Protein Kinases, Protein Phosphatases and Mitochondrial Diseases
*Edited by* WILLIAM S. ALLISON AND ANNE N. MURPHY

VOLUME 458. Complex Enzymes in Microbial Natural Product Biosynthesis, Part A: Overview Articles and Peptides
*Edited by* DAVID A. HOPWOOD

VOLUME 459. Complex Enzymes in Microbial Natural Product Biosynthesis, Part B: Polyketides, Aminocoumarins and Carbohydrates
*Edited by* DAVID A. HOPWOOD

VOLUME 460. Chemokines, Part A
*Edited by* TRACY M. HANDEL AND DAMON J. HAMEL

VOLUME 461. Chemokines, Part B
*Edited by* TRACY M. HANDEL AND DAMON J. HAMEL

VOLUME 462. Non-Natural Amino Acids
*Edited by* TOM W. MUIR AND JOHN N. ABELSON

VOLUME 463. Guide to Protein Purification, 2nd Edition
*Edited by* RICHARD R. BURGESS AND MURRAY P. DEUTSCHER

VOLUME 464. Liposomes, Part F
*Edited by* NEJAT DÜZGÜNEŞ

VOLUME 465. Liposomes, Part G
*Edited by* NEJAT DÜZGÜNEŞ

VOLUME 466. Biothermodynamics, Part B
*Edited by* MICHAEL L. JOHNSON, GARY K. ACKERS, AND JO M. HOLT

VOLUME 467. Computer Methods Part B
*Edited by* MICHAEL L. JOHNSON AND LUDWIG BRAND

VOLUME 468. Biophysical, Chemical, and Functional Probes of RNA Structure, Interactions and Folding: Part A
*Edited by* DANIEL HERSCHLAG

VOLUME 469. Biophysical, Chemical, and Functional Probes of RNA Structure, Interactions and Folding: Part B
*Edited by* DANIEL HERSCHLAG

VOLUME 470. Guide to Yeast Genetics: Functional Genomics, Proteomics, and Other Systems Analysis, 2nd Edition
*Edited by* GERALD FINK, JONATHAN WEISSMAN, AND CHRISTINE GUTHRIE

VOLUME 471. Two-Component Signaling Systems, Part C
*Edited by* MELVIN I. SIMON, BRIAN R. CRANE, AND ALEXANDRINE CRANE

VOLUME 472. Single Molecule Tools, Part A: Fluorescence Based Approaches
*Edited by* NILS G. WALTER

VOLUME 473. Thiol Redox Transitions in Cell Signaling, Part A Chemistry and Biochemistry of Low Molecular Weight and Protein Thiols
*Edited by* ENRIQUE CADENAS AND LESTER PACKER

VOLUME 474. Thiol Redox Transitions in Cell Signaling, Part B Cellular Localization and Signaling
*Edited by* ENRIQUE CADENAS AND LESTER PACKER

VOLUME 475. Single Molecule Tools, Part B: Super-Resolution, Particle Tracking, Multiparameter, and Force Based Methods
*Edited by* NILS G. WALTER

VOLUME 476. Guide to Techniques in Mouse Development, Part A Mice, Embryos, and Cells, 2nd Edition
*Edited by* PAUL M. WASSARMAN AND PHILIPPE M. SORIANO

VOLUME 477. Guide to Techniques in Mouse Development, Part B Mouse Molecular Genetics, 2nd Edition
*Edited by* PAUL M. WASSARMAN AND PHILIPPE M. SORIANO

VOLUME 478. Glycomics
*Edited by* MINORU FUKUDA

VOLUME 479. Functional Glycomics
*Edited by* MINORU FUKUDA

VOLUME 480. Glycobiology
*Edited by* MINORU FUKUDA

VOLUME 481. Cryo-EM, Part A: Sample Preparation and Data Collection
*Edited by* GRANT J. JENSEN

VOLUME 482. Cryo-EM, Part B: 3-D Reconstruction
*Edited by* GRANT J. JENSEN

VOLUME 483. Cryo-EM, Part C: Analyses, Interpretation, and Case Studies
*Edited by* GRANT J. JENSEN

VOLUME 484. Constitutive Activity in Receptors and Other Proteins, Part A
*Edited by* P. MICHAEL CONN

VOLUME 485. Constitutive Activity in Receptors and Other Proteins, Part B
*Edited by* P. MICHAEL CONN

VOLUME 486. Research on Nitrification and Related Processes, Part A
*Edited by* MARTIN G. KLOTZ

VOLUME 487. Computer Methods, Part C
*Edited by* MICHAEL L. JOHNSON AND LUDWIG BRAND

VOLUME 488. Biothermodynamics, Part C
*Edited by* MICHAEL L. JOHNSON, JO M. HOLT, AND GARY K. ACKERS

VOLUME 489. The Unfolded Protein Response and Cellular Stress, Part A
*Edited by* P. MICHAEL CONN

VOLUME 490. The Unfolded Protein Response and Cellular Stress, Part B
*Edited by* P. MICHAEL CONN

VOLUME 491. The Unfolded Protein Response and Cellular Stress, Part C
*Edited by* P. MICHAEL CONN

VOLUME 492. Biothermodynamics, Part D
*Edited by* MICHAEL L. JOHNSON, JO M. HOLT, AND GARY K. ACKERS

VOLUME 493. Fragment-Based Drug Design Tools,
Practical Approaches, and Examples
*Edited by* LAWRENCE C. KUO

VOLUME 494. Methods in Methane Metabolism, Part A
Methanogenesis
*Edited by* AMY C. ROSENZWEIG AND STEPHEN W. RAGSDALE

VOLUME 495. Methods in Methane Metabolism, Part B
Methanotrophy
*Edited by* AMY C. ROSENZWEIG AND STEPHEN W. RAGSDALE

VOLUME 496. Research on Nitrification and Related Processes, Part B
*Edited by* MARTIN G. KLOTZ AND LISA Y. STEIN

VOLUME 497. Synthetic Biology, Part A
Methods for Part/Device Characterization and Chassis Engineering
*Edited by* CHRISTOPHER VOIGT

VOLUME 498. Synthetic Biology, Part B
Computer Aided Design and DNA Assembly
*Edited by* CHRISTOPHER VOIGT

VOLUME 499. Biology of Serpins
*Edited by* JAMES C. WHISSTOCK AND PHILLIP I. BIRD

VOLUME 500. Methods in Systems Biology
*Edited by* DANIEL JAMESON, MALKHEY VERMA, AND HANS V. WESTERHOFF

VOLUME 501. Serpin Structure and Evolution
*Edited by* JAMES C. WHISSTOCK AND PHILLIP I. BIRD

VOLUME 502. Protein Engineering for Therapeutics, Part A
*Edited by* K. DANE WITTRUP AND GREGORY L. VERDINE

Volume 503. Protein Engineering for Therapeutics, Part B
*Edited by* K. Dane Wittrup and Gregory L. Verdine

Volume 504. Imaging and Spectroscopic Analysis of Living Cells
Optical and Spectroscopic Techniques
*Edited by* P. Michael Conn

Volume 505. Imaging and Spectroscopic Analysis of Living Cells
Live Cell Imaging of Cellular Elements and Functions
*Edited by* P. Michael Conn

Volume 506. Imaging and Spectroscopic Analysis of Living Cells
Imaging Live Cells in Health and Disease
*Edited by* P. Michael Conn

Volume 507. Gene Transfer Vectors for Clinical Application
*Edited by* Theodore Friedmann

Volume 508. Nanomedicine
Cancer, Diabetes, and Cardiovascular, Central Nervous System, Pulmonary and Inflammatory Diseases
*Edited by* Nejat Düzgüneş

Volume 509. Nanomedicine
Infectious Diseases, Immunotherapy, Diagnostics, Antifibrotics, Toxicology and Gene Medicine
*Edited by* Nejat Düzgüneş

Volume 510. Cellulases
*Edited by* Harry J. Gilbert

Volume 511. RNA Helicases
*Edited by* Eckhard Jankowsky

Volume 512. Nucleosomes, Histones & Chromatin, Part A
*Edited by* Carl Wu and C. David Allis

Volume 513. Nucleosomes, Histones & Chromatin, Part B
*Edited by* Carl Wu and C. David Allis

Volume 514. Ghrelin
*Edited by* Masayasu Kojima and Kenji Kangawa

Volume 515. Natural Product Biosynthesis by Microorganisms and Plants, Part A
*Edited by* David A. Hopwood

VOLUME 516. Natural Product Biosynthesis by Microorganisms and Plants, Part B
*Edited by* DAVID A. HOPWOOD

VOLUME 517. Natural Product Biosynthesis by Microorganisms and Plants, Part C
*Edited by* DAVID A. HOPWOOD

VOLUME 518. Fluorescence Fluctuation Spectroscopy (FFS), Part A
*Edited by* SERGEY Y. TETIN

VOLUME 519. Fluorescence Fluctuation Spectroscopy (FFS), Part B
*Edited by* SERGEY Y. TETIN

CHAPTER ONE

# FCS in STED Microscopy: Studying the Nanoscale of Lipid Membrane Dynamics

**Veronika Mueller[*], Alf Honigmann[*], Christian Ringemann[*], Rebecca Medda[*], Günter Schwarzmann[†], Christian Eggeling[*,‡,1]**

[*]Department of Nanobiophotonics, Max Planck Institute for Biophysical Chemistry, Göttingen, Germany
[†]Life and Medical Sciences Center (LIMES) Membrane Biology and Lipid Biochemistry Unit, University of Bonn, Bonn, Germany
[‡]Weatherall Institute of Molecular Medicine, University of Oxford, Headley Way, Oxford, United Kingdom
[1]Corresponding author: e-mail address: ceggeli@gwdg.de

## Contents



## Abstract

Details of molecular membrane dynamics in living cells such as lipid–protein interactions or the incorporation of molecules into lipid "rafts" are often hidden to the observer because of the limited spatial resolution of conventional far-field optical microscopy. Fortunately, the superior spatial resolution of far-field stimulated-emission-depletion (STED) nanoscopy allows gaining new insights. Applying fluorescence correlation spectroscopy (FCS) in focal spots continuously tuned down to 30 nm in diameter distinguishes free from anomalous molecular diffusion due to transient binding, as for the

*Methods in Enzymology*, Volume 519
ISSN 0076-6879
http://dx.doi.org/10.1016/B978-0-12-405539-1.00001-4

diffusion of fluorescent phosphoglycero- and sphingolipid analogs in the plasma membrane of living cells. STED-FCS data recorded at different environmental conditions and on different lipid analogs reveal molecular details of the observed nanoscale trapping. Dependencies on the molecular structure of the lipids point to the distinct connectivity of the various lipids to initiate or assist cellular signaling events, but also outline strong differences to the characteristics of liquid-ordered and disordered phase separation in model membranes. STED-FCS is a highly sensitive and exceptional tool to study the membrane organization by introducing a new class of nanoscale biomolecular studies.

## 1. INTRODUCTION

Cellular signaling is known to be linked to the organization of lipids in the plasma membrane, where lipid–lipid or lipid–protein interactions and the formation of cholesterol-assisted lipid nanodomains (or "rafts") (Fig. 1.1) are believed to play a crucial role (Brown & London, 2000; Fielding, 2006; Goswami et al., 2008; Hanzal-Bayer & Hancock, 2007; Jacobson, Mouritsen, & Anderson, 2007; Kusumi, Shirai, Koyama-Honda, Suzuki, & Fujiwara, 2010; Lingwood & Simons, 2010; Simons & Ikonen, 1997). However, the access to a lot of details on the molecular and spatiotemporal characteristics of these interactions is often hindered by the limited spatial resolution of conventional optical microscopy (Eggeling et al., 2009; Hancock, 2006; Jacobson et al., 2007; Lommerse, Spaink, & Schmidt, 2004; Munro, 2003; Shaw, 2006). Using visible light and far-field optics only allows distinguishing features that are approximately more than 200 nm apart (Abbe, 1873; Hell, 2009). The sizes of the lipid–protein complexes or domains are, however, supposed to be on smaller scales (Pike, 2006). The required spatial resolution is provided by other microscopy techniques such as electron microscopy (Fujita et al., 2007), atomic- (Binnig, Quate, & Gerber, 1986) or optical force microscopy (Pralle, Keller, Florin, Simons, & Hoerber, 2000) or near-field optical microscopy (de Bakker et al., 2007; Pohl, Denk, & Lanz, 1984; Wenger et al., 2007), but their use on living cells may be

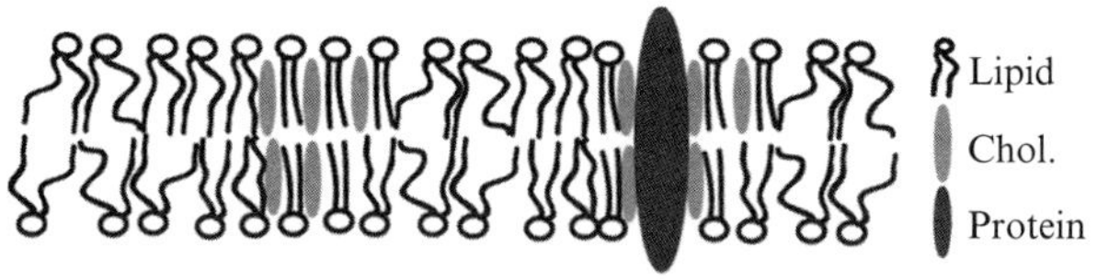


**Figure 1.1** Sketch of membrane heterogeneity: nanodomains may show up as small (10–200 nm), heterogeneous, highly dynamic, cholesterol- (Chol.), and sphingolipid-enriched platforms that compartmentalize cellular processes (Pike, 2006).

challenging. Some of these microscopy techniques are, for example, usually restricted to studies on dead cells or on features close to surfaces, that is, their potentially invasive nature may influence the system under study. Non-invasive far-field optics do not suffer from these restrictions and allow for the direct study on living cells. Far-field approaches such as fluorescence recovery after photobleaching (FRAP) (Feder, Brust-Mascher, Slattery, Baird, & Webb, 1996; Yechiel & Edidin, 1987), fluorescence correlation spectroscopy (FCS) (Fahey et al., 1977; Schwille, Korlach, & Webb, 1999; Wawrezinieck, Rigneault, Marguet, & Lenne, 2005), Förster resonance energy transfer (FRET) (Varma & Mayor, 1998; Zacharias, Violin, Newton, & Tsien, 2002), or single-particle tracking (SPT) (Fujiwara, Ritchie, Murakoshi, Jacobson, & Kusumi, 2002; Saxton & Jacobson, 1997; Schutz, Kada, Pastushenko, & Schindler, 2000) have successfully performed live-cell studies of membrane dynamics and organization. However, they are again diffraction-limited (FCS and FRAP), often are realized at non-endogenous overexpressed concentration of the molecules under study (FRET) or, to reach the desired temporal resolution to follow the fast lipid dynamics, require the use of bright but large and clumsy signal markers, which potentially influence the system under study (SPT).

Previous FCS experiments have demonstrated how some information about nanoscopic movement can be indirectly inferred when using model-based approaches (Schwille et al., 1999), for example, by extrapolating measurement results to the nanoscopic case (Humpolickova et al., 2006; Wawrezinieck et al., 2005). However, these approaches heavily rely on correct quantitative modeling of the investigated system, as the experiments cannot deliver a clear signature of dynamics at length scales below the diffraction limit of far-field optics. Recently, combining FCS with far-field stimulated-emission-depletion (STED) nanoscopy (Eggeling et al., 2009; Kastrup, Blom, Eggeling, & Hell, 2005; Ringemann et al., 2009) allowed for the direct measurements at the length scale of interest and delivered more model-independent results about nanoscopic details. STED nanoscopy offers "diffraction-unlimited" microscopy in the far field by applying stimulated emission to inhibit fluorescence emission everywhere outside small confined regions (Hell, 2009; Hell & Wichmann, 1994). STED-FCS directly revealed nanoscopic details of membrane lipid dynamics. In contrast to a fluorescent phospholipid analog, sphingolipid analogs were strongly trapped in transient, cholesterol-assisted molecular complexes (Eggeling et al., 2009; Mueller et al., 2011). The STED-FCS data were confirmed by fast single-molecule tracking experiments (Sahl, Leutenegger, Hilbert, Hell, & Eggeling, 2010) and by FCS experiments on a near-field optical microscope (Manzo, van Zanten, & Garcia-Parajo, 2011). All experiments applied small organic dye labels, whose influence was (at least for the experiments presented so far) shown to be negligible (Eggeling et al., 2009; Sahl et al., 2010). On- and off-rates in the range of $\sim 1/(10\ \text{ms})$ (Ringemann et al., 2009) were determined and the spatial scale at which the lipids dwell during trapping estimated to be smaller than 20 nm (Eggeling et al., 2009; Sahl et al., 2010).

Phase separation into differently ordered domains of, for example, ternary lipid mixtures of model membranes such as giant unilamellar vesicles (GUV), supported lipid bilayers (SLB) or black lipid bilayers, or of giant plasma membrane vesicles (GPMV) of living cells were often taken as physical basis of lipid "raft" organization (see, e.g., Bacia, Scherfeld, Kahya, & Schwille, 2004; Bagatolli, 2006; Baumgart et al., 2007a; Dietrich et al., 2001; Hac, Seeger, Fidorra, & Heimburg, 2005; London, 2005; Marsh, 2009; Samsonov, Mihalyov, & Cohen, 2001; Sezgin et al., 2012; Veatch & Keller, 2005). It has been shown that the partitioning of a lipid into different phases is highly dependent on its molecular structure, such as lipid type, number of chains and their lengths, and on the type and position of the fluorescent marker, as well as on other environmental parameters such as the temperature (see, e.g., Bacia et al., 2004; Baumgart, Hunt, Farkas, Webb, & Feigenson, 2007b; Wang & Silvius, 2000). However, a lot of debate started on how these model systems can really be applied as a basis for the lipid organization in the living cell. Above all, the molecular composition of the ternary mixtures usually applied to induce phase separation in model membranes is still rather simple to realistically mimic the molecular versatility of the cellular plasma membrane. Nevertheless, it has been shown that despite its molecular diversity the plasma membrane may also separate into large phases at low temperatures and when separated from the cytoskeleton in GPMVs (Baumgart et al., 2007a; Lingwood, Ries, Schwille, & Simons, 2008).

This chapter gives an overview over the STED-FCS studies of lipid dynamics in the plasma membrane of living cells so far. We present different but consistent FCS analysis approaches to distinguish free from trapped diffusion and to quantify its extent. Trapping characteristics of various fluorescent lipid analogs are disclosed, differing in the number of chains, the type or the position of the fluorescent marker, and the head-group structure. In addition, studies on the influence of the temperature, the cell type, and the applied laser intensity are presented. The STED-FCS experiments outline the distinct importance of specific molecular structures such as the hydroxyl-containing head groups or the ceramide (or sphingosine) group close to the water–lipid interface, which both facilitate transient bonding. Detailed cholesterol-depletion experiments highlight differences between head- and ceramide-group mediated trapping interactions. However, no correlation between phase separation in model membrane systems and the nanoscopic trapping has been observed. STED-FCS is an exceptional tool to determine molecular specificities of various lipids and their distinct connectivity to certain cellular activities.

# 2. STED-FCS OF LIPID MEMBRANE DYNAMICS

## 2.1. Cellular labeling

As a far-field fluorescence microscopy technique, STED-FCS requires the use of fluorescent labels. Therefore, for the STED-FCS experiments of lipid membrane dynamics measurements, different fluorescent lipid analogs were usually incoroporated into the plasma membrane of living mammalian cells by incubation with lipid–BSA complexes (Martin & Pagano, 1994).

Several control experiments indicated proper membrane incorporation of the lipid analogs.

**i.** Figure 1.2A shows exemplary confocal scanning images of the cells after incorporation with either an Atto647N-labeled phosphoethanolamine (PE or PE-h) or an Atto647N-labeled sphingomyelin (SM or SM-c). Both fluorescent lipids were distributed homogeneously in the plasma membrane. This homogeneity became obvious in the central cell body part of the PE labeled cells. Due to their increased flatness, the filopodia appeared brighter, as both the upper and lower membrane was imaged in this case. Further, the SM analog took part in membrane trafficking. It was found in several parts of the endoplasmatic reticulum, the Golgi apparatus and perinuclear vesicles, a characteristic of the endocytotic pathway of ceramide-structured lipids.

**ii.** FRAP experiments performed on the fluorescent lipid analogs in the plasma membrane of the living PtK2 cells indicated that almost all signal ($\approx$90%) recovered within minutes after photobleaching a $15 \times 15$ $\mu m^2$ large part of the cell (Fig. 1.2B). Depending on the incubation time and temperature, some lipids were already internalized during the FRAP experiments and did not recover. The background signal in-between cells due to, for example, fluorescent lipids unspecifically adsorbed

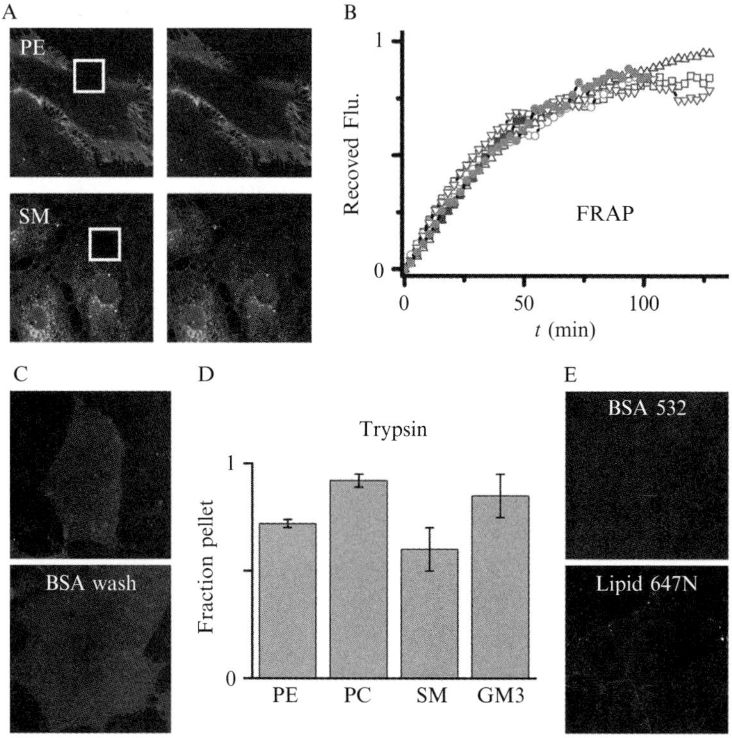


**Figure 1.2** Cellular labeling with fluorescent lipid analogs. (A) Confocal scanning images ($70 \times 70\ \mu m^2$) of living PtK2 cells labeled with an Atto647N-labeled phosphoethanolamine (PE, upper panel) and sphingomyelin (SM, lower panel) after (left panels) photobleaching a $15 \times 15\ \mu m^2$ large area (white box) and after recovery of signal (right panels). (B) FRAP curves of Atto647N-labeled PE and SM on different cells or on different parts of the cell. (C) BSA washing: Confocal scanning images ($80 \times 80\ \mu m^2$) of the plasma membrane of living PtK2 cells incorporating Atto647N-labeled SM before (upper) and after (lower) washing with BSA. (D) Trypsin treatment: Fraction of signal in the cell pellet, that is, fraction of fluorescent lipid analogs (PE: Atto647N-labeled phosphoethanolamine, PC: Atto647N-labeled phosphocholine, SM: Atto647N-labeled sphingomyelin, GM3: Atto647N-labeled GM3) still incorporated after trypsin treatment. (E) Confocal scanning images ($80 \times 80\ \mu m^2$) of living PtK2 cells labeled with Atto647N-labeled PE incorporated by a complex with Atto532-labeled BSA: Atto532 (upper image) and Atto647N fluorescence (lower image). (See Color Insert.)

on the microscope cover glass did not recover at all. Recovery in the membrane was very similar for both PE and SM and for observations on different cells or different parts of the cell. Analysis of the FRAP experiment resulted in a diffusion constant of $D \approx 0.4 \pm 0.2\ \mu m^2/s$, which corresponded well to previous studies (Bacia et al., 2004; Fujiwara et al., 2002; Lenne et al., 2006; Nishimura, Vrljic, Klein, McConnell, &

Moerner, 2006; Schutz et al., 2000), but did not hint to any differences between PE and SM diffusion.

**iii.** Despite the confocal scanning images and the FRAP measurements, the lipid analogs could still be unspecifically bound to the plasma membrane. Back exchange experiments, i.e., washing of the PtK2 cells with BSA after incorporation should remove these unspecifically bound lipids. The fraction of signal lost after BSA 'washing' was low for most of the lipids. Figure 1.2C exemplifies confocal scanning images of the plasma membrane of living PtK2 cells labeled with Atto647N-tagged SM before and after washing with BSA. The washing hardly removed fluorescent SM. Using STED-FCS, similar dynamical characteristics of the lipid analogs on BSA-treated and untreated cells were observed. As a consequence, one may conclude a strong membrane affinity of the fluorescent lipid analogs. We have, for example, shown previously that the dye Atto647N behaves like a lipophilic anchor (Eggeling et al., 2009). In contrast, lipids labeled at the acyl chain with a very polar and charged dye such as Atto532 hardly anchored in the membrane (Eggeling et al., 2009).

**iv.** Washing by BSA may not have removed those fluorescent lipid analogs unspecifically bound to (surface) membrane proteins. Treatment of the PtK2 cells with trypsin after incorporation should remove these unspecifically bound lipids. For various kinds of the fluorescent lipid analogs, Fig. 1.1D exemplarily shows that most of the lipids remained in the cell pellet after trypsin treatment. Using STED-FCS, similar dynamical characteristics of the lipid analogs on trypsin-treated compared to non-treated cells were observed. One can thus safely assume that the lipid analogs were properly incorporated into the plasma membrane and that the observed dynamics were not due to unspecifically bound molecules.

**v.** Finally, control experiments showed a proper release of BSA after incorporation. Confocal scanning images of PtK2 cells incubated with complexes of Atto647N-labeled lipids and Atto532-labeled BSA revealed a properly labeled plasma membrane (Atto647N channel) but no Atto532 fluorescence (Fig. 1.1E). Consequently, the dynamics of the incorporated lipids could not be biased by residually bound BSA.

**vi.** Very randomly the STED-FCS measurements experienced the diffusion of larger and brighter particles. This was accounted to cell debris, internalized vesicles, or larger clusters due to the small fraction of fluorescent lipid analogs still unspecifically bound to the membrane or membrane proteins. Usually, FCS data was recorded only for a short period of time (10–15 s) to avoid distortion of the correlation data by these bright particles. On the other hand, affected FCS data was discarded.

## 2.2. STED nanoscopy

The concept of STED nanoscopy allows creating fluorescence interrogation observation volumes in the far-field at nanometric scales (Hell, 2009; Hell & Wichmann, 1994). In its usual implementation, an additional laser is added to a conventional point-scanning far-field microscope which inhibits fluorescence emission everywhere but at the center of the of the exciting laser focus (Fig. 1.3A). The wavelength of this second laser is tuned to the red edge of the fluorophore's emission spectrum and induces the stimulated de-excitation of the fluorophore's excited (and fluorescent) electronic state. By detecting only the spontaneous (and not the stimulated emission), the registered signal is efficiently decreased and completely switched off by increasing the intensity of the STED laser. The introduction of a phase plate into the STED beam realizes a distortion of its wavefront and an intensity distribution of the focus that features one or several local zeros, such as a doughnut-shaped intensity distribution (Fig. 1.3A). Consequently, increasing the intensity of the STED beams drives the area in which fluorescence emission is still allowed to smaller and smaller subdiffraction scales (Fig. 1.3B). The observation area of the STED microscope is therefore tuned by the intensity of the STED laser, which is one of the central features of STED-FCS.

In the STED-FCS experiments on membranes so far, the diffraction-limited fluorescence excitation spot was overlaid with the doughnut-shaped intensity distribution of the STED light, rendering an observation area of subdiffraction size along all lateral directions (Fig. 1.3B). The size of the observation area can be dynamically tuned by adjusting the STED beam intensity. Figure 1.3B shows the tuning of the observation area of the STED nanoscope by the STED intensity by plotting the diameter-square $d^2$ of the observation areas versus the time-averaged STED beam power $P_{\text{STED}}$. The diameters $d$ were estimated by scanning approximately 20 nm large fluorescent crimson beads for various $P_{\text{STED}}$ and by determining their full-width-at-half-maximum diameters $d$ in the resulting images. The resulting diameters could be described by the STED resolution scaling (Abbe, 1873; Harke et al., 2008; Hell, 2004).

$$\frac{d^2}{{d_0}^2} = \frac{1}{1 + P_{\text{STED}}/P_{\text{SAT}}} \qquad [1.1]$$

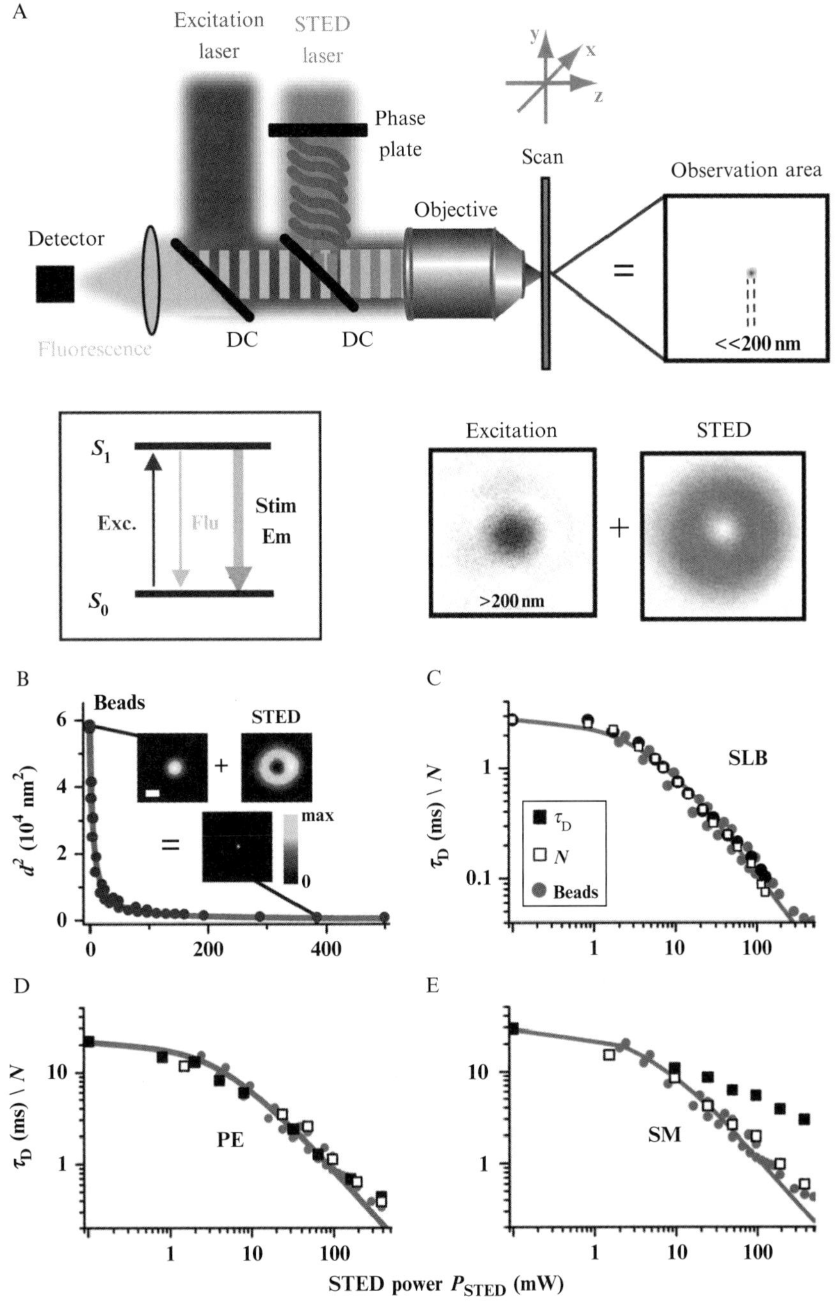


**Figure 1.3** Principle of STED nanoscopy and STED-FCS. (A) Besides the fluorescence excitation laser, a second STED laser is introduced into a conventional microscope (such as a confocal microscope), whose wavefront is (by introducing a phase plate) altered in

$d_0$ is the diameter of the confocal recordings. The saturation power $P_{SAT}$ is a characteristic of the fluorophore in use and given for a certain spatial distribution of the STED light (Harke et al., 2008). With a maximum time-averaged laser power $P_{STED} = 500$ mW in the sample space, diameters of the observation area could be tuned between $d_0 \approx 250$ nm of the diffraction-limited case down to $d \approx 30$ nm, which corresponded to a 70-fold reduced area.

## 2.3. STED-FCS: The principle

FCS analyses characteristic fluctuations $\delta F(t) = \langle F(t) \rangle - F(t)$ in the fluorescence signal $F(t)$ over the time $t$ about an average value $\langle F(t) \rangle$ by calculating the second-order autocorrelation function $G(t_c) = 1 + \langle \delta F(t)\ \delta F(t+t_c) \rangle / \langle F(t) \rangle^2$ (Magde, Webb, & Elson, 1972). Here, $t_c$ represents the correlation lag time. Triangular brackets indicate time averages. Fluctuations in the fluorescence signal are, for example, caused by characteristic variations in the concentration of fluorescent molecules, which diffuse in and out of the effective observation volume, and by transitions into and out of a dark (triplet) state. Following earlier work, the autocorrelation function taking into account diffusion dynamics, the dark (triplet) state population, and other kinetics causing changes in the fluorescence brightness can be approximated by (Elson & Magde, 1974; Widengren, Mets, & Rigler, 1995)

$$G(t_c) = 1 + (1/N) G_D(t_c) G_T(t_c) G_K(t_c) \qquad [1.2]$$

such a way that, for example, a doughnut-like intensity distribution of the focused beam is realized. The overlay of the excitation and STED foci inhibits fluorescence emission everywhere but at the focal center leaving a subdiffraction sized area where emission is still allowed: the new observation area. Inhibition of fluorescence emission by the STED laser is realized by stimulated emission, where the fluorophore in its excited state $S_1$ is de-excited to its ground state $S_0$ more efficiently than the spontaneous fluorescence. (B) The overlay of the foci of the excitation (upper left inset) and of the STED light (upper right inset) create observation areas of subdiffraction size (lower inset); scale bar = 200 nm. The time-averaged STED power $P_{STED}$ tunes the observation area $d^2$. The diameter $d$ has been inferred from scanning fluorescent crimson beads (gray circles). Fitting of Eq. (1.1) (gray line) to the data results in $P_{SAT} = 4$ mW. (C–E) Relative decrease of the transit time $\tau_D$ (black squares) and the particle number $N$ (white squares) of the STED-FCS measurements on SLBs (C), Atto647N-labeled PE in living PtK2 cells (D), and Atto647N-labeled SM in living PtK2 cells (E) in comparison to the decrease of the observation areas $d^2$ of the bead measurements (gray circles). Most data can be described by the STED resolution scaling law (Eq. 1.1 with $P_{SAT} = 4$ mW, gray line), confirming the STED effect and in the case of $\tau_D$ indicating free Brownian diffusion. Deviation in the case of SM depicts strong anomalous subdiffusion. (See Color Insert.)

where $N$ is the particle number (i.e., the mean number of fluorescent molecules in the observation volume, which is proportional to the concentration divided by the observation volume (or area for two-dimensional samples)), $G_D(t_c)$ the correlation term covering diffusion, $G_T(t_c) = 1 + T_{1eq}/(1 - T_{1eq})\exp(-t_c/\tau_T)$ the correlation term covering the dark (triplet) state population (with the equilibrium fraction $T_{1eq}$ of molecules in the dark triplet state, and the triplet correlation time $\tau_T$, characterized by the triplet population and depopulation kinetics), and $G_K(t_c) = 1 + K\exp(-t_c/\tau_K)$ is an additional kinetic term with amplitude $K$ and correlation time $\tau_K$. At the excitation intensities applied, $T_{1eq}$ and $\tau_T$ of the labels were usually around 0.1 and 5 μs, respectively and thus fixed throughout the analysis. Furthermore, the analysis of all FCS data recorded for lipid dynamics of Atto647N-labeled lipids in living cells had to include an additional kinetic term with amplitude $K = 0.05$–$0.1$ and correlation time $\tau_K = 50$–$150$ μs. This kinetic term might stem from the population of an additional dark state or conformational fluctuations of the dye-lipid system leading to changes in the fluorescence brightness. Its amplitude increased with hydrophobic environment as observed by measurements on model membranes at different humidity.

The simplest model to describe FCS data recorded for the two-dimensional diffusion of freely Brownian diffusing lipids in membranes is

$$G_D(t_c) = (1 + (t_c/\tau_D))^{-1} \quad \text{with} \quad \tau_D = d^2/(8\ln 2D) \qquad [1.3]$$

Here, $\tau_D$ is the average transit time of a fluorophore with the apparent diffusion coefficient $D$ through the observation area of diameter $d$. In STED-FCS, FCS data of the lipid diffusion is recorded for different STED powers $P_{STED}$, that is, for different diameters $d$. For a proper interpretation of the resulting $\tau_D(d)$ dependency, an accurate calibration of the STED experiment is required, that is, an accurate knowledge of the diameter $d(P_{STED})$.

The crimson bead calibration measurements yielded a saturation power $P_{SAT} = 4$ mW (Fig. 1.3B). However, this may be different for other fluorophores such as the organic dyes Atto647N used as a fluorescent label throughout most of our experiments. FCS was used to analyze the $d(P_{STED})$-tuning characteristics of the lipid analogs. From FCS, one can extract the average transit time $\tau_D$ through the focal spot as well as the average particle number $N$ in the observation area. Values of $N$ in general (Eq. 1.2) and of $\tau_D$ only for free Brownian diffusion (Eq. 1.3) should scale linearly with $d^2$ (Wawrezinieck et al., 2005) and thus should also obey the STED resolution scaling:

$$\tau_D/\tau_{D0} = N/N_0 = \frac{1}{1 + P_{STED}/P_{SAT}} \qquad [1.4]$$

where $\tau_{D0}$ and $N_0$ are the transit time and the particle number from the confocal measurement. To avoid any bias from out-of-axis uncorrelated background signal such as from internalized lipids, $N$ has been determined by fluorescence-intensity-distribution analysis (FIDA) (Kask, Palo, Ullmann, & Gall, 1999) rather than by FCS. FIDA allowed for a straightforward consideration of such background noise (Kastrup et al., 2005; Ringemann et al., 2009). Figure 1.3C shows the transit time $\tau_D$ and particle number $N$ extracted from correlation data recorded for an Atto647N-labeled PE in SLB at different STED powers $P_{STED}$. A one-component SLB model system on plasma-cleaned cover glass was chosen since it should ensure free Brownian diffusion of the lipids (Kahya, Scherfeld, Bacia, Poolman, & Schwille, 2003). Both $\tau_D$ and $N$ obeyed Eq. (1.4) and scaled with $P_{STED}$ in accordance to the crimson bead measurements with $P_{SAT} = 4$ mW. The concordant decrease of bead image size and of the lipid's $\tau_D$ and $N$ and the accordance to the STED resolution scaling confirmed that the reduction truly resulted from the STED effect and not from other photophysical or -chemical characteristics such as photobleaching, trapping, or heating (Eggeling et al., 2009). The same held true for the fluorescent PE lipid in the plasma membrane of the living PtK2 cells (Fig. 1.3D), indicating close to normal diffusion for the PE. The situation was different for an Atto647N-labeled SM (Fig. 1.3E). While the particle number $N$ decreased with $P_{STED}$ in accordance to the beads and Eq. (1.4), which again confirmed the STED effect, the transit time $\tau_D$ clearly deviated from the expected behavior. Values of $\tau_D$ were significantly larger than predicted for free Brownian diffusion, indicating anomalous subdiffusion.

## 2.4. STED-FCS: Anomalous subdiffusion

The diffusion term includes the time it takes for a lipid to diffuse through the observation area regarding the inhomogeneous intensity profile of the laser focus (or of the detection area) as well as possible heterogeneity in diffusion due to, for example, obstacles in the molecular diffusion pathway such as impermeable barriers, domain incorporation (where diffusion is slowed down), or binding to less mobile or immobile membrane compounds. Deviations from normal Brownian diffusion may be analyzed using the model of anomalous subdiffusion (Bouchaud & Georges, 1990; Feder et al., 1996; Saxton, 1994; Schwille et al., 1999; Wachsmuth, Waldeck, & Langowski, 2000). Due to the anomalous diffusion, the mean-square displacement $\langle r \rangle^2$ of the molecule's diffusion is not linear with time but follows a power law in time $t$ (Bouchaud & Georges, 1990; Feder et al., 1996; Saxton, 1994), $\langle r \rangle^2 = 4Dt^{\alpha}$, with an apparent (averaged) diffusion coefficient $D$ and an anomalous diffusion exponent $\alpha$. Denoted anomaly throughout, the inverse $(1/\alpha)$ describes the degree of hindered diffusion. While diffusion is free for $(1/\alpha) = 1$ and follows Brownian motion characterized by $D$, $(1/\alpha) > 1$ characterizes anomalous subdiffusion; the larger $(1/\alpha)$ the more hindered the diffusion. In FCS, anomalous diffusion is treated by introducing the anomalous diffusion exponent $\alpha$ into the term expressing the diffusion dynamics (Schwille et al., 1999; Wachsmuth et al., 2000).

$$G_{\mathrm{D}}(t_{\mathrm{c}}) = \left(1 + (t_{\mathrm{c}}/\tau_{\mathrm{D}})^{a}\right)^{-1} \quad \text{with} \quad \tau_{\mathrm{D}} = d^2/(8 \ln 2D) \qquad [1.5]$$

$\tau_{\mathrm{D}}$ is the average transit time of a fluorophore with the apparent diffusion coefficient $D$ through the focal detection area of diameter $d$. For free Brownian diffusion, $D$ is constant for different $d$ and thus $\tau_{\mathrm{D}}$ scales linearly with the focal area $d^2$. This is different for anomalous diffusion, where $D$ varies with $d^2$ depending on the spatial and temporal characteristic of the process causing the anomaly (e.g., time span and size of a trap or dimension of an obstacle).

Figure 1.4A depicts exemplary FCS data of the confocal recordings of the PE and SM analog in the plasma membrane of living PtK2 cells. The inflection point of the correlation curves gives an estimate of the average transit time, which was slightly larger for the SM than for the PE lipid. A fit of Eq. (1.5) to the data resulted in transit times $\tau_{\mathrm{D}} \approx 20$ ms for PE and $\approx 30$ ms for SM. However, for both PE and SM the anomaly $(1/\alpha)$ was close to one, delivering no indication of heterogeneous diffusion. The diffraction-limited confocal recording with a $>200$ nm large observation area could not clearly explore the cause of the decreased mobility and could not distinguish between slower but free diffusion and anomalous subdiffusion due to, for example, nanoscopic obstacles or traps, because they averaged over such nanoscopic details of molecular diffusion. In contrast, nanoscale observation areas created by STED directly provided the desired details, as depicted in Fig. 1.4B. Correlation data recorded for $P_{\mathrm{STED}} = 160$ mW ($d \approx 40$ nm) showed a clear difference between PE and SM diffusion. While the PE curve could still be described with $(1/\alpha) \approx 1$, the SM data rendered $(1/\alpha) \approx 1.5$ and a significantly slower transit time $\tau_{\mathrm{D}}$ compared to PE.

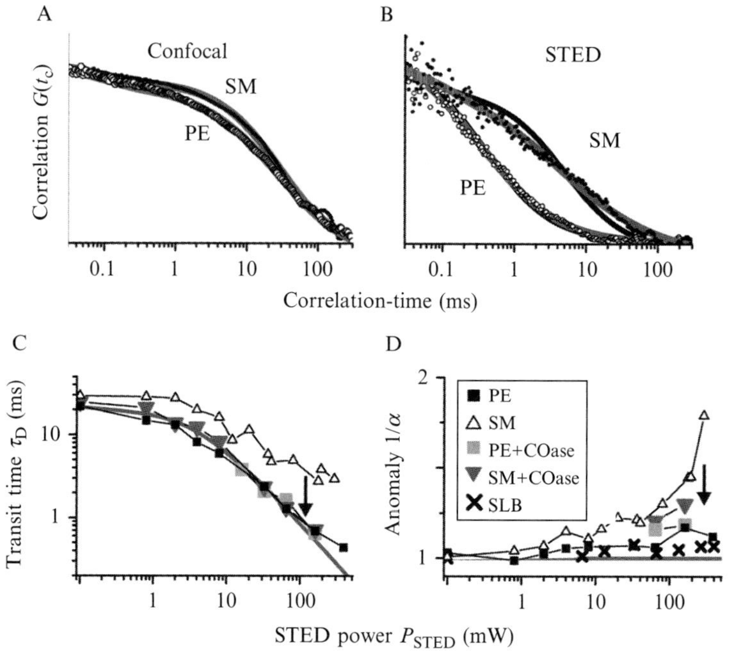


**Figure 1.4** STED-FCS reveals anomalous diffusion of SM in the plasma membrane of living PtK2 cells. (A and B) Exemplary autocorrelation data for confocal (A) and STED ($P_{STED} = 160$ mW, $d \approx 40$ nm) (B) recording. The confocal and the Atto647N-labeled PE STED-FCS data but not the Atto647N-labeled SM STED-FCS data could be described by an anomaly $(1/\alpha) = 1$, indicating normal diffusion (black lines, Eq. 1.5). An anomaly of $(1/\alpha) \approx 1.5$ was necessary to fit the SM STED-FCS data (gray line). (C and D) Dependence of the average transit time $\tau_D$ (C) and anomaly $(1/\alpha)$ (D) determined (Eq. 1.5) from correlation data of PE (black squares) and SM (open triangles), and of PE (gray squares) and SM (gray triangles) after cholesterol depletion by COase treatment on increasing STED power $P_{STED}$, that is, decreasing observation area. For comparison, values are also given for free diffusion on SLB (cross). PE diffused almost normal, while diffusion of SM was hindered, depending on cholesterol (arrow). The difference in PE and SM diffusion became obvious only for $P_{STED} > 10$ mW, that is, for small observation areas tuned by STED.

The dynamical tuning of the observation spot by STED allowed for a more detailed view on this anomaly. Transit times $\tau_D$ (Fig. 1.4C) and anomalies $(1/\alpha)$ (Fig. 1.3D) determined from PE and SM correlation data (Eq. 1.5) for increasing STED power $P_{STED}$ showed an increasing difference between PE and SM diffusion with decreasing observation area, that is the distinction of diffusion heterogeneity due to nanoscopic obstacles improved with spatial resolution. Obviously, only ensuring a significantly reduced transient time of free diffusion allowed distinguishing between normal and anomalous diffusion. Strikingly, the difference between SM and PE diffusion almost diminished after depletion of cholesterol from the plasma membrane by cholesterol oxidase (COase). After COase treatment SM revealed the same transit times as PE (Fig. 1.4C), which may be an indication for 'raft'-like interactions (Pike, 2006).

The anomaly values of SM as well as of PE after COase treatment were slightly larger ($(1/\alpha) > 1.2$) than for normal PE diffusion ($(1/\alpha) \approx 1.1$–$1.15$) (Fig. 1.4D). Besides possible incomplete cholesterol depletion, such slight anomalous behavior may be explained by changes in the phase behavior of the membrane due to cholesterol depletion, for example, by the creation of solid like regions in the plasma membrane (see, e.g., Nishimura et al., 2006). Nevertheless, a significant change in the diffusion coefficient of SM or PE following the cholesterol depletion were not observed (Fig. 1.4C).

Both, the intensity profile of the laser spot, which establishes the fluorescence emission profile (at the excitation intensities applied we can safely assume a linear dependency of the fluorescence emission on the laser intensity) and the detection profile of the microscope (given by the point-spread function of the microscope objective and the transmission function of the confocal pinhole) determine the spatial profile of the effective focal spot. Equation (1.5) assumes a Gaussian-intensity profile, which is a good approximation for confocal FCS experiments at low laser intensities (Aragon & Pecora, 1975; Enderlein, Gregor, Patra, Dertinger, & Kaupp, 2005). For very small observation areas created by STED, this assumption slightly deviated from the actual shape of the focal spots (Kastrup et al., 2005). Consequently, the analysis of the correlation data may show an anomaly artifact. However, this bias hardly influenced the correlation analysis in the case of hindered diffusion because the focal passage was dominated by trapping, which, for example, resulted in values of $1/\alpha > 1.4$. These values exceeded by far the artifact due to non-Gaussian spot profiles where $1/\alpha \approx 1.1$ (compare SLB data in Fig. 1.4D).

## 2.5. STED-FCS: Quantification of anomalous subdiffusion

There are further ways to parameterize anomalous subdiffusion than just the anomaly parameter ($1/\alpha$). Having calibrated the STED nanoscope for the lipid marker Atto647N by SLB or bead measurements and therefore knowing the dependency of the diameter $d$ of the observation area versus STED power $P_{STED}$ (Fig. 1.3), one can now plot the average transit time $\tau_D$ (obtained from fitting Eq. (1.5) to the correlation data) against the focal area $d^2$ (Fig. 1.5A). The plot exhibits the $\tau_D(d^2)$ dependencies expected from

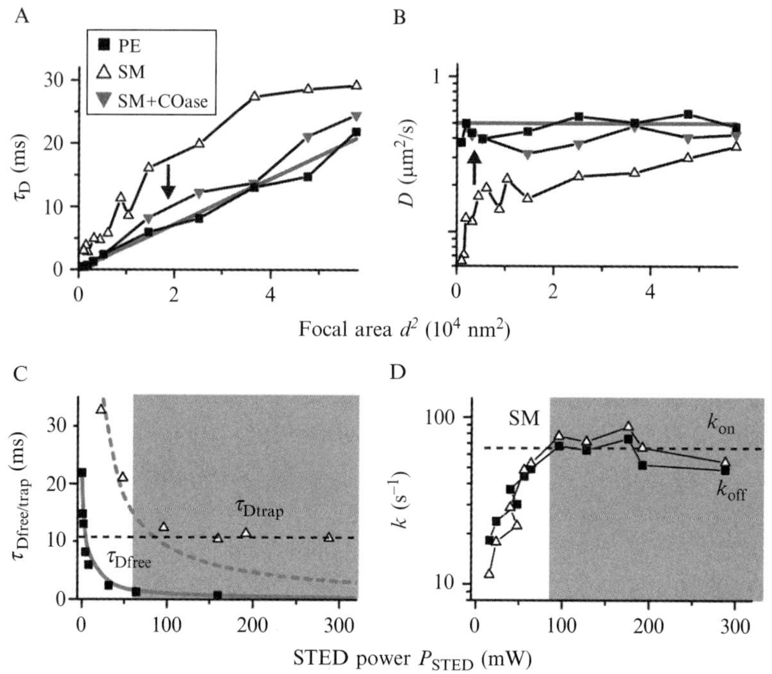


**Figure 1.5** Different visualization of anomalous subdiffusion. (A and B) Anomaly model (Eq. 1.5): Dependence of the average transit time $\tau_D$ (A) and apparent diffusion coefficient $D$ (B) of PE (black squares), SM (open triangles), and SM after cholesterol depletion by COase treatment (gray triangles) on the observation area $d^2$ formed by STED. Constraint diffusion of SM became obvious by a nonlinear dependence of $\tau_D$ and a nonconstant $D$. (C) Model of two-diffusion-modalities (Eq. 1.6): transit times $\tau_{Dfree}$ (black squares) and $\tau_{Dtrap}$ (open triangles) of SM for increasing STED powers $P_{STED}$, that is, decreasing observation area. The clear deviation of $\tau_{Dtrap}$ from normal free diffusion (gray dotted line, Eq. 1.6 with $\tau_{Dtrap0} = 240$ ms and $P_{SAT} = 4$ mW) indicated transient binding of SM with trapping times of approximately 10 ms (horizontal dotted line). (D) Binding model (Eq. 1.7): on- $k_{on}$ and off-rates $k_{off}$ of transient SM binding for increasing STED powers $P_{STED}$, that is, decreasing spot size, revealing $k_{on} \approx k_{off} \approx 60$–$80\ s^{-1}$ (horizontal dotted line). Both the two-diffusion-modality and the binding model could accurately be used only for $P_{STED} > 60$–$70$ mW (gray shaded areas), that is, for observation areas with $d < 60$–$70$ nm where $\tau_{Dfree} < 1$–$1.5$ ms.

theory (Wawrezinieck et al., 2005): linear relationships for the almost free diffusion of PE and of SM after COase treatment and the nonlinear characteristic of SM predicted for transient incorporation into molecular complexes or nanodomains. Meshwork-based or hopping-like diffusion due to compartmentalization of the plasma membrane by the underlying cytoskeleton network would result in an exponential-like increase of $\tau_D$ with $d^2$, which was not observed in this case.

The relation $\tau_D(d^2)$ further allows to calculate the apparent diffusion coefficient $D = d^2/(8 \ln 2\tau_D)$ (Eq. 1.5) for every $d$. A constant value $D(d)$ is to be expected for normal free diffusion, while an enhanced hindrance in diffusion results in a reduced mobility and thus in reduced values of $D$. A decrease of $D$ for small observation areas $d^2$ reveals transient incorporation into molecular complexes or nanodomains, while an increase of $D$ for small observation areas $d^2$ may be explained by meshwork-based or hopping-like diffusion (Mueller et al., 2011). Figure 1.5B clearly confirms that only the reduction of $d$ far beyond the diffraction limit realized an accurate disclosure of hindered diffusion due to transient trapping with strong indications for SM and slight indications for PE.

The $\tau_D(d^2)$ and $D(d^2)$ dependencies showed that heterogeneity in SM diffusion was caused by transient formation of cholesterol-assisted molecular complexes, such as lipid– protein binding or lipid shells. During trapping, the lipid dwelled only within areas of <20 nm diameter. This knowledge allows using other analysis approaches of the STED-FCS data to quantify the kinetic parameters of the trapping interaction.

(i) *Two-diffusion-modalities*: On one hand, two dissimilar modalities of focal transits may be assumed: one modality with transit time $\tau_{Dfree}$, where the lipid happens to diffuse freely, and the other modality with a more complex transit time $\tau_{Dtrap}$ stemming from hindered diffusion.

$$G_D(t_c) = A_{free}(1 + t_c/\tau_{Dfree})^{-1} + A_{trap}\left(1 + t_c/\tau_{Dtrap}\right)^{-1} \qquad [1.6]$$

$A_{free}$ and $A_{trap}$ are the amplitudes associated with the transit times $\tau_{Dfree}$ and $\tau_{Dtrap}$. While $\tau_{Dfree}$ is given by free Brownian diffusion (with free diffusion constant $D_{free}$ excluding anomalous subdiffusion) and thus follows the linear dependence on the effective focal area $\tau_{Dfree} \sim d^2$, $\tau_{Dtrap}$ reveals the combined transit time in the case of hindered lipid diffusion. With $A_{free} + A_{trap} = 1$, the value of $A_{trap}$ correlates with the amount of trapping events. $\tau_{Dfree}$ is usually fixed to values $\tau_{Dfree} = d^2/(8 \ln 2D_{free})$ expected for a value of $D_{free}$ that was estimated from measurements at large $d$. Instead of following the resolution scaling by STED (Eq. 1.4) as for $\tau_{Dfree}$, values of $\tau_{Dtrap}$ leveled off for large $P_{STED}$ (Fig. 1.5C). This characteristic behavior excluded diffusing and temporally stable lipid–protein complexes or domains but rather demonstrated highly localized trapping on a timescale of $\sim$10 ms. A clear separation of the two-diffusion-modalities $\tau_{Dfree}$ and $\tau_{Dtrap}$ and, therewith, an accurate determination of the trapping time from the two-diffusion-modality model was only possible for tiny observation areas as well (Fig. 1.5C gray shaded area).

(ii) *Binding model*: A molecular trapping mechanism causing the anomalous diffusion may further be validated when analyzing the correlation data with a model introducing transient binding of the diffusing lipid to a fixed or comparatively slow moving particle (Michelman-Ribeiro et al., 2009; Ringemann et al., 2009). One then can introduce an effective encounter rate $k_{on}$ and a dissociation rate $k_{off}$ of this complex. If the trapping time $1/k_{off}$ is much longer than the average time the freely diffusing molecule would spend in the observation focus, the diffusion is reaction-dominated and we can describe the correlation function by

$$G_D(t_c) = (1 - B)(1 + t_c/\tau_{Dfree})^{-1} + B\exp(-k_{off}t_c) \qquad [1.7]$$

where $\tau_{Dfree}$ is the average transit time for free diffusion (with free diffusion constant $D_{free}$) and $B = k_{off}/(k_{off} + k_{on})$ is the fraction of bound molecules (Michelman-Ribeiro et al., 2009; Ringemann et al., 2009). $\tau_{Dfree}$ is usually fixed to values $\tau_{Dfree} = d^2/(8 \ln 2D_{free})$ expected for a value of $D_{free}$ that was estimated from measurements at large $d$. Figure 1.5D shows values of $k_{on}$ and $k_{off}$ determined by fitting Eq. (1.7) to the correlation data of SM for increasing $P_{STED}$. Note that Eq. (1.7) is only valid for reaction-dominated diffusion, that is, for the case that the transit time of free diffusion is much shorter than the trapping time $1/k_{off}$. This was only the case for $P_{STED} > 60$–$70$ mW, that is, for observation areas with $d < 60$–$70$ nm and $\tau_{Dfree} < 1$–$1.5$ ms (gray shaded area). The values of $k_{on} \approx k_{off} \approx 60$–$80$ s$^{-1}$ in this range were in line with the dwell times $\tau_{Dtrap} \approx 10$ ms of trapped molecular transits estimated from the two-diffusion-modality model.

# 3. LIPID MEMBRANE DYNAMICS

## 3.1. Anomaly of PE diffusion

At STED powers $P_{STED} > 100$ mW, that is, at observation areas of diameter $d < 100$ nm diffusion of PE showed a slight deviation from normal behavior with values of $\tau_D$ deviating from the ideal STED resolution scaling (Fig. 1.4C), values of $(1/\alpha) > 1.1$ (Fig. 1.4D), and slightly decreasing values of the diffusion coefficient $D$ for small focal spots $d^2$ (Fig. 1.5B). While only a minor part of this behavior may be ascribed to an analysis artifact (compare discussion to Eq. 1.5), the slight decrease of $D$ for small observation areas $d^2$ rather indicated transient trapping for PE also. However, the observations may not be explained by a hopping-like diffusion, as has been reported for other phosphoglycerolipids in different cells (Fujiwara et al., 2002). In that case, one should have observed an increase of $D$ for small observation areas $d^2$. In a previous analysis, on- and off-rates $k_{on} \approx 190\ s^{-1}$ and $k_{off} \approx 800\ s^{-1}$ of PE trapping, respectively, were determined (Ringemann et al., 2009). Due to the short trapping duration of $\sim 1$ ms, diffusion was not reaction-dominated anymore, which banned the application of Eq. (1.7) and required a more sophisticated analysis (Michelman-Ribeiro et al., 2009; Ringemann et al., 2009). Applying the model of two-diffusion-modalities (Eq. 1.6) resulted in rather inaccurate values of $\tau_{Dtrap} < 2$ ms, but nevertheless revealed a fraction $A_{free} > 80\%$ of free molecular transits as compared to a fraction $A_{free} = 40\%$ of hindered transits in the case of SM (Table 1.1). The same was observed in fast single-molecule studies (Sahl et al., 2010).

## 3.2. Consistency of the different analysis procedures

As outlined above, three different approaches could be applied to parameterize deviations from normal, free diffusion due to nanoscale trapping: (i) An anomaly model (Eq. 1.5) yielding parameters for the apparent diffusion coefficient $D$ and the anomaly $(1/\alpha)$, (ii) a model for two-diffusion-modalities (Eq. 1.6) yielding the average fraction $A_{trap}$ of molecular transits characterized by constrained diffusion with an average dwell time $\tau_{Dtrap}$, and (iii) a binding model (Eq. 1.7) yielding the on- and off-rates $k_{on}$ and $k_{off}$. Figure 1.6 shows that all three analysis methods quantified the extent of nanoscale trapping equally well: values of $A_{trap}$, $\tau_{Dtrap}$, $(1/k_{off})$, of the equilibrium binding constant $K = k_{on}/k_{off}$, and of $(1/\alpha)$ of the STED recordings and of the ratio $D_{STED}/D_{Conf}$ of the apparent diffusion coefficients $D_{Conf}$ and $D_{STED}$ determined for the confocal and STED recordings, respectively, correlate well. These values were gathered for different lipids, as outlined further on (Table 1.1). Strikingly, $K$ tended to 0 for $A_{trap} \to 0$, $(1/k_{off})$ to 0 for $\tau_{Dtrap} \to 0$, and $D_{STED}/D_{Conf}$ and $(1/\alpha)$ to 1 for $K \to 0$, as expected for normal free diffusion.

## 3.3. Environmental parameters

Using STED-FCS, one may also check the sensitivity of SM trapping on some environmental parameters. Figure 1.7 shows values of the apparent diffusion coefficient $D$ and of the anomaly $(1/\alpha)$ determined from STED recordings at $P_{STED} \approx 160$ mW ($d \approx 40$ nm) versus different incubation times by the cholesterol-depleting additives COase and β-cyclodextrin (β-CD) (A), versus increasing excitation power $P_{exc}$ (B), and at different sample temperatures $T$ (C). As expected for the cholesterol-assisted interaction, $D$ increased and $(1/\alpha)$ decreased with increasing cholesterol depletion, reaching

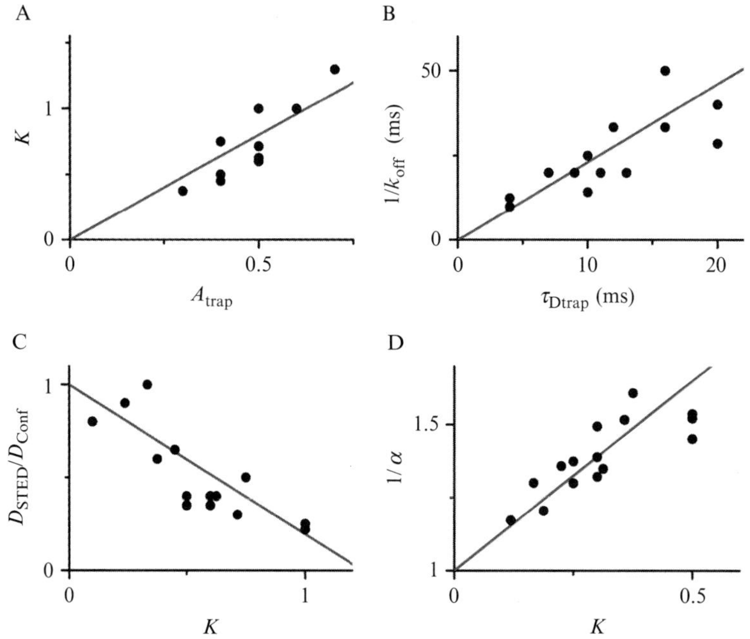


**Figure 1.6** Consistency of different analysis procedures. Correlation of (A) the equilibrium constant $K = k_{on}/k_{off}$ from the binding model (Eq. 1.7) and the amplitude $A_{trap}$ of molecular transits with constraint diffusion from the two-diffusion-modality model (Eq. 1.6), (B) the off-rate $1/k_{off}$ from the binding model (Eq. 1.7) and the transit time $\tau_{Dtrap}$ from the two-diffusion-modality model (Eq. 1.6), (C) the ratio $D_{STED}/D_{Conf}$ of diffusion coefficients determined for the STED and confocal recordings from the anomaly model (Eq. 1.5) and the equilibrium constant $K = k_{on}/k_{off}$ from the binding model (Eq. 1.7), and (D) the anomaly ($1/\alpha$) determined for the STED recordings from the anomaly model (Eq. 1.5) and the equilibrium constant $K = k_{on}/k_{off}$ from the binding model (Eq. 1.7). Data taken from measurements on different fluorescent lipid analogs.

levels of normal free diffusion after approximately 30–50 min. At the experimental conditions, cholesterol depletion by β-CD was slightly slower than that by COase. Following the procedure to determine on- and off-rates of complex formation (Eq. 1.7), the decrease of anomaly with cholesterol depletion seemed to be caused by a faster off-rate. For example, $k_{off}$ increased from $\approx$60–80 $s^{-1}$ to $>$150 $s^{-1}$ after 30 min of COase treatment, whereas $k_{on}$ stayed approximately constant ($\approx$70 $s^{-1}$).

Larger excitation powers $P_{exc}$ result in an increased photobleaching of the fluorescent marker, especially for slowly moving particles (Eggeling, Widengren, Rigler, & Seidel, 1998). The apparent transit time of photobleached fluorophores is shortened, which lowers the observed average transit time $\tau_D$ and increases the apparent diffusion constant $D$ when applying

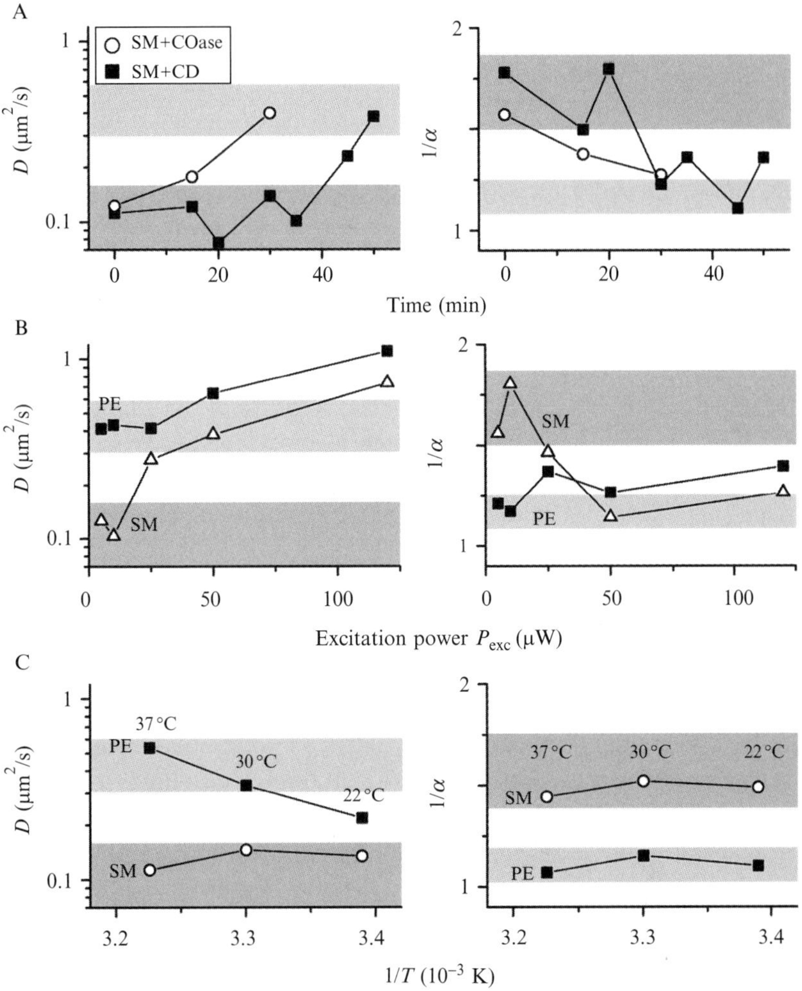


**Figure 1.7** Dependence of the molecular trapping of SM on environmental parameters. Diffusion coefficient $D$ (left panels) and anomaly ($1/\alpha$) (right panels) of the STED recordings ($P_{STED} = 160$ mW, $d \approx 40$ nm) of (A) SM for different incubation times of the cholesterol-depleting additives COase (open circles) or β-CD (black squares), (B) SM (open triangles) and PE (black squares) for increasing excitation power $P_{exc}$ and thus increasing levels of photobleaching, and (C) SM (open triangles) and PE (black squares) for different sample temperatures $T$. Shaded areas indicate usual value ranges of PE (light gray) and SM diffusion (dark gray).

Eq. (1.5) (Eggeling et al., 1998). We have shown previously (Ringemann et al., 2009) that for a bias-free analysis of the FCS data much larger excitation powers may be applied in the case of STED compared to conventional confocal recordings, as molecules move much faster through the reduced observation area. Large $P_{exc}$ led to a more pronounced bleaching of the trapped SM lipids as compared to the quickly moving PE lipids. Consequently, the dwell time and the fraction of trapped SM molecules were increasingly biased at large $P_{exc}$, weakening the observed trapping by lowering $D$ and $(1/\alpha)$: diffusion of PE and SM was almost similar at $P_{exc} > 30$ μW (Fig. 1.7B). In order to perform accurate measurements, the excitation power is usually set to below 10 μW. Another option would be scanning of the laser spots during recording of the correlation data (Ries & Schwille, 2006; Ruan, Cheng, Levi, Gratton, & Mantulin, 2004).

While the apparent diffusion coefficient $D$ of PE increased with temperature $T$, that of SM stayed constant (Fig. 1.7C). The Arrhenius behavior confirmed the (almost) free diffusion characteristics of the PE lipids in the plasma membrane. In contrast, the insensitivity of the SM dynamics on $T$ demonstrated that temperature did not influence the formation of the lipid complexes, at least not in the measured temperature range of 22–37 ° C.

### 3.4. Molecular dependence

To explore the molecular details responsible for the nanoscale trapping (Mueller et al., 2011), values of the apparent diffusion coefficient $D$ and the anomaly $(1/\alpha)$ (Eq. 1.5), of the fraction $A_{trap}$ and the dwell time $\tau_{Dtrap}$ (Eq. 1.6), and of the rate constants $k_{on}$ and $k_{off}$ and the equilibrium constant $K = k_{on}/k_{off}$ of the transient complex (Eq. 1.7) of various lipids for conventional diffraction-limited and for STED recordings were determined (Fig. 1.8 and Table 1.1). The lipids differed in their head-group structure, in the number, the length and the saturation degree of the acyl chains, and the type and position of the label (Fig. 1.8A). As outlined above, anomalous subdiffusion due to transient molecular trapping resulted in decreased values of $D$ and increased values of $(1/\alpha)$ for the STED recordings as compared to the confocal ones.

At first, it is confirmed that the diffusion characteristics of our fluorescent lipid analogs were not dye-induced. For example, the difference in PE and SM dynamic remained unchanged when changing the labeling position of the Atto647N marker, that is, when either labeling at the head group (-h) or by replacing the acyl chain (-c), when applying another lipophilic marker (Kolmakov et al., 2010), or when both, PE and SM, were head-group labeled with a very polar dye Atto532 (Atto532-PE-h and Atto532-SM-h). From these similarities, one may conclude a minimal influence of the dye label on the lipid dynamics of PE and SM. Further, the Lyso derivatives of PE and

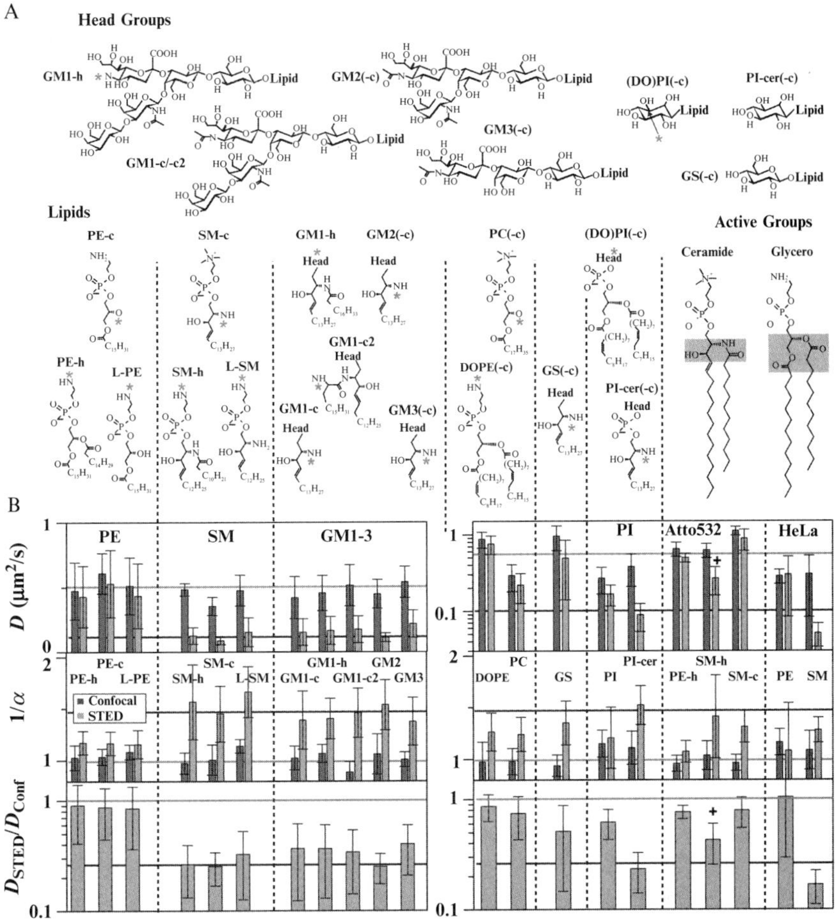


**Figure 1.8** Molecular dependence of nanoscale trapping. (A) Structures of the different fluorescent lipid analogs used in the STED-FCS studies. Upper panel lists the different head groups and right panel indicates the ceramide and the glycerol group of the sphingo- and phosphoglycerolipids, respectively (shaded area). The star indicates the position of the dye. In the studies, all lipids were labeled with Atto647N unless otherwise stated. PE (DPPE), phosphatidylethanolamine; PC (DSPC), phosphatidylcholine; PI, phosphatidylinositol; DOPE, unsaturated phosphatidylethanolamine; SM, sphingomyelin; GS, galactosylceramide; PI-Cer, ceramide phosphorylinositol; and the gangliosides GM1, GM2, and GM3. -c, labeling via acyl-chain replacement; -h, head-group labeling; L-, Lyso derivatives. (B) Diffusion coefficient $D$ (upper panel) and anomaly ($1/\alpha$) (middle panel) of the confocal (dark gray columns) and STED recordings (light gray columns; $P_{STED} = 160$ mW, $d \approx 40$ nm) and the ratio $D_{STED}/D_{Conf}$ (lower panel) of diffusion coefficients determined for the STED and confocal recordings for the different lipids. Decreased values of $D$ and increased values of ($1/\alpha$) for the STED compared to the confocal recordings indicate anomalous subdiffusion due to transient trapping.

SM (L-) showed the same diffusion behavior as their double-chained derivatives, respectively, that is, even without the long acyl chain, trapping of SM was unchanged. The same held true for the different kinds of gangliosides GM1, GM2, and GM3. No matter whether labeled at the polar head group (-h) (using Atto647N or polar dyes such as Atto532 (Eggeling et al., 2009; Polyakova et al., 2009)) or by acyl-chain replacement (-c) (using Atto647N), nor for the different structures of the head group (compare GM1, GM2, and GM3), the gangliosides were trapped and characterized by anomalous subdiffusion. Even the labeling by Atto647N at one of the long acyl chains (-c2, i.e., GM1-c2 carried the dye anchor in addition to the two native chains) did not change the dynamical behavior of the lipid.

In conclusion, the trapping was negligible for all PE analogs and dominant for all SM or ganglioside analogs. A common difference between the structures of the PE and of the SM and gangliosides is the ceramide (or sphingosine) structure of the SM and gangliosides close to the water–lipid interface, which is absent in the PE lipid and may facilitate hydrogen bonds to (endogenous) membrane components such as cholesterol, other (sphingo)lipids or proteins (compare Fig. 1.8A) (Pascher, 1976). Thus, on the first view the ceramide structure seemed to be the only key element in the cholesterol-assisted nanoscale trapping observed in the STED-FCS studies.

The STED-FCS studies of other fluorescent lipid analogs confirmed the above assumption (Fig. 1.8). The unsaturated PE derivative DOPE as well as another phosphoglycerolipid phosphocholine (PC) showed nearly normal diffusion similar to PE. Carrying the ceramide moiety, galactosylsphingosine (GS) behaved similar to the other gangliosides. On the other hand, trapping of the phosphoinositol (PI) analogs was only as pronounced as that of SM, if the ceramide moiety was artificially introduced (PI-cer compared to PI; the ceramide is absent in endogenous PI). Further, the difference between PE and SM was not specific for PtK2 cells, but also pronounced in the human HeLa cell line (Fig. 1.8) or in mammalian BHK (baby hamster kidney) cells (data not shown).

Yet, slight differences between the various phosphoglycero- and sphingo- or gangliosides arose (Table 1.1 and Fig. 1.8). (i) Diffusion of unsaturated DOPE was somewhat faster ($D_{\mathrm{free}} = 1.2\ \mu m^2/s$) and that of PC somewhat slower ($D_{\mathrm{free}} = 0.35\ \mu m^2/s$) than of PE ($D_{\mathrm{free}} = 0.5\ \mu m^2/s$). One can only speculate on reasons for this difference: On one hand, the unsaturated chains may induce an increased mobility of DOPE; on the other hand, the larger head group or the slightly increased chain length of PC ($C_{17}$ compared to $C_{15}$ for PE) might slow down its diffusion. When using

**Table 1.1** Parameters of trapping interactions of different fluorescent lipid analogs in the plasma membrane of living cells as extracted from the STED-FCS analysis

| Lipid | $D$ ($\mu m^2/s$) | $D_{STED}/D_{Conf}$ | $Recov_{COase}$ (%) | $A_{trap}$ | $\tau_{Dtrap}$ (ms) | $k_{on}$ ($s^{-1}$) | $k_{off}$ ($s^{-1}$) | $K$ |
|---|---|---|---|---|---|---|---|---|
| Phospholipids | | 0.9 | 0 | <0.2 | – | 190* | 800* | 0.2 |
| PE | 0.5 | | | | | | | |
| DOPE | 1.2 | | | | | | | |
| PC | 0.35 | | | | | | | |
| SM | 0.5 | 0.25 | 100 | 0.6 | 10 | 70 | 70 | 1 |
| +COase | 0.5 | 1 | | <0.2 | – | ≈60 | >150 | <0.35 |
| SM (HeLa) | 0.3 | 0.2 | 100 | 0.7 | 12 | 40 | 30 | 1.1 |
| +COase | 0.3 | 1 | | <0.2 | – | – | – | – |
| SM (Atto532) | 1.1 | 0.8 | | 0.2 | 7 | 5 | 50 | 0.1 |
| GM1 | 0.5 | 0.35 | 10 | 0.5 | 11 | 30 | 50 | 0.6 |
| +COase | 0.5 | 0.4 | | 0.5 | 9 | 30 | 50 | 0.6 |
| GM2 | 0.5 | 0.22 | 20 | 0.6 | 16 | 30 | 30 | 1 |
| +COase | 0.5 | 0.5 | | 0.4 | 16 | 15 | 20 | 0.75 |
| GM3 | 0.5 | 0.35 | 0 | 0.4 | 13 | 25 | 50 | 0.5 |
| +COase | 0.5 | 0.3 | | 0.5 | 20 | 25 | 35 | 0.7 |
| GS | 1 | 0.4 | 30 | 0.5 | 4 | 50 | 80 | 0.6 |
| +COase | 1 | 0.65 | | 0.4 | 4 | 45 | 100 | 0.45 |
| PI | 0.25 | 0.6 | <0 | 0.3 | 10 | 15 | 40 | 0.4 |
| +COase | 0.25 | 0.4 | | 0.4 | 12 | 15 | 30 | 0.5 |
| PI-cer | 0.5 | 0.25 | 20 | 0.5 | 25 | 35 | 35 | 1 |
| +COase | 0.5 | 0.35 | | 0.5 | 20 | 20 | 30 | 0.7 |

*Values determined by a more sophisticated analysis (see text).
$D$, apparent diffusion coefficient of confocal recordings (Eq. 1.5).
$D_{STED}/D_{Conf}$, ratio of the apparent diffusion coefficients $D_{Conf}$ and $D_{STED}$ determined for the confocal and STED recordings, respectively—the smaller $D_{STED}/D_{Conf}$, the larger the extent of trapping.
$Recov_{Chol} = \Delta D(STED)/\Delta D(conf)$ relating the changes of the apparent diffusion coefficient following COase treatment as observed for the confocal ($D$(conf)) and STED recordings ($D$(STED)). If we assume that $\Delta D$(conf) was mainly due to a change in free diffusion, $Recov_{Chol}$ would quantify the change in binding affinity due to cholesterol depletion. "Recover" in the main text.
$A_{trap}$, $\tau_{Dtrap}$, molecular fraction $A_{trap}$ and the dwell time $\tau_{Dtrap}$ of trapped transits (Eq. 1.6).
$k_{on}$, $k_{off}$, and $K = k_{on}/k_{off}$: on- and off-rate constants and equilibrium constant of the transient lipid complexes (Eq. 1.7).

a $C_{11}$- instead of a $C_4$-linker to tag Atto647N to PC, diffusion was further slowed down ($D_{free}=0.15\ \mu m^2/s$) (data not shown). (ii) Free diffusion of PE and SM in HeLa cells was slightly slower ($D_{free}=0.3\ \mu m^2/s$) than in PtK2 cells ($D_{free}=0.5\ \mu m^2/s$), and, although the trapping was similar ($A_{trap}\approx 60$–70% and $K=k_{on}/k_{off}\approx 1$), slightly lower on- and off-rates of the transient complex were determined ($k_{on}$, $k_{off}\approx 40$, 30 $s^{-1}$ compared to $k_{on}$, $k_{off}\approx 70$, 70 $s^{-1}$ in PtK2 cells). (iii) After labeling at the water–lipid interface (i.e., by acyl-chain replacement), trapping of the Atto532-SM analog was less pronounced ($A_{trap}\approx 20\%$, $\tau_{Dtrap}\approx 7$ ms, and $K\approx 0.1$) and diffusion much faster ($D_{free}\approx 1\ \mu m^2/s$) than for the Atto647N-SM or the head-group labeled Atto532-SM analog. This difference may stem from a less tight membrane anchoring of the polar Atto532 sphingolipid analog compared to the lipid derivatives labeled by lipophilic markers such as Atto647N, which seemed to be a proper label for acyl-chain replacement (Kolmakov et al., 2010). (iv) Trapping of the ganglioside GM1 was slightly less pronounced ($K\approx 0.6$) than for SM ($K\approx 1$). The polar sugar head group of GM1 is its most distinct difference to SM, seemingly influencing the trapping behavior: the relatively large size of the head group may on one hand destabilize complexes, but its numerous hydroxyl-groups may on the other hand facilitate molecular interactions via hydrogen bonding. In the case of the GM1, we determined slightly lowered off- ($k_{off}\approx 50\ s^{-1}$) and more significantly lowered on-rates of transient trapping ($k_{on}\approx 30\ s^{-1}$) compared to SM ($k_{off}\approx k_{on}\approx 70\ s^{-1}$). (v) The influence of the size and the structure of the sugar head group became more pronounced when studying the slight differences between the different gangliosides. Missing one galactose residue, trapping of GM2 was more pronounced ($K\approx 1$) than of GM1 ($K\approx 0.6$; we determined slightly longer traps of GM2: $k_{off}\approx 30\ s^{-1}$ compared to $\approx 50\ s^{-1}$ for GM1). However, the size of the sugar head group could not be the only ruler: Missing even one galactose and one *N*-acetylgalactosamine residue, the observed dynamics of GM3 were again almost indistinguishable from GM1. Further, carrying only one galactose group, free diffusion of GS was observed twice faster ($D_{free}=1\ \mu m^2/s$ compared to $0.5\ \mu m^2/s$ for the other gangliosides), and this lipid analog had a similar binding constant than GM1 ($K\approx 0.6$) but with a faster on- ($k_{on}\approx 50\ s^{-1}$) and off-rate ($k_{off}\approx 80\ s^{-1}$). Distinct structural differences of the head group must be responsible for these variations in dynamic properties and thus may reveal molecule-specific binding affinities. It is known that gangliosides facilitate distinct molecular pathways such as endocytosis and toxicity of the cholera toxin mediated by binding to GM1 (Chinnapen, Chinnapen, Saslowsky, & Lencer, 2007), cellular binding of the VP1 protein of simian virus 40 via GM1 (Tsai et al., 2003), or cell growth modulated by the head-group mediated binding of GM3 to the epidermal growth factor (EGF) receptor (Bremer, Schlessinger, & Hakomori, 1986; Kawashima, Yoon, Itoh, & Nakayama, 2009). (v) A slight influence of the head group was also observed when studying in detail the trapping dynamics of the PI-cer. PI-cer and SM differed only in their head group (SM: phosphocholine, PI-cer: phosphoinositol). While both showed similar binding constants ($K \approx 1$), slightly lowered on- and off-rates were observed in the case of PI-cer ($k_{on} \approx k_{off} \approx 35\ s^{-1}$ compared to $k_{on} \approx k_{off} \approx 70\ s^{-1}$ of SM). (vi) Strikingly, PI without the ceramide group also revealed trapping ($A_{trap} \approx 30\%$ and $k_{on}$, $k_{off} \approx 15$, $40\ s^{-1}$, $K \approx 0.4$), which was much more pronounced than that of the phosphoglycerolipids (e.g., $A_{trap} < 20\%$ and $K \approx 0.2$ for PE or DOPE). This characteristic indicates purely head-group mediated interactions. Hindered diffusion of PI is not surprising, as trapping of glycosyl–phosphatidyl–inositol (GPI) anchored proteins, where the lipid anchor has a similar structure as PI, has been reported several times before (see, e.g., Fielding, 2006; Hanzal-Bayer & Hancock, 2007; Jacobson et al., 2007).

Our observations confirmed that the polar head group with its numerous hydroxyl-groups provides a basis for the formation of molecular complexes, for example, by hydrogen bonding.

## 3.5. Dependence on COase treatment

The differences in dynamical characteristics of the various lipid analogs became more pronounced when comparing the diffusion after cholesterol depletion by COase treatment (Mueller et al., 2011). In Fig. 1.9, we plotted the difference of the apparent diffusion coefficients $\Delta D = D_{COase} - D$ and of the anomaly $\Delta(1/\alpha) = (1/\alpha)_{COase} - (1/\alpha)$ observed before COase treatment ($D$, $(1/\alpha)$) and after COase treatment for 30 min ($D_{COase}$, $(1/\alpha)_{COase}$) using the conventional confocal and the STED ($P_{STED} \approx 160$ mW, $d \approx 40$ nm) measurement modes. We also plotted a percentage value "Recover" $= \Delta D(\text{STED})/\Delta D(\text{conf})$ relating the changes of the apparent diffusion coefficient following COase treatment as observed for the confocal ($\Delta D$(conf)) and STED recordings ($\Delta D$(STED)). If one assumes that $\Delta D$(conf) was mainly due to a change in free diffusion, "Recover" would quantify the change in binding affinity due to COase treatment. Table 1.1 lists values of "Recover," of the trapping amplitude $A_{trap}$ and time $\tau_{Dtrap}$ (Eq. 1.6),

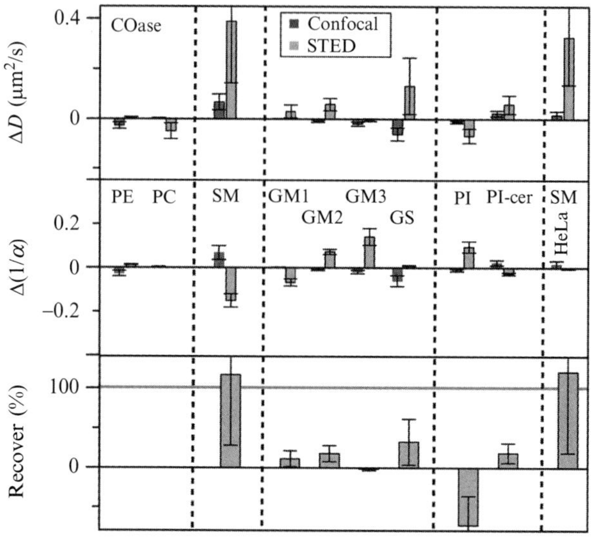


**Figure 1.9** Dependence of nanoscale trapping of different fluorescent lipid analogs on COase treatment. Difference $\Delta D$ (upper panel) and $\Delta(1/\alpha)$ (middle panel) of apparent diffusion coefficients $D$ and of anomaly $(1/\alpha)$ with and without cholesterol depletion by COase for the confocal (dark grey columns) and STED recordings (light gray columns; $P_{STED} = 160$ mW, $d \approx 40$ nm), and "Recover"-values of recovery of free diffusion after cholesterol depletion (lower panel). Larger values of $\Delta D$, negative values of $\Delta(1/\alpha)$, and large "Recover"-values indicate a pronounced cholesterol dependence.

as well as of the on- and off-rates $k_{on}$ and $k_{off}$ (Eq. 1.7) observed for COase-treated cells.

Clear differences in the dependence on COase treatment of the lipids' membrane dynamic were observed. (i) The phosphoglycerolipids such as PE and PC did not show a significant effect after COase treatment. If possible to observe at all, the slight trapping of the phosphoglycerlipids seemed not to be facilitated by cholesterol. (ii) Whether in PtK2 or HeLa cells, SM diffused almost freely after COase treatment ("Recover" $\approx$ 100%). This also resulted in the slight increase of the apparent diffusion coefficient $D$ observed in the confocal recordings (compare also Fig. 1.3B). Cholesterol seemed to significantly stabilize the transient complexes of SM. For example, the off-rate $k_{off}$ of SM binding increased from 70 s$^{-1}$ for non-treated to >150 s$^{-1}$ for COase-treated cells. (iii) Recovery of free diffusion after COase treatment was much less pronounced for all of the gangliosides ("Recover" < 0–30%, $A_{trap} > 0.4$). For example, GM1 diffusion was only very little influenced by COase treatment with "Recover" $\approx$ 10% and no observable changes in $A_{trap}$, $\tau_{Dtrap}$, $k_{on}$, and $k_{off}$. (iv) Trapping of the

PI-ceramide (PI-cer) seemed also to be much less dependent on cholesterol ("Recover" ≈ 20%) than of SM ("Recover" ≈ 100%). (v) Strikingly, trapping of the PI without a ceramide group did not at all depend on cholesterol. We even observed a slight increase in heterogeneity of PI diffusion following COase treatment ("Recover" < 0%).

One can safely conclude that the differences of the dependence on COase treatment of lipid membrane dynamics are a result of the differences of the molecular structure of the lipids' head groups. While sole ceramide-based interactions are fully assisted by cholesterol (see SM), hydroxyl-groups of the ganglioside or PI lipids' polar head group facilitate molecular complexes that are not mediated by cholesterol. This fact became most obvious when comparing PI and PI-cer. Having a ceramide group (PI-cer) the phosphoinositol showed a slight dependence on cholesterol, while even a slight increase of anomalous diffusion was observed for the phopsphoinositol without a ceramide group (PI). The observations may not be surprising, as cholesterol is only present in the membrane bilayer, that is, only within the interaction radius of the ceramide group, which is located at the water–lipid interface.

## 4. CONCLUSIONS

STED- FCS was applied to reveal molecular details of constraint lipid diffusion in the plasma membrane of living cells. From these experiments, it was observed that constraint diffusion was caused by transient, often cholesterol-assisted molecular complexes, but not by meshwork-based or hopping-like diffusion due to compartmentalization of the plasma membrane. Most probably, binding occurred to other membrane constituents such as membrane proteins or other lipids, whether of the same or of other molecular type remained unclear. While an entire complex of several lipids and proteins may have been of larger spatial scales, the trapped molecules dwelled only within <20 nm diameter areas. Direct observation of nanoscopic lipid binding events was only possible in observation areas tuned significantly below the diffraction limit as nicely supplied by the STED nanoscope. Multiple ways of analyzing FCS data recorded in observation areas with diameters as small as 30 nm rendered a consistent picture of trapping. One could even determine on- and off-rates of the transient binding, which were in the range of 1/(10 ms) for sphingolipids.

The synthesis and study of different fluorescent phosphoglycero-, sphingo-, ganglioside, and phosphoinositol lipid analogs enabled revealing characteristics of the nanoscale trapping. No significant dependence of the nanoscale trapping on molecular or experimental parameters such as the number or structure of the alkyl and acyl chains, the temperature, and the labeling conditions were found. At least for the current conditions neither the label type nor the label position had any significant influence except if labeled with a polar dye at the acyl chain. This independence was on one hand surprising because it revealed that only the polarity but not the bulkiness of the dye label had an influence on at least the observed trapping dynamics. On the other hand, the observations may contradict observations made on model membranes or GPMV, where lipids partitioned differently into liquid disordered and ordered phases depending on the abovementioned molecular and experimental parameters. The findings clearly indicated a different molecular basis for nanoscopic live-cell trapping than for large-scale phase separation in model membranes (Sezgin et al., 2009). STED-FCS measurements in the different phases of model membranes may highlight this difference. It will be important to investigate, whether the slowed-down diffusion in the liquid-ordered phase is caused by anomalous subdiffusion.

The STED-FCS measurements on various lipids showed that binding within nanoscopic traps was most probably related to hydrogen bonding to hydroxyl-groups of the lipids' head group or to the ceramide (or sphingosine) moiety close to the water–lipid interface. Phosphoglycerolipids showed only very weak interactions, while lipids carrying the ceramide group distinctively interacted with other membrane components on a timescale of $\sim$ 10 ms. Yet, slight differences appeared between the trapping of the ceramide lipids, which became very pronounced after depleting cholesterol from the plasma membrane using COase treament. A clear dependence of the nanoscale trapping on COase treatment was only observed for lipids without hydroxyl-containing head groups, such as SM, for which an almost complete recovery of free diffusion after COase treatment could be observed. The membrane dynamics of lipids with head groups carrying numerous hydroxyl-groups such as gangliosides (GM1, etc.) were much less dependent on COase treatment. Most strikingly, trapping of a phosphoinositol lipid analog having a ceramide group was lowered after COase treatment, while that of a phosphoinositol lipid analog without a ceramide group was even slightly increased. At least for the analyzed lipids, one could thus differentiate between three classes of molecular lipid interactions: (i) weak interactions of phosphoglycerolipids, (ii) cholesterol-assisted binding mediated by the ceramide group, and (iii) hydroxyl-head-group assisted cholesterol-independent binding. Depending on the head group's size and on the number of hydroxyl-groups in the head group, small differences within these binding classes showed up.

The STED-FCS measurements disclosed that the lipids did not explore an area >20 nm in diameter during trapping, that is, they interacted on nanoscopic spatial scales and formed a transient molecular-sized complex. However, since one observed the dynamics of single molecules only, one cannot rule out that further lipid or protein molecules were (temporarily) included in such a complex. Therefore, the question arises how the observed nanoscopic interactions relate to the most common picture of membrane heterogeneity: lipid nanodomains or "rafts" (Hanzal-Bayer & Hancock, 2007; Jacobson et al., 2007; Lingwood & Simons, 2010; Simons & Ikonen, 1997). The lipid raft concept was introduced in 1997 (Simons & Ikonen, 1997) and postulated tightly packed sphingolipid–cholesterol–protein assemblies that could function in membrane trafficking and signaling. Cholesterol-dependent, liquid-ordered phases in model membranes are often taken as model systems for the tight molecular packing in lipid "rafts" (Hanzal-Bayer & Hancock, 2007; Jacobson et al., 2007; Lingwood & Simons, 2010; Simons & Ikonen, 1997), even though the order parameter of the phases in model membranes is expected to differ substantially from those observed in cell membranes (Kaiser et al., 2009). Altogether, the raft picture is still diffuse, another reason being the lack in spatial resolution of conventional optical microscopy.

Using the superior resolution of STED nanoscopy, stable and immobile complexes were not detected before, as has, for example, been reported by other methods for lipid-anchored proteins previously (e.g., Goswami et al., 2008). Previous STED images of fluorescent SM had rather highlighted a homogeneous distribution of these heterogeneously diffusing lipids. This homogeneous distribution followed from the fact that the complexes were highly dynamic and that at a certain time point about half of the labeled lipids moved freely and the other half were trapped (Eggeling et al., 2009).

A decrease of the trapping of SM upon decreasing the levels of cholesterol as well as of endogenous SM (Mueller et al., 2011) indicates not only the involvement of cholesterol but also of other SM lipids in the molecular complexes of SM. However, the interactions of GM1 (often referred to as a lipid "raft marker") showed much less dependence on levels of endogenous SM or cholesterol than SM, that is, the observed molecular complexes built up by GM1 differ from those assisted by SM and cholesterol. In contrast to general observations made for the phase partitioning in model membranes (e.g., Bacia et al., 2004; Baumgart, Hunt, Farkas, Webb, & Feigenson, 2007b; Shaw et al.,

2006; Wang & Silvius, 2000), the STED-FCS experiments demonstrated hardly any influence by the dye label (Mueller et al., 2011; Sezgin et al., 2012). However, the fluorescent analogs of the lipids used in the STED-FCS experiments did not enter the liquid-ordered phase in model membranes (Honigmann, Walter, Erdmann, Eggeling, & Wagner, 2010; Sezgin et al., 2012) (probably due to the presence of the rather bulky dye label). Consequently, we assume that the observed transient trapping with its rather strong binding to other membrane constituents follows a different molecular mechanism than that of the weak interactions responsible for the formation of ordered phases in model membranes. Because of the low-order partitioning of the applied fluorescent lipid analogs, the observation of ordered nanodomains in living cell membranes was infeasible, and we do not rule out their existence.

If we picture the STED-FCS observations in the context of lipid "rafts," we may support a current view that (sphingo)lipids rafts may establish "fluctuating nanoscale assemblies of sphingo lipid, cholesterol, and proteins that can be stabilized to coalesce, forming platforms that function in membrane signaling and trafficking" (Lingwood & Simons, 2010). Here, the STED-FCS experiments may highlight the fluctuating nanoscale assemblies, which then seem to be highly diverse and strongly depending on the lipid structure. It remains to be shown whether these fluctuating nanoscale assemblies may be stabilized to coalesce to maybe more tightly packed domains (Joly, 2004; Lingwood & Simons, 2010). A fluorescent lipid analog that partitions into the liquid-ordered phase of model membranes (like its natural counterpart) may be able to report on this coalescence (Honigmann, Mueller, Hell, & Eggeling, 2012). Other currently available liquid-order markers such as Laurdan (Parasassi, Krasnowska, Bagatolli, & Gratton, 1998), fluorescent cholesterol analogues (Shaw et al., 2006), Nile Red derivatives (Kucherak et al., 2010), or phosphoglycero- and GM1 analogues labeled with NBD (Wan, Kiessling, & Tamm, 2008) or Alexa 488 (Burns, Frankel, & Buranda, 2005) are, however, not compatible with the present STED-FCS setup.

The observed nanoscopic molecular connectivity of lipids may play an important role in cellular signaling (Lingwood, Ries, Schwille, & Simons, 2008). For example, the molecular structure and thus the distinct molecular specificity of the different lipids may regulate or mediate specific signaling pathways. For example, a sphingolipid binding domain peptide was found to bind specifically to SM (Hebbar et al., 2008). Ganglioside interactions may facilitate several molecular pathways such as endocytosis and toxicity of the cholera toxin mediated by binding to GM1 (Chinnapen et al., 2007), cellular binding of the VP1 protein of simian virus 40 via GM1 (Tsai et al., 2003), or cell growth modulated by the head-group mediated binding of GM3 to the EGF receptor (Bremer et al., 1986; Kawashima et al., 2009; Coskun et al., 2012). Whereas the head-group or ceramide structure may provide molecular specificity to a dedicated cellular pathway, certain lipid-mediated cellular signaling events were found to be less efficient for different saturation degrees or significantly shortened acyl chains of the same lipid type (see, e.g., Ewers et al., 2009). Therefore, aside the molecular connectivity the strength of the membrane anchoring may also play an important role in the complex mechanisms of cellular signaling. Also, the favor of a lipid for the inner or outer plasma membrane leaflet (e.g., in contrast to PE and PI, SM or gangliosides are more probable found in the outer leaflet, while cholesterol is present in both leaflets, see, e.g., Mondal, Mesmin, Mukherjee, & Maxfield, 2009) or the influence of membrane curvatures (for example in caveolae) may play a role in the lipid's dynamic. Much more details of these mechanisms may be discovered when, for example, using STED-FCS to unravel the dependence of the nanoscopic trapping on the saturation degree or acyl-chain length, to study the membrane dynamics close to caveolae markers, or to monitor specific lipid–protein interactions during activation of the cell. As the results show, STED-FCS is a highly sensitive and exceptional tool to study nanoscale membrane organization, which introduces a new approach to determine their cellular functionality and molecular interdependencies.

## ACKNOWLEDGMENTS

We thank V. N. Belov and S. Polyakova (MPI Göttingen) for synthesis of the fluorescent lipids, M. Leutenegger, C. v. Middendorff, A. Schönle, and L. Kastrup (MPI Göttingen) for fruitful discussion, T. Gilat, and E. Rothermel (MPI Göttingen) for helpful assistance, and S.W. Hell (MPI Göttingen) for outstanding support.

## REFERENCES

Abbe, E. (1873). Beiträge zur Theorie des Mikroskops und der mikroskopischen Wahrnehmung. *Archiv für Mikroskopische Anatomie*, *9*, 413–468.

Aragon, S. R., & Pecora, R. (1975). Fluorescence correlation spectroscopy and brownian rotational diffusion. *Biopolymers*, *14*, 119–137.

Bacia, K., Scherfeld, D., Kahya, N., & Schwille, P. (2004). Fluorescence correlation spectroscopy relates rafts in model and native membranes. *Biophysical Journal*, *87*, 1034–1043.

Bagatolli, L. A. (2006). To see or not to see: Lateral organization of biological membranes and fluorescence microscopy. *Biochimica et Biophysica Acta*, *1758*, 1541–1556.

Baumgart, T., Hammond, A. T., Sengupta, P., Hess, S. T., Holowka, D. A., Baird, B. A., et al. (2007). Large-scale fluid/fluid phase separation of proteins and lipids in giant plasma

membrane vesicles. *Proceedings of the National Academy of Sciences of the United States of America, 104*, 3165–3170.
Baumgart, T., Hunt, G., Farkas, E. R., Webb, W. W., & Feigenson, G. W. (2007). Fluorescence probe partitioning between Lo/Ld phases in lipid membranes. *Biochimica et Biophysica Acta, 1768*, 2182–2194.
Binnig, G., Quate, C. F., & Gerber, C. (1986). Atomic force microscope. *Physical Review Letters, 56*, 930–933.
Bouchaud, J. P., & Georges, A. (1990). Anomalous diffusion in disordered media: Statistical mechanisms, models, and physical applications. *Physics Reports, 195*, 127–193.
Bremer, E. G., Schlessinger, J., & Hakomori, S. (1986). Ganglioside-mediated modulation of cell growth. *The Journal of Biological Chemistry, 261*, 2434–2440.
Brown, D. A., & London, E. (2000). Structure and function of sphingolipid- and cholesterol-rich membrane rafts. *The Journal of Biological Chemistry, 275*, 17221–17224.
Burns, A. R., Frankel, D. J., & Buranda, T. (2005). Local mobility in lipid domains of supported bilayers characterized by atomic force microscopy and fluorescence correlation spectroscopy. *Biophysical Journal, 89*, 1081–1093.
Chinnapen, D. J. F., Chinnapen, H., Saslowsky, D., & Lencer, W. (2007). Rafting with cholera toxin: Endocytosis and tracking from plasma membrane to ER. *FEMS Microbiology Letters, 266*, 129–137.
Coskun, Ü., Grzybek, M., Drechsler, D., & Simons, K. (2011). Regulation of human EGF receptor by lipids. *PNAS, 108*, 9044–9048.
de Bakker, B. I., de Lange, F., Cambi, A., Korterik, J. P., van Dijk, E. M. H. P., van Hulst, N. F., et al. (2007). Nanoscale organization of the pathogen receptor DC-SIGN mapped by single-molecule high-resolution fluorescence microscopy. *Chemphyschem, 8*, 1473–1480.
Dietrich, C., Bagatolli, L. A., Volovyk, Z. N., Thompson, N. L., Levi, M., Jacobson, K., et al. (2001). Lipid rafts reconstituted in model membranes. *Biophysical Journal, 80*, 1417–1428.
Eggeling, C., Ringemann, C., Medda, R., Schwarzmann, G., Sandhoff, K., Polyakova, S., et al. (2009). Direct observation of the nanoscale dynamics of membrane lipids in a living cell. *Nature, 457*, 1159–1162 U1121.
Eggeling, C., Widengren, J., Rigler, R., & Seidel, C. A. M. (1998). Photobleaching of fluorescent dyes under conditions used for single-molecule detection: Evidence of two-step photolysis. *Analytical Chemistry, 70*, 2651–2659.
Elson, E. L., & Magde, D. (1974). Fluorescence correlation spectroscopy I. Conceptual basis and theory. *Biopolymers, 13*, 1–27.
Enderlein, J., Gregor, I., Patra, D., Dertinger, T., & Kaupp, U. B. (2005). Performance of fluorescence correlation spectroscopy for measuring diffusion and concentration. *Chemphyschem, 6*, 2324–2336.
Ewers, H., Römer, W., Smith, A. E., Bacia, K., Dmitrieff, S., Chai, W., et al. (2009). GM1 structure determines SV40-induced membrane invagination and infection. *Nature Cell Biology, 12*, 11–18.
Fahey, P. F., Koppel, D. E., Barak, L. S., Wolf, D. E., Elson, E. L., & Webb, W. W. (1977). Lateral diffusion in planar lipid bilayers. *Science, 195*, 305–306.
Feder, T. J., Brust-Mascher, I., Slattery, J. P., Baird, B. A., & Webb, W. W. (1996). Constrainted diffusion or immobile fraction on cell surfaces: A new interpretation. *Biophysical Journal, 70*, 2767–2773.
Fielding, C. J. (2006). *Lipid rafts and caveolae*. Weinheim: Wiley-VCH.
Fujita, A., Cheng, J. L., Hirakawa, M., Furukawa, K., Kusunoki, S., & Fujimoto, T. (2007). Gangliosides GM1 and GM3 in the living cell membrane form clusters susceptible to cholesterol depletion and chilling. *Molecular Biology of the Cell, 18*, 2112–2122.
Fujiwara, T., Ritchie, K., Murakoshi, H., Jacobson, K., & Kusumi, A. (2002). Phospholipids undergo hop diffusion in compartmentalized cell membrane. *The Journal of Cell Biology, 157*, 1071–1081.

Goswami, D., Gowrishankar, K., Bilgrami, S., Ghosh, S., Raghupathy, R., Chadda, R., et al. (2008). Nanoclusters of GPI-anchored proteins are formed by cortical actin-driven activity. *Cell*, *135*, 1085–1097.

Hac, A. E., Seeger, H. M., Fidorra, M., & Heimburg, T. (2005). Diffusion in two-component lipid membranes—A fluorescence correlation spectroscopy and monte carlo simulation study. *Biophysical Journal*, *88*, 317–333.

Hancock, J. F. (2006). Lipid rafts: Contentious only from simplistic standpoints. *Nature Reviews. Molecular Cell Biology*, 7, 457–462.

Hanzal-Bayer, M. F., & Hancock, J. F. (2007). Lipid rafts and membrane traffic. *FEBS Letters*, *581*, 2098–2104.

Harke, B., Keller, J., Ullal, C. K., Westphal, V., Schoenle, A., & Hell, S. W. (2008). Resolution scaling in STED microscopy. *Optics Express*, *16*, 4154–4162.

Hebbar, S., Lee, E., Manna, M., Steinert, S., Kumar, G. S., Wenk, M., et al. (2008). A fluorescent sphingolipid binding domain peptide probe interacts with sphingolipids and cholesterol-dependent raft domains. *Journal of Lipid Research*, *49*, 1077–1089.

Hell, S. W. (2004). Strategy for far-field optical imaging and writing without diffraction limit. *Physics Letters A*, *326*, 140–145.

Hell, S. W. (2009). Microscopy and its focal switch. *Nature Methods*, *6*, 24–32.

Hell, S. W., & Wichmann, J. (1994). Breaking the diffraction resolution limit by stimulated-emission—Stimulated-emission-depletion fluorescence microscopy. *Optics Letters*, *19*, 780–782.

Honigmann, A., Walter, C., Erdmann, F., Eggeling, C., & Wagner, R. (2010). Characterization of horizontal lipid bilayers as a model system to study lipid phase separation. *Biophysical Journal*, *98*, 2886–2894.

Honigmann, A., Mueller, V., Hell, S. W., & Eggeling, C. (2012). STED microscopy detects and quantifies liquid phase separation in lipid membranes using a new far-red emitting fluorescent phosphoglycerolipid analogue. *Faraday Discussion* http://dx.doi.org/10.1039/C2FD20107K.

Humpolickova, J., Gielen, E., Benda, A., Fagulova, V., Vercammen, J., vandeVen, M., et al. (2006). Probing diffusion laws within cellular membranes by Z-scan fluorescence correlation spectroscopy. *Biophysical Journal*, *91*, L23–L25.

Jacobson, K., Mouritsen, O. G., & Anderson, G. W. (2007). Lipid rafts: At a crossroad between cell biology and physics. *Nature Cell Biology*, *9*, 7–14.

Joly, E. (2004). Hypothesis: Could the signalling function of membrane microdomains involve a localized transition of lipids from liquid to solid state? *BMC Cell Biology*, *5*, 3.

Kahya, N., Scherfeld, D., Bacia, K., Poolman, B., & Schwille, P. (2003). Probing lipid mobility of raft-exhibiting model membranes by fluorescence correlation spectroscopy. *The Journal of Biological Chemistry*, *278*, 28109–28115.

Kaiser, H.-J., Lingwood, D., Levental, I., Sampaio, J. L., Kalvodova, L., Rajendran, L., et al. (2009). Order of lipid phases in model and plasma membranes. *Proceedings of the National Academy of Sciences of the United States of America*, *106*, 16645–16650.

Kask, P., Palo, K., Ullmann, D., & Gall, K. (1999). Fluorescence-intensity distribution analysis and its application in biomolecular detection technology. *Proceedings of the National Academy of Sciences of the United States of America*, *96*, 13756–13761.

Kastrup, L., Blom, H., Eggeling, C., & Hell, S. W. (2005). Fluorescence fluctuation spectroscopy in subdiffraction focal volumes. *Physical Review Letters*, *94*, 178104.

Kawashima, N., Yoon, S.-J., Itoh, K., & Nakayama, K. (2009). Tyrosine kinase activity of epidermal growth factor receptor is regulated by GM3 binding through carbohydrate to carbohydrate interactions. *The Journal of Biological Chemistry*, *284*, 6147–6155.

Kolmakov, K., Belov, V., Bierwagen, J., Ringemann, C., Mueller, V., Eggeling, C., et al. (2010). Red-emitting rhodamine dyes for fluorescence microscopy and nanoscopy. *Chemistry*, *16*, 158–166.

Kucherak, O. A., Oncul, S., Darwich, Z., Yushchenko, D. A., Arntz, Y., Didier, P., et al. (2010). Switchable nile red-based probe for cholesterol and lipid order at the outer leaflet of biomembranes. *Journal of the American Chemical Society*, *132*, 4907–4916.

Kusumi, A., Shirai, Y. M., Koyama-Honda, I., Suzuki, K. G. N., & Fujiwara, T. K. (2010). Hierarchical organization of the plasma membrane: Investigations by single-molecule tracking vs. fluorescence correlation spectroscopy. *FEBS Letters*, *584*, 1814–1823.

Lenne, P. F., Wawrezinieck, L., Conchonaud, F., Wurtz, O., Boned, A., Guo, X. J., et al. (2006). Dynamic molecular confinement in the plasma membrane by microdomains and the cytoskeleton meshwork. *The EMBO Journal*, *25*, 3245–3256.

Lingwood, D., Ries, J., Schwille, P., & Simons, K. (2008). Plasma membranes are poised for activation of raft phase coalescence at physiological temperature. *Proceedings of the National Academy of Sciences of the United States of America*, *105*, 10005–10010.

Lingwood, D., & Simons, K. (2010). Lipid rafts as a membrane-organizing principle. *Science*, *327*, 46–50.

Lommerse, P. H. M., Spaink, H. P., & Schmidt, T. (2004). In vivo plasma membrane organization: Results of biophysical approaches. *Biochimica et Biophysica Acta*, *1664*, 119–131.

London, E. (2005). How principles of domain formation in model membranes may explain ambiguities concerning lipid raft formation in cells. *Biochimica et Biophysica Acta*, *1746*, 203–220.

Magde, D., Webb, W. W., & Elson, E. (1972). Thermodynamic fluctuations in a reacting system—Measurement by fluorescence correlation spectroscopy. *Physical Review Letters*, *29*, 705–708.

Marsh, D. (2009). Cholesterol-induced fluid membrane domains: A compendium of lipid-raft ternary phase diagrams. *Biochimica et Biophysica Acta*, *1788*, 2114–2123.

Martin, O. C., & Pagano, R. C. (1994). Internalization and sorting of a fluorescent analogue of glucosylceramide to the Golgi apparatus of human skin fibroblasts: Utilization of endocytic and nonendocytic transport mechanisms. *The Journal of Cell Biology*, *125*, 769–781.

Michelman-Ribeiro, A., Mazza, D., Rosales, T., Stasevich, T. J., Boukari, H., Rishi, V., et al. (2009). Direct measurement of association and dissociation rates of DNA binding in live cells by fluorescence correlation spectroscopy. *Biophysical Journal*, *97*, 337–346.

Mondal, M., Mesmin, B., Mukherjee, S., & Maxfield, F. R. (2009). Sterols are mainly in the cytoplasmic leaflet of the plasma membrane and the endocytic recycling compartment in CHO cells. *Molecular Biology of the Cell*, *20*, 581–588.

Mueller, V., Ringemann, C., Honigmann, A., Schwarzmann, G., Medda, R., Leutenegger, M., et al. (2011). STED nanoscopy reveals molecular details of cholesterol- and cytokeleton-modulated lipid interactions in living cells. *Biophysical Journal*, *101*, 1651–1660.

Munro, S. (2003). Lipid rafts: Elusive or illusive? *Cell*, *115*, 377–388.

Nishimura, S. Y., Vrljic, M., Klein, L. O., McConnell, H. M., & Moerner, W. E. (2006). Cholesterol depletion induces solid-like regions in the plasma membrane. *Biophysical Journal*, *90*, 927–938.

Parasassi, T., Krasnowska, E. K., Bagatolli, L., & Gratton, E. (1998). Laurdan and prodan as polarity-sensitive fluorescent membrane probes. *Journal of Fluorescence*, *8*, 365–373.

Pascher, I. (1976). Molecular arrangements in sphingolipids conformation and hydrogen-bonding of ceramide and their implication on membrane stability and permeability. *Biochimica et Biophysica Acta*, *455*, 433–451.

Pike, L. J. (2006). Rafts defined: A report on the Keystone symposium on lipid rafts and cell function. *Journal of Lipid Research*, *47*, 1597–1598.

Pohl, D. W., Denk, W., & Lanz, M. (1984). Optical stethoscopy—Image recording with resolution lambda/20. *Applied Physics Letters*, *44*, 651–653.

Polyakova, S., Belov, V., Yan, S. F., Eggeling, C., Ringemann, C., Schwarzmann, G., et al. (2009). New GM1 ganglioside derivatives for selective single and double labelling of the natural glycosphingolipid skeleton. *European Journal of Organic Chemistry, 2009*, 5162–5177.

Pralle, A., Keller, P., Florin, E. L., Simons, K., & Hoerber, J. K. H. (2000). Sphingolipid-cholesterol diffuse as small entities in the plasma membrane of mammalian cells. *The Journal of Cell Biology, 148*, 997–1007.

Ries, J., & Schwille, P. (2006). Studying slow membrane dynamics with continuous wave scanning fluorescence correlation spectroscopy. *Biophysical Journal, 91*, 1915–1924.

Ringemann, C., Harke, B., Middendorff, C. V., Medda, R., Honigmann, A., Wagner, R., et al. (2009). Exploring single-molecule dynamics with fluorescence nanoscopy. *New Journal of Physics, 11*, 103054.

Ruan, Q. Q., Cheng, M. A., Levi, M., Gratton, E., & Mantulin, W. W. (2004). Spatial-temporal studies of membrane dynamics: Scanning fluorescence correlation spectroscopy (SFCS). *Biophysical Journal, 87*, 1260–1267.

Sahl, S. J., Leutenegger, M., Hilbert, M., Hell, S. W., & Eggeling, C. (2010). Fast molecular tracking maps nanoscale dynamics of plasma membrane lipids. *Proceedings of the National Academy of Sciences of the United States of America, 107*, 6829–6834.

Samsonov, A. V., Mihalyov, I., & Cohen, F. S. (2001). Characterization of cholesterol-sphingomyelin domains and their dynamics in bilayer membranes. *Biophysical Journal, 81*, 1486–1500.

Saxton, M. J. (1994). Anomalous diffusion due to obstacles—A Monte-Carlo study. *Biophysical Journal, 66*, 394–401.

Saxton, M. J., & Jacobson, K. (1997). Single particle tracking: Applications to membrane dynamics. *Annual Review of Biophysics, 26*, 373–399.

Schutz, G. J., Kada, G., Pastushenko, V. P., & Schindler, H. (2000). Properties of lipid microdomains in a muscle cell membrane visualized by single molecule microscopy. *The EMBO Journal, 19*, 892–901.

Schwille, P., Korlach, J., & Webb, W. W. (1999). Fluorescence correlation spectroscopy with single-molecule sensitivity on cell and model membranes. *Cytometry, 36*, 176–182.

Sezgin, E., Levental, I., Grzybek, M., Schwarzmann, G., Mueller, V., Honigmann, A., et al. (2012). Partitioning, diffusion, and ligand binding of raft lipid analogs in model and cellular plasma membranes. *Biochimica et Biophysica Acta—Biomembranes, 1818*, 1777–1784.

Shaw, A. S. (2006). Lipid rafts: Now you see them, now you don't. *Nature Immunology, 7*, 1139–1142.

Shaw, J. E., Epand, R. F., Epand, R. M., Li, Z., Bittman, R., & Yip, C. M. (2006). Correlated fluorescence-atomic force microscopy of membrane domains: Structure of fluorescence probes determines lipid localization. *Biophysical Journal, 90*, 2170–2178.

Simons, K., & Ikonen, E. (1997). Functional rafts in cell membranes. *Nature, 387*, 569–572.

Tsai, B., Gilbert, J. M., Stehle, T., Lencer, W., Benjamin, T. L., & Rapaport, T. A. (2003). Gangliosides are receptors for murine polyoma virus and SV40. *The EMBO Journal, 22*, 4346–4355.

Varma, R., & Mayor, S. (1998). GPI-anchored proteins are organized in submicron domains at the cell surface. *Nature, 394*, 798–801.

Veatch, S. L., & Keller, S. L. (2005). Seeing spots: Complex phase behavior in simple membranes. *Biochimica et Biophysica Acta, 1746*, 172–185.

Wachsmuth, M., Waldeck, W., & Langowski, J. (2000). Anomalous diffusion of fluorescent probes inside living cell nuclei investigated by spatially-resolved fluorescence correlation spectroscopy. *Journal of Molecular Biology, 298*, 677–689.

Wan, C., Kiessling, V., & Tamm, L. K. (2008). Coupling of cholesterol-rich lipid phases in asymmetric bilayers. *Biochemistry, 47*, 2190–2198.

Wang, T. Y., & Silvius, J. R. (2000). Different sphingolipids show differential partitioning into sphingolipid/cholesterol-rich domains in lipid bilayers. *Biophysical Journal, 79*, 1478–1489.

Wawrezinieck, L., Rigneault, H., Marguet, D., & Lenne, P. F. (2005). Fluorescence correlation spectroscopy diffusion laws to probe the submicron cell membrane organization. *Biophysical Journal, 89*, 4029–4042.

Wenger, J., Conchonaud, F., Dintinger, J., Wawrezinieck, L., Ebbesen, T. W., Rigneault, H., et al. (2007). Diffusion analysis within single nanometric apertures reveals the ultrafine cell membrane organization. *Biophysical Journal, 92*, 913–919.

Widengren, J., Mets, U., & Rigler, R. (1995). Fluorescence correlation spectroscopy of triplet-states in solution—A theoretical and experimental-study. *The Journal of Physical Chemistry, 99*, 13368–13379.

Yechiel, E., & Edidin, M. (1987). Micrometer-scale domains in fibroblast plasma-membranes. *The Journal of Cell Biology, 105*, 755–760.

Zacharias, D. A., Violin, J. D., Newton, A. C., & Tsien, R. Y. (2002). Partitioning of lipid-modified monomeric GFPs into membrane microdomains of live cells. *Science, 296*, 913–916.

CHAPTER TWO

# Analyzing Förster Resonance Energy Transfer with Fluctuation Algorithms

**Suren Felekyan, Hugo Sanabria, Stanislav Kalinin, Ralf Kühnemuth, Claus A.M. Seidel**[1]

Institut für Physikalische Chemie, Lehrstuhl für Molekulare Physikalische Chemie, Heinrich-Heine-Universität, Düsseldorf, Germany

[1]Corresponding author: e-mail address: cseidel@hhu.de

## Contents



## Abstract

Fluorescence correlation spectroscopy (FCS) in combination with Förster resonance energy transfer (FRET) has been developed to a powerful statistical tool, which allows for the analysis of FRET fluctuations in the huge time of nanoseconds to seconds. FRET–FCS utilizes the strong distance dependence of the FRET efficiency on the donor (D)–acceptor (A) distance so that it developed to a perfect method for studying structural fluctuation in biomolecules involved in conformational flexibility, structural dynamics, complex formation, folding, and catalysis. Structural fluctuations thereby

*Methods in Enzymology*, Volume 519
ISSN 0076-6879
http://dx.doi.org/10.1016/B978-0-12-405539-1.00002-6

result in anticorrelated donor and acceptor signals, which are analyzed by FRET–FCS in order to characterize underlying structural dynamics. Simulated and experimental examples are discussed. First, we review experimental implementations of FRET–FCS and present theory for a two-state interconverting system. Additionally, we consider a very common case of FRET dynamics in the presence of donor-only labeled species. We demonstrate that the mean relaxation time for the structural dynamics can be easily obtained in most of cases, whereas extracting meaningful information from correlation amplitudes can be challenging. We present a strategy to avoid a fit with an underdetermined model function by restraining the D and A brightnesses of the at least one involved state, so that both FRET efficiencies and both rate constants (i.e., the equilibrium constant) can be determined. For samples containing several fluorescent species, the use of pulsed polarized excitation with multiparameter fluorescence detection allows for filtered FCS (fFCS), where species-specific correlation functions can be obtained, which can be directly interpreted. The species selection is achieved by filtering using fluorescence decays of individual species. Analytical functions for species auto- and cross-correlation functions are given. Moreover, fFCS is less affected by photophysical artifacts and often offers higher contrast, which effectively increases its time resolution and significantly enhances its capability to resolve multistate kinetics. fFCS can also differentiate between species even when their brightnesses are the same and thus opens up new possibilities to characterize complex dynamics. Alternative fluctuation algorithms to study FRET dynamics are also briefly reviewed.

## 1. INTRODUCTION

The use of fluorescence correlation spectroscopy (FCS) to investigate conformational flexibility, structural dynamics in biomolecules, complex formation, folding, and catalysis has grown in popularity since its first publication more than 40 years ago (Ehrenberg & Rigler, 1974; Magde, Elson, & Webb, 1972, 1974). The combination of FCS with Förster resonance energy transfer (FRET) (Braslavsky et al., 2008; Förster, 1948; Sisamakis, Valeri, Kalinin, Rothwell, & Seidel, 2010) takes advantage of the ultimate sensitivity of fluorescence and the strong distance dependence of the FRET process. An excited fluorophore (Donor, D) transfers energy to another fluorophore (Acceptor, A) when they are in close proximity. FRET is particularly useful for probing interdye distances between 20 and 100 Å. For the case of perfect fluorophores and detection, the efficiency of FRET, $E$, is defined as a yield of acceptor fluorescence $F_A$ normalized to the total fluorescence of donor and acceptor:

$E = F_A/(F_D + F_A)$ (see definition for real experimental conditions by Eq. 2.17 in Section 2.1). The FRET efficiency is related to the interdye distance $R_{DA}$. In this relation, $R_{DA}$ is normalized by a dye pair-specific (dipolar coupling) constant referred to as Förster radius $R_0$: $E = 1/(1 + (R_{DA}/R_0)^6)$ (for details, see Lakowicz, 2006; Sisamakis et al., 2010). From these equations, it can be seen that fluctuations in $R_{DA}$ result in anticorrelated fluctuations of the D and A fluorescence intensities, which can be quantified by FRET–FCS as summarized in this review.

The use of FRET between two fluorophores has become increasingly popular (Clegg, 2009; Clegg et al., 1992; De Angelis, 1999; Sun, Wallrabe, Seo, & Periasamy, 2011; Szollosi, Damjanovich, & Matyus, 1998; Vogel, Thaler, & Koushik, 2006; Wu & Brand, 1994). As an alternative to native fluorophores, such as tryptophan (Lakowicz, 2006), the green fluorescent protein (Giepmans, Adams, Ellisman, & Tsien, 2006), a wide range of small and photostable organic fluorophores (Gonçalves, 2009), and chemically activated nanoparticles (Bruchez, Moronne, Gin, Weiss, & Alivisatos, 1998; Curutchet, Franceschetti, Zunger, & Scholes, 2008) have been introduced. A prerequisite for this method is to covalently attach a donor (D) and acceptor (A) dye as FRET pair at specific sites of the biomolecule. The following functional moieties are popular for covalent coupling: amino or mercapto groups as well as various bioorthogonal functional groups in nonnatural amino acids of genetically engineered proteins (Wu & Schultz, 2009).

Over the past decades, a large variety of sophisticated FRET-based spectroscopic assays has been developed in order to study biological reactions and the characteristics of freely diffusing or immobilized biomolecules *in vitro* and in living cells (Giepmans et al., 2006; Lippincott-Schwartz, Snapp, & Kenworthy, 2001; Sun, Day, & Periasamy, 2011; Weidtkamp-Peters et al., 2009). The successful application of FRET on the ensemble level (Lakowicz, 2006) led to efforts to exploit the same phenomenon on the single-molecule level in smFRET by Ha et al. (1996), so that many questions on supramolecular assemblies and biological systems were studied such as (1) generation of quantitative structural models with respect to distances, domain orientation, conformational arrangement, and dynamics (Knight, Mekler, Mukhopadhyay, Ebright, & Levy, 2005; Margittai, Widengren, et al., 2003; Mekler et al., 2002; Muschielok et al., 2008; Sindbert et al., 2011; Woźniak, Schröder, Grubmüller, Seidel, & Oesterhelt, 2008); (2) binding equilibria (Kask et al., 2000; Sun,

Wallrabe, et al., 2011; Wang, Guo, Golding, Cox, & Ong, 2009) and on/off rates (Gansen et al., 2009; Rothwell et al., 2003); (3) complex stoichiometries (Bader et al., 2011; Chen & Müller, 2007; Eggeling, Kask, Winkler, & Jager, 2005; Ulbrich & Isacoff, 2007); (4) folding of biomolecules (Michalet, Weiss, & Jager, 2006; Nettels, Gopich, Hoffmann, & Schuler, 2007; Schuler & Eaton, 2008); (5) arrangement and function of complex molecular machines (Blanchard, 2009; Majumdar et al., 2007; Mekler et al., 2002); (6) catalytic reaction cycles (Diez et al., 2004; Joo et al., 2006; Mickler, Hessling, Ratzke, Buchner, & Hugel, 2009; Pandey et al., 2009).

In view of this wide areas of applications, the anticorrelated fluctuations of the D and A fluorescence intensity (sensitization of A fluorescence is linked with D quenching) were utilized to study dynamics of structure and composition of biomolecular complexes under equilibrium conditions (Felekyan, Kalinin, Sanabria, Valeri, & Seidel, 2012; Gurunathan & Levitus, 2010; Hanson & Yang, 2008; Henzler-Wildman et al., 2007; Kudryavtsev et al., 2012; Levitus, 2010; Margittai, Widengren, et al., 2003; Müller, Zaychikov, Bräuchle, & Lamb, 2005; Nettels et al., 2007; Price, Aleksiejew, & Johnson, 2011; Price, DeVore, & Johnson, 2010; Sahoo & Schwille, 2011; Tan & Yang, 2011). In this review, we give a survey on the fundamental theory and instrumentation for FRET–FCS. Considering molecular mixtures with increasing complexity, we discuss the potential and the pitfalls of FCS combined with FRET. Many of these limitations can be overcome by multiparameter fluorescence detection (MFD) including the measurement of fluorescence lifetimes and polarization. MFD can be then combined with filter techniques to arrive at filtered FCS (fFCS), which is perfectly suited to perform selective analyses in complex mixtures. Pulsed interleaved excitation of D and A can be used to further increase specificity (Kudryavtsev et al., 2012; Müller et al., 2005; Sahoo & Schwille, 2011). Finally, we provide a short record on successful and promising applications.

## 1.1. Definition of the correlation functions

In FCS, the fluctuating fluorescence arises from fluorescent molecules diffusing through a confocal volume. This signal is used to obtain information about dynamic processes at the molecular level (Elson & Magde, 1974; Magde et al., 1972, 1974) by analyzing the fluctuations using the normalized correlation function $G(t_c)$:

$$
\begin{aligned}
{}^{A,B}G(t_c) &= \frac{\langle {}^{A}S(t)\cdot{}^{B}S(t+t_c)\rangle}{\langle {}^{A}S(t)\rangle\langle {}^{B}S(t)\rangle} = \frac{\langle(\langle {}^{A}S(t)\rangle + \delta^{A}S(t))(\langle {}^{B}S(t)\rangle + \delta^{B}S(t+t_c))\rangle}{\langle {}^{A}S(t)\rangle\langle {}^{B}S(t)\rangle} \\
&= 1 + \frac{\langle \delta^{A}S(t)\cdot\delta^{B}S(t+t_c)\rangle}{\langle {}^{A}S(t)\rangle\langle {}^{B}S(t)\rangle},
\end{aligned}
\qquad [2.1]
$$

where $t_c$ is the correlation time, ${}^{A,B}S(t)$ represents the detected intensity signal (number of detected photons per time interval) at channels A or B, and $\delta^{A,B}S(t)$ corresponds to the fluctuations from the time average signal denoted as $\langle {}^{A,B}S(t)\rangle$. The autocorrelation function is defined when the detected channels are the same A=B, if A≠B, then ${}^{A,B}G(t_c)$ is called cross-correlation.

For a single molecular species, the following analytical form of the correlation function was derived (Rigler & Mets, 1992; Schaefer, 1973; Thompson, 1991).

$$G(t_c) = 1 + \frac{1}{N} G_{\text{diff}}(t_c), \qquad [2.2]$$

where

$$G_{\text{diff}}(t_c) = \left(1 + \frac{t_c}{t_{\text{diff}}}\right)^{-1} \left(1 + \left(\frac{\omega_0}{z_0}\right)^2 \frac{t_c}{t_{\text{diff}}}\right)^{-1/2}. \qquad [2.3]$$

This model assumes a three-dimensional Gaussian-shaped volume element with spatial distribution of the detection probabilities: $W(x,y,z) = \exp\left(-2(x^2+y^2)/\omega_0^2\right)\exp\left(-2z^2/z_0^2\right)$. The $1/e^2$ radii in $x$ and $y$ or in $z$ direction are denoted by $\omega_0$ and $z_0$, respectively. The characteristic diffusion time $t_{\text{diff}}$ can be used to estimate the diffusion coefficient $D$, $t_{\text{diff}} = \omega_0^2/4D$, and represents the average time for the molecule inside the confocal volume. The amplitude of the correlation is scaled with the reciprocal of the average number of fluorescent particles $N$ in the confocal volume.

On top of the diffusion, several physical mechanisms can give rise to additional fluorescence fluctuations. For example, triplet kinetics is a reversible mechanism intrinsic of fluorescent markers where after the excitation process the electronic state intercrosses from a singlet to a dark triplet state. This first-order reaction contributes with a bunching term $G_T(t_c)$ to the correlation function as

$$G(t_c) = 1 + \frac{1}{N} G_{\text{diff}}(t_c)[1 + G_T(t_c)], \qquad [2.4]$$

with

$$G_T(t_c) = \frac{T}{(1-T)} \exp\left(-\frac{t_c}{t_T}\right). \qquad [2.5]$$

From the equilibrium fraction of molecules in the triplet state, T, and the triplet correlation time, $t_T$, the rate constants for intersystem crossing, $k_{ISC}$, and triplet decay, $k_T$, can be derived (Widengren, Mets, & Rigler, 1995; Widengren, Rigler, & Elson, 2001; Widengren, Rigler, & Mets, 1994).

Other kinetic reactions can be studied with FCS as long as they influence fluorescence. Since the pioneering work of Webb, Magde, and Elson, FCS has been widely used to study molecular interactions. Ever since, the question at hand has been: how to design experiments that generate fluctuations in fluorescence at timescales faster than diffusion? This approach has been extensively pursued with the help of FRET. The fluctuations in the FRET signal due, for example, to intra- or intermolecular transitions add additional kinetic terms to the correlation function. If the triplet time $t_T$ is faster than the reaction time $t_R$ ($t_T << t_R$) and assuming that triplet and reaction are decoupled, Eq. (2.4) can be modified to account for an additional kinetic term such as

$$G(t_c) = 1 + \frac{1}{N} G_{\text{diff}}(t_c)[1 + G_T(t_c)][1 + G_K(t_c)]. \qquad [2.6]$$

For a two-state system, the kinetic reaction term $G_K(t_c)$ is given by

$$G_K(t_c) = A_K \exp\left(-\frac{t_c}{t_R}\right). \qquad [2.7]$$

Assuming $t_T << t_R$, Eq. (2.6) can be equivalently presented as

$$G(t_c) \cong 1 + \frac{1}{N} G_{\text{diff}}(t_c)\left[1 + A_K \exp\left(-\frac{t_c}{t_R}\right) + (1 + A_K)\frac{T}{1-T} \exp\left(-\frac{t_c}{t_T}\right)\right], \qquad [2.8]$$

since $t_c/t_R + t_c/t_T \cong t_c/t_T$. The relaxation time $t_R$ and the amplitude of the kinetic term $A_K$ contain information on kinetic rate constants and molecular fractions. A general solution exists which allows one to model $G_K(t_c)$ for arbitrarily complex systems (Enderlein & Gregor, 2005; Gregor & Enderlein, 2007; Nettels et al., 2007), which makes use of a matrix formalism. In this case, $G_K(t_c)$ is generally given by a sum of $n-1$ exponential terms for an $n$-state system.

Most of the formalism contained herein considers the case where the kinetic reaction and triple kinetics are decoupled and therefore the factor $[1+G_T(t_c)]$ can be treated separately as shown in Eq. (2.6).

## 1.2. Heterogeneous mixtures

Considering a mixture of $n$ species, with corresponding brightnesses $Q^{(i)}$, diffusion constants $D^{(i)}$, and fractions $x^{(i)}$, with the index $i=1,\ \ldots, n$ needs to be taken into account for writing the autocorrelation function of the mixture (Kim, Heinze, Bacia, Waxham, & Schwille, 2005; Schwille, 2001):

$$G(t_c)=1+\frac{1}{N}\frac{\sum_i^n x^{(i)}\left(Q^{(i)}\right)^2 G_{\mathrm{diff}}^{(i)}(t_c)}{\left(\sum_i^n x^{(i)}Q^{(i)}\right)^2}, \qquad [2.9]$$

where

$$G_{\mathrm{diff}}^{(i)}(t_c)=\left(1+\frac{t_c}{t_D^{(i)}}\right)^{-1}\left(1+\left(\frac{\omega_0}{z_0}\right)^2\frac{t_c}{t_{\mathrm{diff}}^{(i)}}\right)^{-1/2} \qquad [2.10]$$

has the same meaning as Eq. (2.3), but defined for each species ($i$) in the mix. Each species has a concentration $c(i)$, diffusion time $t_{\mathrm{diff}}^{(i)}$, brightness $Q^{(i)}$, and molecular fraction $x^{(i)}=(c^{(i)}/\sum_i c^{(i)})$. Please note that no simple analytical expression can then be found for the case of distinct diffusion terms together with a kinetic reaction term. We restrict the theory in Section 2 of this review to the case of equal diffusion times.

## 1.3. Hardware

Various realizations of equipment needed in FCS have been described in the past. Depending on the requirements of the experiment different levels of sophistication can be implemented. In the following, we briefly review confocal setups to measure FRET–FCS ranging from low-cost hardware-correlator-based solutions to elaborate multidetector single-photon counting devices.

Separation of FRET-based effects from fluctuations due to rotational and translational diffusion or other photophysical processes in general requires a cross-correlation of donor and acceptor signal in addition to donor and acceptor autocorrelation. Consequently, the minimum setup consists of two single-photon sensitive detectors (SPADs or PMTs) attached to electronics capable of either performing real-time correlations with nanosecond resolution or steaming data to hard drive for subsequent offline software correlation. The optical setup in Fig. 2.1 consists of a high numerical aperture objective, a dichroic beamsplitter to separate fluorescence from the exciting laser light, a confocal pinhole, and a second dichroic beamsplitter to divide the fluorescence into donor and acceptor channel, followed by two band-pass filters. Hardware correlators with ns-resolution are offered commercially (ALV GmbH, correlator.com), and recently, low-price field-programmable gate array (FPGA)-based alternatives became available (Kalinin, Kuhnemuth, Vardanyan, & Seidel, 2012; Mocsár, et al., 2012). Here simultaneous auto- and cross-correlation of multiple inputs can be performed in real time.

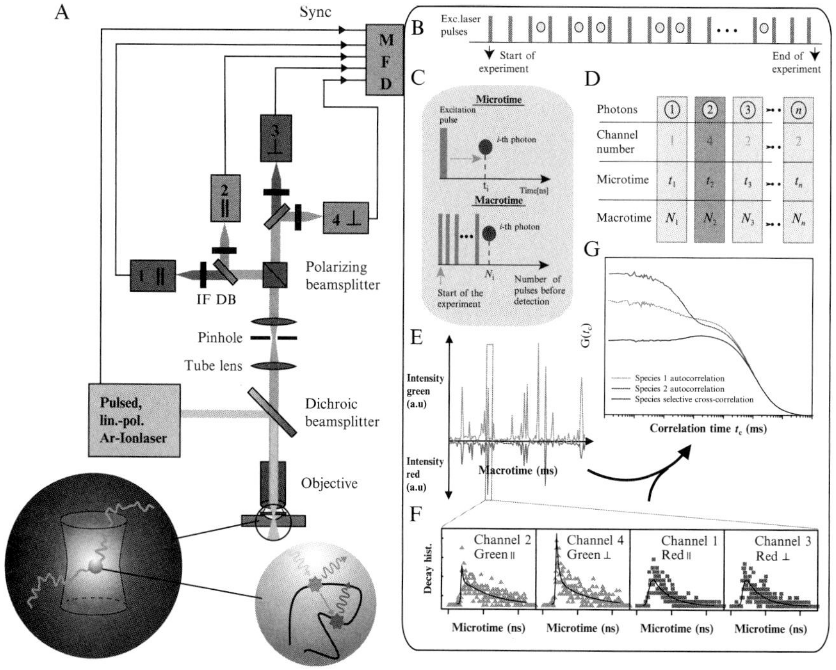


**Figure 2.1** Typical four-detector setup for FRET–FCS by multiparameter fluorescence detection (MFD) and flowchart of subsequent data processing. The identical optical hardware (A) can be employed for hardware-correlator-based solutions. (B–D) Assignment of the detection parameters (detector number, microtime (time after laser pulse), and macrotime (number of laser pulses since experiment start)). (E) Generation of intensity traces. (F) Generation of selective decay histograms for each detection channel. (G) Species selective FCS using filters obtained from decay pattern. (For color version of this figure, the reader is referred to the online version of this chapter.)

To study processes extending into the time regime of detector dead time (<100 ns) and/or detector afterpulsing (<1–2 μs), an expansion of the concept by dividing donor and acceptor signal onto two detectors each is needed. In this way, it is possible to have virtually dead time and artifact free autocorrelation of donor and acceptor signal. Performing this splitting with a polarizing beamsplitter adds also the advantage to measure fluorescence anisotropy at the same time (see Fig. 2.1A). MFD (Eggeling et al., 2001; Kühnemuth & Seidel, 2001; Widengren et al., 2006) takes advantage of this fact to study more complex systems. In an MFD experiment, the photon stream from the four detectors (Fig. 2.1A) is fed into the time-correlated single-photon counting (TCSPC) electronics with photon time stamping (Fig. 2.1B). Each event is tagged with its channel id, microtime (time after laser or external clock pulse) and macrotime (number of laser or clock pulses since experiment start) (Fig. 2.1C and D). To this end, single-photon registration and complete data storage are mandatory to allow for the exertion of advanced software correlation algorithms to generate full correlation curves from a single measurement with correlation times ranging from picoseconds to minutes.

Software correlation of the recorded photons is performed either after generation of multichannel scaler (MCS) traces or, in most cases more efficiently, directly via a "single-photon MCS trace" (all details are described in Felekyan et al., 2005) (Fig. 2.1E and G). The main difference of a "single-photon MCS trace" from conventional MCS traces is the fact that it contains only the time information of the registered photons instead of intensity information for equally spaced time bins. In this way, the same information is stored more efficiently, equivalent to only saving the filled bins and not the more frequent empty bins. Our fast correlation algorithm published elsewhere (Felekyan et al., 2005) makes use of this fact.

Software correlation also opens up the possibility for selective FCS of a sample containing several fluorescent species. There are in principle two ways to achieve this (Fig. 2.1B–G): (1) analysis of highly diluted samples (p*M*) necessary for the detection of single molecules, which allow for specific burst selection and subsequent subensemble FCS; or (2) the use of pulsed polarized excitation allows for fFCS where species selection is achieved by filtering of species-specific fluorescence decays (Felekyan et al., 2012). The fluorescence decay patterns for species selective fFCS are obtained from single-molecule subensembles or multiexponential fits to ensemble TCSPC data (Fig. 2.1F). More details are given in Section 3.

## 1.4. Timescales studied by FRET

The beautiful and static insights into protein structure obtained from X-ray diffraction result in the impression that biomolecules possess absolute functional specificity and a single, fixed structure. This perspective is in conflict with the ability of proteins to adopt different functions and structures. Proteins and other such biomolecules are often flexible with a range of motions spanning from picoseconds for localized vibrations to seconds for concerted global conformational rearrangements.

This suggests an intimate relationship between dynamics and molecular function, where biomolecular function is determined by the thermodynamic stability and kinetic accessibility of their functional states (Henzler-Wildman & Kern, 2007). Figure 2.2 illustrates that fluorescence spectroscopy is well suited to follow all biomolecular processes slower than the fluorescence lifetime (for organic dyes usually a few nanoseconds). There are various distinct fluorescence techniques in combination with single-molecule detection which allow one to resolve dynamic biomolecular heterogeneity over timescales ranging from picoseconds to hours. TCSPC can be applied to measure fluorescence lifetime decays (for organic dyes usually a few nanoseconds), which provide ns-snapshots of the molecular environment, so that the presence detection of longer lived distributions is possible. MFD using the confocal microscope of Fig. 2.1 can be applied to study photon bursts (burstwise) due to transits of freely diffusing FRET-labeled molecules at picomolar concentrations (Sisamakis et al., 2010). As the first step in burst analysis, efficient analysis algorithms are used to determine all possible fluorescence parameters (Kühnemuth & Seidel, 2001; Widengren et al., 2006) such as FRET efficiency and fluorescence lifetime of the donor in the presence of an acceptor. These two parameters can be plotted against each other in a 2D MFD-FRET diagram, which counts the frequency of bursts with certain fluorescence properties time averaged over the burst duration. We can distinguish two scenarios for molecular kinetics. In case of an interchange slower than the characteristic diffusion time (upper panel in light red), two species are immediately detected as quasi-static populations (more details on the MFD analysis are given in Section 4). The molecules must be immobilized (Roy, Hohng, & Ha, 2008; Tan & Yang, 2011) or encapsulated in liposomes (Boukobza, Sonnenfeld, & Haran, 2001; Cisse, Okumus, Joo, & Ha, 2007) to monitor their slow FRET interchange in a time trace. As the distribution detected at the nanosecond timescale is still present the same in the milliseconds time regime, it is clear that the species cross-correlation curve of fFCS does not show any amplitude for dynamics

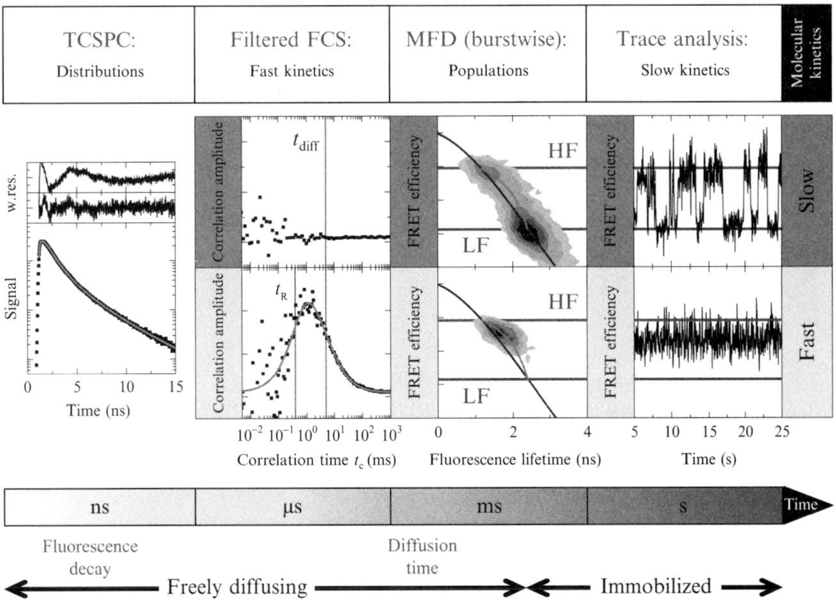


**Figure 2.2** Fluorescence timescales to study molecular kinetics using FRET. Either freely diffusing or immobilized biomolecules can be studied using FRET over multiple decades in time to resolve dynamic biomolecular heterogeneity. Time resolved fluorescence decays are better described not a by a fixed donor–acceptor distance but a distribution of interdye distances. Lifetime and polarization information, particular for each photon, can be used in combination with FCS, filtered FCS, where interconversion of states can be monitored. If relaxation is much slower than the diffusion through the detection volume, the quasi-static populations do not correlate. However, when the relaxation time is faster than diffusion time of molecules, an anticorrelation can be observed due to the interconversion between FRET states. In burstwise analysis results are presented as 2D MFD-FRET diagrams. Theoretical static and dynamic FRET lines serve as reference for position of the populations. On the top row, slow kinetics would show a smear over the dynamic FRET line, where the ends of the distributions lie on the static FRET line. The lower row in burstwise analysis represents the case of faster kinetics; the smear is gone but the single distribution lies on top of the dynamic FRET line. For slow kinetic processes, immobilized biomolecules or encapsulation in liposomes can be traced over seconds showing interconversion of FRET states. When the transitions are fast for trace analysis this alternating behavior in FRET efficiencies is lost. (For color version of this figure, the reader is referred to the online version of this chapter.)

instead a flat baseline is obtained. In the second case of fast interchange, we see the opposite (lower panel in light yellow). In 2D MFD-FRET, we see a single broad peak. Therefore, it is clear that the FRET interchange must take place somehow between nanoseconds and milliseconds. This is supported by fFCS curve, which is a strong anticorrelation peak in submillisecond time range. No further dynamics is observed in the second time range.

To conclude, a set of fluorescence fluctuation techniques in combination FRET exist, which allows us to follow structural dynamics from nanoseconds to seconds. In this context, fluorescence correlation analysis is a powerful statistical tool that allows for the analysis of FRET fluctuations without presumptions.

## 2. FRET AND FCS

FRET can be used to produce additional fluctuations on the fluorescence signal that can be detected by FCS. However, some experimental constraints limit the possible cases to be studied as discussed in the next sections. The following flow chart (Table 2.1) describes a standard procedure to perform FRET–FCS. Special care, from alignment to data acquisition and analysis, has to be taken. For example, some experimental parameters need to be determined for a typical MFD setup (e.g., molecular brightness, diffusion time, and if necessary *G*-factor for anisotropy, a spectral detection efficiencies for green and red detectors $g_G$ and $g_R$, respectively).

We start from the simplest case and increase in level of complexity in implementation and data analysis. For data analysis, we can consider the following cases.

**Case 1** Only FRET molecules (Section 2.1): A set of analytical functions is given. If these relationships do not describe the data satisfactorily, this implies that some assumptions are not held. One likely possibility is that a small fraction of donor-only molecules is in solution and Case 2 better describes the experiments. Furthermore, care needs to be taken on data fitting and warnings are mentioned at the end of each section.

**Case 2** FRET molecules mixed with donor-only (Section 2.2): For this case, the analytical solution of the correlation functions is more complicated and depends also on properties of the donor-only label sample. Special care in fitting is advised.

**Case 3** FRET–fFCS: FRET molecules mixed with donor-only at picomolar concentration (Section 3): This case applies only if a TCSPC collection board with photon time stamping is in place. fFCS can be used with several advantages summarized in Section 5.3. fFCS can be used in two variations:

- **3.1** Scatter filter and color auto- and cross-correlations (Section 3.1);
- **3.2** Interconversion between states by species selective auto- and cross-correlation (Section 3.2).

**Table 2.1** Steps for FRET–FCS experiments

| |
|---|
| 1. Align detectors in both spectral ranges and pinhole (maximize count rate) using appropriate dye solutions (n*M* concentrations) |
| 2. Measure Raman scattered light from water/buffer; if pulsed excitation and TCSPC data are used, get the *scatter* decay pattern |
| 3. Measure GREEN dye (n*M* concentration); assure proper diffusion time, $t_{\mathrm{diff}}$, get *G*-factor (if MFD setup); calculate counts-per-molecule (CPM) |
| 4. Measure GREEN dye (p*M* concentration); check single-molecule steady state anisotropy (if MFD setup) |
| 5. Measure RED dye (n*M* concentration); calculate CPM, compare diffusion time for green and red detectors |
| 6. Measure reference FRET sample (static); calibrate detection efficiency |
| 7. Measure dynamic FRET system (n*M* or p*M* concentration) |
| 8. Measure donor-labeled sample (n*M* or p*M* concentration) |
| 9. Analyze data (select which case is more appropriate)<br>9.1. Only FRET molecules (Section 2.1)<br>9.2. FRET molecules mixed with a fraction of donor-only sample (Section 2.2)<br>9.3. FRET–fFCS: FRET molecules mixed with a fraction of donor-only sample at picomolar concentration (Section 3; only possible with TCSPC data)<br>9.3.1. Scatter filter and color auto- and cross-correlations (Section 3.1)<br>9.3.2. Interconversion between states by species selective auto- and cross-correlation (Section 2.2) |

## 2.1. Only FRET molecules

For the simplest case, let us assume fluorescent molecules labeled with a donor and an acceptor dye, a FRET pair emitting in the green and red spectral range, respectively. The molecules can undergo a reversible conformational transition (i.e., open–closed, unfolding–folding, free–bound) that can be described with a first-order two-state dynamics model. For general purposes, we label these states as [1] and [2], and the equilibrium reaction is written as

$$[1] \underset{k_{2,1}}{\overset{k_{1,2}}{\rightleftharpoons}} [2], \qquad [2.11]$$

with the equilibrium constant $K = (k_{2,1}/k_{1,2})$ and the relaxation time $t_R$ measured by fluctuation analysis:

$$t_R = \left(k_{1,2} + k_{2,1}\right)^{-1}, \qquad [2.12]$$

where $k_{1,2}$ and $k_{2,1}$ are the kinetic rate constants.

If only FRET-labeled molecules are present in solution, "DA" (Donor, D and Acceptor, A) the signal detected by the green ($G$) and red ($R$) channels is given by

$$^{\mathrm{G,R}}S(t) = {}^{\mathrm{G,R}}S^{(\mathrm{DA})}. \quad [2.13]$$

For a two-color, two detection channel setup, two autocorrelation functions, one for each detector, green and red signals, respectively, and a cross-correlation function between green and red channels can be calculated (Eq. 2.14).

$$\begin{aligned}
^{\mathrm{G,G}}G(t_c) &= 1 + \frac{\langle \delta^{\mathrm{G}}S(t)\cdot\delta^{\mathrm{G}}S(t+t_c)\rangle}{\langle {}^{\mathrm{G}}S(t)\rangle\langle {}^{\mathrm{G}}S(t)\rangle},\\
^{\mathrm{R,R}}G(t_c) &= 1 + \frac{\langle \delta^{\mathrm{R}}S(t)\cdot\delta^{\mathrm{R}}S(t+t_c)\rangle}{\langle {}^{\mathrm{R}}S(t)\rangle\langle {}^{\mathrm{R}}S(t)\rangle},\\
^{\mathrm{G,R}}G(t_c) &= 1 + \frac{\langle \delta^{\mathrm{G}}S(t)\cdot\delta^{\mathrm{R}}S(t+t_c)\rangle}{\langle {}^{\mathrm{G}}S(t)\rangle\langle {}^{\mathrm{R}}S(t)\rangle}.
\end{aligned} \quad [2.14]$$

In the following, we present an analytical solution where the detection volume element is described by a three-dimensional Gaussian, as found to be appropriate for modern confocal detection. The analytical correlation function for two-level interconverting systems has been derived and used frequently in the past (Al-Soufi et al., 2005; Gurunathan & Levitus, 2009; Price et al., 2011, 2010; Slaughter, Allen, Unruh, Bieber Urbauer, & Johnson, 2004; Torres & Levitus, 2007). Therefore, we present only the final results to be used by experimentalists. Assuming $t_{\mathrm{T}} << t_{\mathrm{R}}$ and small triplet population (low excitation power), one can neglect the contribution of triplet kinetics to the correlation functions and write the model solutions for Eq. (2.14) as

$$\begin{aligned}
^{\mathrm{G,G}}G(t_c) &= 1 + \frac{1}{N}G_{\mathrm{diff}}(t_c)\left[1 + {}^{\mathrm{G,G}}AC(t_c)\right],\\
^{\mathrm{R,R}}G(t_c) &= 1 + \frac{1}{N}G_{\mathrm{diff}}(t_c)\left[1 + {}^{\mathrm{R,R}}AC(t_c)\right],\\
^{\mathrm{G,R}}G(t_c) &= 1 + \frac{1}{N}G_{\mathrm{diff}}(t_c)\left[1 + {}^{\mathrm{G,R}}CC(t_c)\right],
\end{aligned} \quad [2.15]$$

where $N = N^{(\mathrm{DA})}$ is the number of FRET molecules in the detection volume element and $G_{\mathrm{diff}}(t_c)$ is the diffusion term assuming equal diffusion coefficients, $D^{(1)} = D^{(2)}$. The kinetic reaction terms are defined by

$$
\begin{aligned}
{}^{G,G}AC(t_c) &= \frac{\left({}^{G}Q^{(1)} - {}^{G}Q^{(2)}\right)^2}{\left(K\cdot{}^{G}Q^{(1)} + {}^{G}Q^{(2)}\right)^2} K \exp\left(-\frac{t_c}{t_R}\right),\\
{}^{R,R}AC(t_c) &= \frac{\left({}^{R}Q^{(2)} - {}^{R}Q^{(1)}\right)^2}{\left({}^{R}Q^{(2)} + K\cdot{}^{R}Q^{(1)}\right)^2} K \exp\left(-\frac{t_c}{t_R}\right),\\
{}^{G,R}CC(t_c) &= -\frac{\left({}^{G}Q^{(1)} - {}^{G}Q^{(2)}\right)}{\left(K\cdot{}^{G}Q^{(1)} + {}^{G}Q^{(2)}\right)} \frac{\left({}^{R}Q^{(2)} - {}^{R}Q^{(1)}\right)}{\left({}^{R}Q^{(2)} + K\cdot{}^{R}Q^{(1)}\right)} K \exp\left(-\frac{t_c}{t_R}\right),
\end{aligned}
\qquad [2.16a]
$$

where it is assumed that the molecular brightness in the green channel of the state [1] is larger than the molecular brightness of state [2] (${}^{G}Q^{(1)} > {}^{G}Q^{(2)}$). The anticorrelated fluctuation observed in ${}^{G,R}G$ is a consequence of the fact that in the case of FRET, the brightnesses in the red channel have the opposite relation ${}^{R}Q^{(1)} < {}^{R}Q^{(2)}$. We can write Eq. (2.16a) in a more compact way by

$$
\begin{aligned}
{}^{G,G}AC(t_c) &= {}^{G,G}AC(0) \exp\left(-\frac{t_c}{t_R}\right),\\
{}^{R,R}AC(t_c) &= {}^{R,R}AC(0) \exp\left(-\frac{t_c}{t_R}\right),\\
{}^{G,R}CC(t_c) &= -\left({}^{G,G}AC(0)\cdot{}^{R,R}AC(0)\right)^{1/2} \exp\left(-\frac{t_c}{t_R}\right).
\end{aligned}
\qquad [2.16b]
$$

The pre-exponentials at correlation time $t_c = 0$, ${}^{G,G}AC(0)$, ${}^{R,R}AC(0)$, and ${}^{G,R}CC(0)$ are the amplitudes of the corresponding kinetic reaction terms, which depend on the molecular brightnesses $Q^{(i)}$ of each FRET state of the molecule. The molecular brightness corresponds to the observed photon count rate per molecule, ${}^{G,R}Q^{(i)} = ({}^{G,R}F^{(i)}/N^{(i)})$, where ${}^{G,R}F^{(i)}$ is the total fluorescence of $N^{(i)}$ molecules of species $i$. The molecular brightness is proportional to the product of the focal excitation irradiance $I_0$, the extinction coefficient $\varepsilon^{(i)}$, fluorescence quantum yield $\Phi_F^{(i)}$, and spectral-dependent detection efficiencies $g_G$ or $g_R$ for green and red detectors, respectively, as ${}^{G,R}Q^{(i)} \propto I_0\varepsilon^{(i)}\Phi_F^{(i)} g_{G,R}$ (Eggeling et al., 2001; Fries, Brand, Eggeling, Köllner, & Seidel, 1998). The FRET efficiency $E^{(i)}$ of species $i$ is related to the molecular brightness by the relationship

$$E^{(i)} = \frac{{}^{\mathrm{R}}Q^{(i)} - \alpha \cdot {}^{\mathrm{G}}Q^{(i)}}{{}^{\mathrm{R}}Q^{(i)} - \alpha \cdot {}^{\mathrm{G}}Q^{(i)} + {}^{\mathrm{G}}Q^{(i)} \frac{\Phi_{\mathrm{FA}}}{\Phi_{\mathrm{FD(0)}}} \frac{g_{\mathrm{R}}}{g_{\mathrm{G}}}}, \qquad [2.17]$$

where $\Phi_{\mathrm{FD(0)}}$ is the quantum yield of the donor in absence of acceptor, $\Phi_{\mathrm{FA}}$ is the quantum yield of the acceptor, and $\alpha$ is the spectral cross talk into the red channel.

The corresponding fractions of the number of molecules per species are defined as

$$N^{(1)} = N \frac{k_{2,1}}{k_{1,2} + k_{2,1}}, \quad N^{(2)} = N \frac{k_{1,2}}{k_{1,2} + k_{2,1}}, \quad \text{and } K = \frac{N^{(1)}}{N^{(2)}}. \qquad [2.18]$$

An example for the contribution of the kinetic reaction term and the dynamic range in the correlation amplitudes for all three cases in Eqs. (2.16a) and (2.16b) are shown in Fig. 2.3A.

In the following, we demonstrate the use of the analytical functions by simulating an experiment where FRET-labeled molecules interconvert from state [1] to state [2] with kinetic rates $k_{1,2} = 5\ \mathrm{ms}^{-1}$, $k_{2,1} = 3\ \mathrm{ms}^{-1}$, and characteristic diffusion time $t_{\mathrm{diff}} = 1$ ms. The basis for the Brownian dynamics simulations is described in Felekyan et al. (2012).

The computed correlations ${}^{\mathrm{G,G}}G(t_c)$, ${}^{\mathrm{R,R}}G(t_c)$, and ${}^{\mathrm{G,R}}G(t_c)$ for the simulated experiment are shown in Fig 2.3B. One approach to fit these correlation functions is to fit Eqs. (2.15)–(2.16b) with pre-exponential factors ${}^{\mathrm{G,G}}AC(0)$, ${}^{\mathrm{R,R}}AC(0)$, and ${}^{\mathrm{G,R}}CC(0)$ as free parameters (Table 2.2). This yields the correct relaxation time, but the amplitudes are not well defined. However, before proceeding to more complex analysis, one needs to verify that ${}^{\mathrm{G,R}}CC(t_c) = -\left({}^{\mathrm{G,G}}AC(0) \cdot {}^{\mathrm{R,R}}AC(0)\right)^{1/2}$ is applicable. If this relationship is not satisfied, the most likely explanation is that there is a small fraction of donor-only in the mixed solution and Section 2.2 is more suitable for in depth analysis.

The detailed analysis of the data shown in Fig. 2.3B requires a proper consideration of the molecular brightnesses. In order to correctly account for them, the auto- and cross-correlations are fit globally using the model functions (2.15) and (2.16a). Thereafter, three different scenarios for the fit are possible:

**i.** *Overdetermined*: This ideal case assumes that all brightnesses are known by prior knowledge. This is possible if single isolated states are measured independently. Under this assumption, the rest of the parameters are recovered with good accuracy.

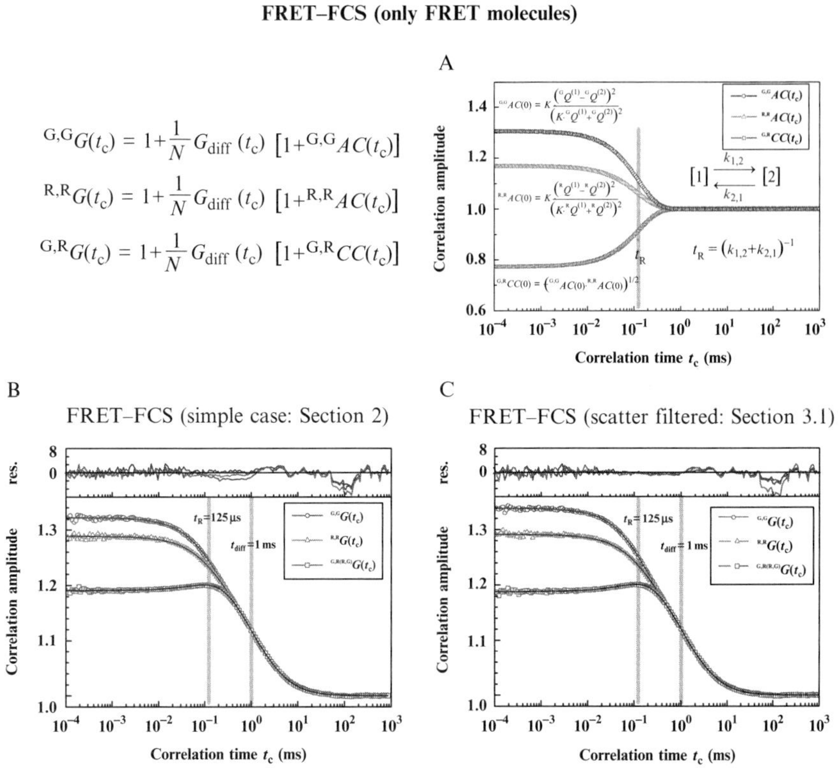


**Figure 2.3** FRET–FCS for DA-labeled molecules. Top left shows the minimum set of analytical functions needed to correctly describe a two-state kinetic scheme. (A) Dynamic terms of the model functions shown in Eq. (2.15). The used parameters are $k_{1,2}=5\ \text{ms}^{-1}$, $k_{2,1}=3\ \text{ms}^{-1}$ for rate constants and 1 ms diffusion time. The molecular brightnesses $^{G}Q^{(1)}=75$ kHz, $^{G}Q^{(2)}=25$ kHz, $^{R}Q^{(1)}=14.46$ kHz, and $^{R}Q^{(2)}=40.38$ kHz. (B) Correlation functions from simulated two-state dynamic molecules labeled with Donor and Acceptor (DA). Global fits of Eqs. (2.15) and (2.16a) are overlaid and residuals are shown on top. The following parameters were used for simulation: The average number of molecules in the detection volume $\langle N\rangle$ was 4.0. The average green and red background count rates were $\langle B_g\rangle=1.25$ kHz and $\langle B_r\rangle=0.5$ kHz, respectively. An estimated 1.5% of cross talk signal was accounted for. The molecular brightnesses are the same as in (A). The ratio of detection efficiencies ($g_G/g_R$) was 0.75, the D quantum yield $\Phi_{FD(0)}=0.8$, A quantum yield $\Phi_{FA}=0.32$, donor-only lifetime $\tau_{D(0)}=4.0$ ns. The kinetic rate constants are the same as in (A) resulting in a relaxation time of $t_R=0.125$ ms. The comparison of simulated and fitted values is shown in Tables 2.2 and 2.3. (C) Scatter filtered correlation curves from the same simulation as in panel (B). Details on this procedure are given in Section 3.1. (For interpretation of the references to color in this figure legend, the reader is referred to the online version of this chapter.)

**Table 2.2** Results of formal FRET–FCS analysis by Eqs. (2.15)–(2.16b) to simulated data shown in Figs. 2.3 and 2.4

| Parameter | Simulated values | ${}^{G,G}G(t_c)^a$ | ${}^{R,R}G(t_c)^a$ | Average ${}^{G,R}G(t_c)$ and ${}^{R,G}G(t_c)^a$ |
|---|---|---|---|---|
| *100% double-labeled molecules (Section 2.1)* | | | | |
| $\langle N\rangle$ | 4.0 | 4.08 | 4.06 | 4.07 |
| $t_{diff}$ (ms) | 1.0 | | 1.012 | |
| Pre-exponentials | 0.306 0.168 0.226 | 0.318 | 0.172 | 0.223 |
| $t_R$ (ms) | 0.125 | | 0.125 | |
| *80% double-labeled + 20% donor-only molecules (Section 2.2)* | | | | |
| $\langle N\rangle_{FCS}$ | 5 | 4.32 | 4.16 | 6.24 |
| $t_{diff}$ (ms) | 1.0 | | 0.99 | |
| Pre-exponentials | 0.133 0.167 0.220 | 0.134 | 0.165 | 0.224 |
| $t_R$ (ms) | 0.125 | | 0.125 | |

$\langle N\rangle$ and pre-exponentials were free whereas $t_{diff}$ and $t_R$ were global fit parameters.
[a]Additional parameters were kept fixed during fitting: Offset = 1, $\omega_0 = 0.55$ μm, $z_0 = 2$ μm, $t_{diff}^{(D)}/t_{diff}^{(DA)} = 1$.

**ii.** *Partially determined*: A more realistic case is when, at least, the FRET efficiency $E^{(i)}$ of one state is known, and therefore the corresponding molecular brightnesses in green and red channels are determined. Additionally, a good estimate for the second efficiency is suggested. The determination of one FRET efficiency and the corresponding brightnesses can be achieved by shifting equilibrium to one of the states and measuring fluorescence lifetimes by TCSPC or by fluorescence intensity distribution analysis (FIDA) (see Section 5.2). After fitting, one should verify that the dye properties do not change between two FRET states, which is formally described by Eq. (2.19) below. By adding these constraints, fits are stabilized and reliability increases.

**iii.** *Determined and scatter filtered*: as in (ii). Additional filtering to reduce influence of uncorrelated scatter signal improves accuracy of the amplitudes. A more detailed description can be found in Section 3.

Care must be taken not to let all parameters free, because the system is *undetermined*, that means more variables than experimental observables are in the system of equations that describe FRET–FCS. Fitting an undetermined system would lead to multiple solutions, where all but one are wrong, unless initial guesses are in close proximity to the real parameters.

For the partially determined case, the following relationship for total corrected brightness ${}^{\mathrm{tot}}Q$ also needs to hold true,

$$ {}^{\mathrm{G}}Q^{(i)} + \beta.{}^{\mathrm{R}}Q^{(i)} = {}^{\mathrm{tot}}Q = \mathrm{const.}, \qquad [2.19] $$

where the $\beta$ factor is defined as

$$ \beta = \frac{1}{\frac{\Phi_{\mathrm{FA}}}{\Phi_{\mathrm{FD(0)}}}\frac{g_{\mathrm{R}}}{g_{\mathrm{G}}} - \alpha}. $$

This relationship comes from the fact that the brightness and FRET efficiencies are related (Eq. 2.17), independently of the state in which the molecule is found, that is, it is necessary that dye spectra and quantum yields $\Phi_{\mathrm{FD(0)}}$ and $\Phi_{\mathrm{FA}}$ do not change between states.

Table 2.3 summarizes the recovered parameters considering the discussed global fitting scenarios. First column shows the simulated parameters.

In conclusion, when only FRET (DA)-labeled molecules are present in solution, we can state the following:

**i.** The reaction time $t_{\mathrm{R}}$ can be easily extracted only when $t_{\mathrm{T}} << t_{\mathrm{R}}$. This is possible even by fitting only the color cross-correlation function: formal Eqs. (2.15)–(2.16b) and detailed Eqs. (2.15–2.16a) can be used. But only the detailed equations allow one to extract the information of the amplitudes, which is crucial to determine equilibrium constant. If one is interested only in the relaxation time, Eqs. (2.15)–(2.16b) are recommended.

**ii.** The amplitudes of the kinetic reaction terms ${}^{\mathrm{G,G}}AC(t_c)$, ${}^{\mathrm{R,R}}AC(t_c)$, and ${}^{\mathrm{G,R}}CC(t_c)$ depend on two factors:

 **a.** the ratio of the brightnesses $Q^{(1)}/Q^{(2)}$ of the states;

 **b.** the equilibrium constant K.

**iii.** Global fit of FRET–FCS equations improves the stability of variable parameters.

## 2.2. FRET molecules mixed with donor-only sample

Having in an experiment a 100% clean solution of only functional DA molecules is, in the majority of cases, unrealistic. It is almost inevitable to have a small fraction of donor-only molecules. To consider the donor-only contribution, the average number of molecules observed in the confocal

**Table 2.3** Alternative results of detailed FRET–FCS analysis by Eqs. (2.15) and (2.16a) applied to the same averaged color auto- and cross-correlation curves of a simulated FCS experiment (details in caption of Fig. 2.3)

| | | FRET–FCS (only FRET molecules) | | Scatter filtered |
|---|---|---|---|---|
| | Simulated values | Overdetermined[a] | Partially determined[a] | Partially determined[a] |
| $\langle N \rangle$ | 4.0 | 4.05 | 4.05 | 4.01 |
| $t_{diff}$ (ms) | 1.0 | 1.002 | 0.996 | 1.005 |
| ${}^{G}Q^{(1)}$ (kHz) | 75 | Fixed | Fixed | Fixed |
| ${}^{G}Q^{(2)}$ (kHz) | 25 | Fixed | 25.02 | 22.58 |
| ${}^{R}Q^{(1)}$ (kHz) | 14.46 | Fixed | Fixed | Fixed |
| ${}^{R}Q^{(2)}$ (kHz) | 40.38 | Fixed | 40.95 | 40.40 |
| $K$ | 0.6 | 0.605 | 0.59 | 0.600 |
| $t_R$ (ms) | 0.125 | 0.125 | 0.124 | 0.125 |
| $E^{(1)\ b}$ | 0.25 | Fixed | Fixed | Fixed |
| $E^{(2)\ b}$ | 0.75 | fixed | 0.75 | 0.77 |
| $k_{1,2}\ (\text{ms}^{-1})^b$ | 5 | 4.98 | 5.10 | 5.00 |
| $k_{2,1}\ (\text{ms}^{-1})^b$ | 3 | 3.02 | 3.00 | 3.00 |

[a]Additional parameters were kept fixed during fitting: Offset$=1$, $\omega_0=0.55$ μm, $z_0=2$ μm, ${}^{G}Q^{(D)}=100$ kHz, cross talk $\alpha=0.015$, $x^{(D)}=0$, $t_{diff}{}^{(D)}/t_{diff}{}^{(DA)}=1$.
[b]Not a fit parameter: $k_{1,2}=\frac{1}{(1+K)t_R}$; $k_{2,1}=K\cdot k_{1,2}$; $E^{(i)}=\frac{{}^{R}Q^{(i)}-\alpha\cdot{}^{G}Q^{(i)}}{{}^{R}Q^{(i)}-\alpha\cdot{}^{G}Q^{(i)}+{}^{G}Q^{(i)}\frac{\Phi_{FA}}{\Phi_{FD(0)}}\frac{g_R}{g_G}}$.

volume element at any given time is given by the sum of donor-only molecules $N^{(D)}$ and the number of FRET molecules $N^{(DA)}$,

$$N=N^{(DA)}+N^{(D)}. \qquad [2.20]$$

One can express the fraction of donor-only molecules $x^{(D)}$ as

$$x^{(D)}=\frac{N^{(D)}}{N^{(D)}+N^{(DA)}}. \qquad [2.21]$$

The brightnesses of donor-only molecules are

$${}^{G}Q^{(D)} \text{ and } {}^{R}Q^{(D)}=\alpha\cdot{}^{G}Q^{(D)} \qquad [2.22]$$

for green and red channels, respectively. $\alpha$ is the spectral cross talk to the red detection channels, given by the experimental setup. The correlation functions can be written as

$$
\begin{aligned}
{}^{G,G}G(t_c) &= 1 + \frac{1}{N} G_{\text{diff}}(t_c)\left[1 + {}^{G,G}AC'^{(DA)}(t_c)\right],\\
{}^{R,R}G(t_c) &= 1 + \frac{1}{N} G_{\text{diff}}(t_c)\left[1 + {}^{R,R}AC'^{(DA)}(t_c)\right],\\
{}^{G,R}G(t_c) &= 1 + \frac{1}{N} G_{\text{diff}}(t_c)\left[1 + {}^{G,R}CC'^{(DA)}(t_c)\right].
\end{aligned}
\qquad [2.23]
$$

Assuming that all molecules have the same diffusion coefficient, that is, $D^{(D)} = D^{(DA)}$ or $t_{\text{diff}}^{(D)} = t_{\text{diff}}^{(DA)}$, and ${}^{G,R}G_{\text{diff}}^{(D)}(t_c) = {}^{G,R}G_{\text{diff}}^{(DA)}(t_c) = G_{\text{diff}}(t_c)$, the differences between Eqs. (2.15) and (2.23) are the definitions of the kinetic reaction terms. In the latter case, they are defined as

$$
{}^{G,G}AC'^{(DA)}(t_c) = \frac{x^{(D)}\left({}^{G}Q^{(D)}\right)^2 + \frac{\left(1-x^{(D)}\right)}{(K+1)^2}\left[K\cdot{}^{G}Q^{(1)} + {}^{G}Q^{(2)}\right]^2\left[1 + {}^{G,G}AC(t_c)\right]}{\left(x^{(D)}\cdot{}^{G}Q^{(D)} + \frac{\left(1-x^{(D)}\right)}{(K+1)}\left[K\cdot{}^{G}Q^{(1)} + {}^{G}Q^{(2)}\right]\right)^2} - 1,
$$

$$
{}^{R,R}AC'^{(DA)}(t_c) = \frac{x^{(D)}\left(\alpha\cdot{}^{G}Q^{(D)}\right)^2 + \frac{\left(1-x^{(D)}\right)}{(K+1)^2}\left[K\cdot{}^{R}Q^{(1)} + {}^{R}Q^{(2)}\right]^2\left[1 + {}^{R,R}AC(t_c)\right]}{\left(x^{(D)}\cdot\alpha\cdot{}^{G}Q^{(D)} + \frac{\left(1-x^{(D)}\right)}{(K+1)}\left[K\cdot{}^{R}Q^{(1)} + {}^{R}Q^{(2)}\right]\right)^2} - 1,
$$

$$
{}^{G,R}CC'^{(DA)}(t_c) = \frac{x^{(D)}\alpha\left({}^{G}Q^{(D)}\right)^2 + \frac{\left(1-x^{(D)}\right)}{(K+1)^2}\left[K\cdot{}^{G}Q^{(1)} + {}^{G}Q^{(2)}\right]\left[K\cdot{}^{R}Q^{(1)} + {}^{R}Q^{(2)}\right]\left[1 + {}^{G,R}CC(t_c)\right]}{\left(x^{(D)}\cdot{}^{G}Q^{(D)} + \frac{\left(1-x^{(D)}\right)}{(K+1)}\left[K\cdot{}^{G}Q^{(1)} + {}^{G}Q^{(2)}\right]\right)\left(x^{(D)}\cdot\alpha\cdot{}^{G}Q^{(D)} + \frac{\left(1-x^{(D)}\right)}{(K+1)}\left[K\cdot{}^{R}Q^{(1)} + {}^{R}Q^{(2)}\right]\right)} - 1.
\qquad [2.24a]
$$

One can easily show that for $x^{(D)} = 0$, Eq. (2.24a) reduce to Eq. (2.16a). A simple alternative to Eq. (2.24a) is to write the kinetic terms as

$$
\begin{aligned}
{}^{G,G}AC'^{(DA)}(t_c) &= A\exp\left(-\frac{t}{t_c}\right) + B,\\
{}^{R,R}AC'^{(DA)}(t_c) &= C\exp\left(-\frac{t}{t_c}\right) + D,\\
{}^{G,R}CC'^{(DA)}(t_c) &= -(AC)^{1/2}\exp\left(-\frac{t}{t_c}\right) - (BD)^{1/2},
\end{aligned}
\qquad [2.24b]
$$

where $A$, $B$, $C$, and $D$ are arbitrary constants (Price et al., 2010).

To validate the results, we have taken the parameters of the simulation described in the case for only FRET molecules and added a small fraction of donor-only molecules, $N^{(D)}=1$ and $N^{(DA)}=4$, with $x^{(D)}=0.2$. The brightness of the donor-only ${}^{G}Q^{(D)}=100$ kHz. The simulated data were correlated and fit globally using Eqs. (2.23) and (2.24a) as shown in Fig. 2.4. The fit results are listed in Table 2.4.

As in the previous section, the following three scenarios of fits are discussed:

**i.** *Overdetermined*: This can only happen if prior knowledge of states is available. If this is the case, the proper fraction of donor-only can be extracted.

**ii.** *Partially determined*: Same conditions apply as in the previous section; experimental determination of at least the FRET efficiency of one state is required and Eq. (2.19) has to be fulfilled. In addition, donor-only brightness needs to be known, which can be determined in the separate measurement. Absolute brightnesses of the FRET species are not required, but they can be scaled to the donor-only brightness. Finally, to assure proper fitting Eq. (2.19) has to be satisfied.

**FRET–FCS (FRET + Donor-only molecules)**

$$^{G,G}G(t_c) = 1+\frac{1}{N}\,G_{\mathrm{diff}}(t_c)\,\left[1+{}^{G,G}AC'^{(DA)}(t_c)\right]$$

$$^{R,R}G(t_c) = 1+\frac{1}{N}\,G_{\mathrm{diff}}(t_c)\,\left[1+{}^{R,R}AC'^{(DA)}(t_c)\right]$$

$$^{G,R}G(t_c) = 1+\frac{1}{N}\,G_{\mathrm{diff}}(t_c)\,\left[1+{}^{G,R}CC'^{(DA)}(t_c)\right]$$

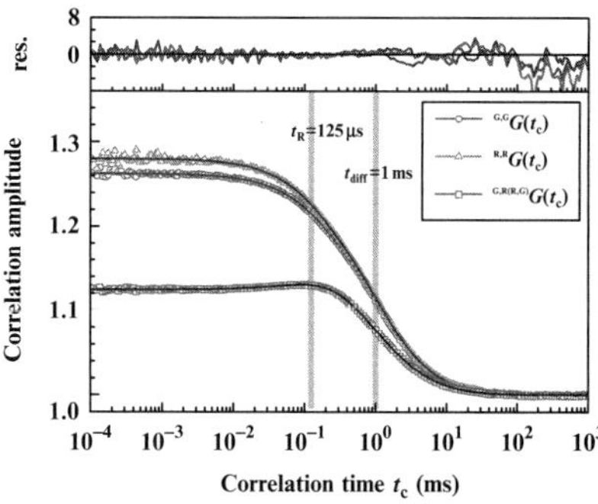


**Figure 2.4** FRET–FCS of FRET molecules mixed with donor-only. Left: The minimum set of analytical functions needed to correctly describe a two-state kinetic scheme in the presence of donor-only. The kinetic terms are described in Eq. (2.24a). Right: The color auto- and averaged (${}^{G,R}G(t_c)$ and ${}^{R,G}G(t_c)$) cross-correlation curves of simulated data for two-state dynamic molecules labeled with Donor and Acceptor (DA) mixed with 20% donor-only molecules. Overlaid fit curves correspond to Eqs. (2.23) and (2.24a). Residuals are shown on top. The parameters for simulations were identical to those in Fig. 2.3 with the exception of the additional donor-only fraction with brightness ${}^{G}Q^{(D)}=100$ kHz. Results of global fits are summarized in Table 2.4. (For color version of this figure, the reader is referred to the online version of this chapter.)

**Table 2.4** Results of detailed FRET–FCS in the presence of donor-only sample in the simulation (for parameters, see caption of Figs. 2.3 and 2.4)

| | | **FRET–FCS (FRET + donor-only molecules)** | | **Scatter filtered** |
|---|---|---|---|---|
| | **Simulated values** | **Overdetermined**[a] | **Partially determined**[a] | **Partially determined**[a] |
| $\langle N \rangle$ | 5.0 | 5.06 | 5.06 | 5.03 |
| $t_{diff}$ (ms) | 1.0 | 0.992 | 0.994 | 0.995 |
| ${}^{G}Q^{(1)}$ (kHz) | 75 | Fixed | Fixed | Fixed |
| ${}^{G}Q^{(2)}$ (kHz) | 25 | Fixed | 24.9 | 22.01 |
| ${}^{R}Q^{(1)}$ (kHz) | 14.46 | Fixed | Fixed | Fixed |
| ${}^{R}Q^{(2)}$ (kHz) | 40.38 | Fixed | 40.89 | 40.77 |
| $x^{(D)}$ | 0.2 | 0.197 | 0.195 | 0.205 |
| $K$ | 0.6 | 0.58 | 0.59 | 0.597 |
| $t_R$ (ms) | 0.125 | 0.125 | 0.126 | 0.126 |
| $E^{(1)}$ [b] | 0.25 | Fixed | Fixed | 0.25 |
| $E^{(2)}$ [b] | 0.75 | Fixed | 0.75 | 0.78 |
| $k_{1,2}$ $(ms^{-1})$[b] | 5 | 5.08 | 5.02 | 4.98 |
| $k_{2,1}$ $(ms^{-1})$[b] | 3 | 2.95 | 2.94 | 2.97 |

Global fit of data shown in Fig. 2.4 using Eqs. (2.23) and (2.24a).
[a]Additional parameters were kept fixed during fitting: Offset = 1, $\omega_0 = 0.55\ \mu m$, $z_0 = 2\ \mu m$, ${}^{G}Q^{(D)} = 100$ kHz, cross talk $\alpha = 0.015$, $t_{diff}{}^{(D)}/t_{diff}{}^{(DA)} = 1$.
[b]Not a fit parameter: $k_{1,2} = \frac{1}{(1+K)t_R}$; $k_{2,1} = K \cdot k_{1,2}$; $E^{(i)} = \frac{{}^{R}Q^{(i)} - \alpha \cdot {}^{G}Q^{(i)}}{{}^{R}Q^{(i)} - \alpha \cdot {}^{G}Q^{(i)} + {}^{G}Q^{(i)} \frac{\Phi_{FA}}{\Phi_{FD(0)}} \frac{g_R}{g_G}}$.

**iii.** *Determined and scatter filtered*: as in (ii). Additional filtering to reduce influence of uncorrelated scatter signal improves accuracy of the amplitudes.

In a similar fashion to the previous section, allowing all parameters to be free in the fitting routine will result in an unstable and *underdetermined* system.

If one is unsure of the existence of a significant fraction of donor-only in the mixture, the FRET–FCS curves can be treated as the simplest case (only FRET molecules) and fit by Eqs. (2.15)–(2.16b). The results are listed in Table 2.2. If donor-only sample is present the pre-exponential amplitudes will not fulfill the relationship ${}^{G,G}CC(0) = -\left({}^{G,G}AC(0) \cdot {}^{R,R}AC(0)\right)^{1/2}$ (see Eq. 2.16b).

In summary, in the presence of a fraction of donor-only labeled species, we can conclude that

**i.** The reaction time $t_R$ can be easily and accurately extracted when $t_T << t_R$.

**ii.** The amplitudes of the kinetic reaction terms ${}^{G,G}AC'^{(DA)}(t_c)$, ${}^{R,R}AC'^{(DA)}(t_c)$, and ${}^{G,R}CC'^{(DA)}(t_c)$ depend on the brightnesses of each state and their concentrations, adding two additional parameters to the already multidimensional fit. Without prior knowledge of the brightness of at least one state (e.g., by TCSPC analysis of single-molecule subensembles), the fits are underdetermined and therefore unstable.

**iii.** Global fit of FRET–FCS, with the addition of donor-only, improves stability of the variable parameters.

## 3. FILTERED FCS

The standard color auto- and cross-correlation analysis for FRET has several limitations. For example, dynamics cannot be detected when there are no significant changes in FRET efficiency between states, and at low sample concentration, the background distorts the amplitudes of the correlations.

The simplest filter procedure is time to amplitude converter (TAC) gating where the photons in the first TAC channels, where most of scatter photons contribute to the decay histograms, are excluded from computation of correlation curves (Eggeling et al., 2001; Lamb, Schenk, Rocker, Scalfi-Happ, & Nienhaus, 2000). This methodology is very subjective in the definition of the TAC gate.

fFCS is based on fluorescence lifetime correlation spectroscopy developed by other authors (Böhmer, Wahl, Rahn, Erdmann, & Enderlein, 2002; Enderlein & Gregor, 2005; Gregor & Enderlein, 2007; Kapusta, Wahl, Benda, Hof, & Enderlein, 2007) but improves its selectivity by adding polarization and spectral information of the fluorescence decays. In this way, fFCS, as described in Felekyan et al. (2012) and Felekyan, Kalinin, Valeri, and Seidel (2009), can differentiate between two species even when their brightnesses are the same. This is possible only if changes in rotational correlation times are significant as indicated by their anisotropies.

We can apply this methodology in two different ways:

**i.** *Background removal to obtain correct correlation amplitudes.* In confocal single-molecule experiments, subnanomolar concentrations are needed. Under these conditions, the signal contains a significant amount of scatter photons. fFCS can be used to extract the true fluorescence signal from background noise. Given that pulsed excitation and fast electronics are used, fFCS allows one to eliminate the influence of scattered background for any FCS input data (Böhmer et al., 2002; Enderlein and Gregor, 2005; Gregor and Enderlein, 2007; Kapusta et al., 2007). fFCS fully recovers the correct $G(0)$ amplitude, which is crucial for any complete analysis of FCS measurements. In Section 3.1, we discuss the use of fFCS for a mixture of FRET-labeled species in the presence of a fraction of donor-only labeled sample at picomolar concentrations.

**ii.** *Selective species auto- and cross-correlation in a mixture.* fFCS was introduced as a tool to extract species-specific auto- and cross-correlation curves from a heterogeneous mixture (Felekyan et al., 2012, 2009). Separation is possible if the species differ significantly in lifetime, anisotropy, or spectra. Therefore, dynamics that lead to changes in FRET efficiencies can be investigated with fFCS and it is possible to determine the kinetics of interconverting species.

To briefly describe fFCS, we need to understand that the contribution to the total signal detected in all channels comes from the sum of three different factors.

$$S(t) = B^{(\text{scatter})} + F^{(\text{D})} + F^{(\text{DA})}, \qquad [2.25]$$

where $B^{(\text{scatter})}$ is the signal coming from Raman (scatter) photons giving rise to the instrument response function. $F^{(\text{D})}$ is the pure fluorescence of the donor-only labeled species, and $F^{(\text{DA})} = F^{(1)} + F^{(2)}$ is the total fluorescence contribution of the FRET-labeled species in state [1], $F^{(1)}$, and species [2], $F^{(2)}$, respectively.

The goal is to compute the correlations that are detected from each species and not the full signal correlation as stated in Eq. (2.1). The species-specific auto- and cross-correlation functions are defined in Eq. (2.26):

$$G^{(i,m)}(t_c) = \frac{\langle F^{(i)}(t) F^{(m)}(t+t_c)\rangle}{\langle F^{(i)}(t)\rangle \langle F^{(m)}(t+t_c)\rangle} = \frac{\left\langle \left(\sum_{j=1}^{dL} f_j^{(i)} S_j(t)\right)\left(\sum_{j=1}^{dL} f_j^{(m)} S_j(t+t_c)\right)\right\rangle}{\left\langle \sum_{j=1}^{dL} f_j^{(i)} S_j(t)\right\rangle \left\langle \sum_{j=1}^{dL} f_j^{(m)} S_j(t+t_c)\right\rangle}, \qquad [2.26]$$

where $(i)$ and $(m)$ are two species in a mixture. A set of filters $f_j^{(i)}$ is introduced, therefore the name of fFCS, that are dependent on the arrival time of each photon after each excitation pulse. The signal $S_j(t)$ obtained via pulsed excitation is recorded at each $j = 1, \ldots, L$ TAC channel via TCSPC.

The signals and filters for all (d) detectors are stacked in their respective arrays with dimensions $dL$ for global minimization according to Felekyan et al. (2012). The prerequisite of fFCS is that each species has a specific fluorescence decay histogram. The filters are defined in such a way that the relative "error" or difference between the photon count per species ($w^{(i)}$) and the weighted histogram $f_j^{(i)} H_j$ is minimized as defined in Eq. (2.27).

$$\left\langle \left( \sum_{j=1}^{dL} f_j^{(i)} H_j - w^{(i)} \right)^2 \right\rangle \to \min., \qquad [2.27]$$

<> brackets represent time averaging.

It is required to express the decay histogram $H_j$ as a linear combination of the conditional probability distributions $p_j^{(i)}$, such as

$$H_j = \sum_{i=1}^{n(=2)} w^{(i)} p_j^{(i)}, \quad \text{with} \sum_{j=1}^{dL} p_j^{(i)} = 1. \qquad [2.28]$$

There are several strategies to obtain the proper decay histograms (Eq. 2.28) for each species in the mixture.

**i.** Individual species can be measured independently at >n$M$ concentrations to minimize contributions from scattered light.

**ii.** The assignment of species can be performed at single-molecule conditions. When analyzing single-molecule bursts, the separation of species can only be achieved if $t_R > t_{diff}$. Dynamic mixing occurs when $t_R < t_{diff}$ and separation of states is not possible. In this last case, subensemble decay analysis of mixed peaks is needed to identify the species-specific fluorescence lifetime components. Synthetic decay patterns can be used in the second step to generate decays for each species based on the fit results.

### 3.1. FRET–fFCS: FRET molecules mixed with donor-only at picomolar concentration

One of the advantages of fFCS is that measurements can be done even at ~p$M$ concentrations. At these concentrations, most of the signal recorded is from scattered photons and only a small fraction of events arrive from the FRET-labeled sample.

As an example, we present a simulation similar to the one presented in Section 2.2, but differs in one parameter, $\langle N \rangle = 0.005$. At this picomolar concentration, the contribution of scatter to our signal is considerable and has to be taken into account; otherwise, the unfiltered correlation amplitudes are a factor ~20 times smaller than the expected ones. Therefore, we generated two filters following Eq. (2.27) where the scatter photons contribute to species [1] and the mixture of FRET-labeled and donor-only molecules is a joined pseudo-species [2].

A set of similar correlation curves as done in Section 2.2 can be computed for the pseudo-species [2] (FRET labeled with donor-only) where the scatter photons have been "filtered." The filters were generated using Eq. (2.27) for a setup with four detectors, two spectral ranges and two polarizations (see Fig. 2.1). The "known" green and red decay patterns, for *scatter* and the mixture, in perpendicular and parallel channels are shown as stacked histograms in Fig. 2.5A. The scatter patterns for green and red detection channels were simulated in the same way.

From these patterns, the corresponding stacked "filters" with length $2L$ ($d=2$) are shown in Fig. 2.5B. The color auto- and cross-correlation functions are shown in Fig. 2.5C. It can be seen directly that the amplitude of the correlation function is at least two orders of magnitude larger when compared to Fig. 2.4. An additional simulation with $\langle N \rangle = 0.05$ shows a decrease of exactly one order of magnitude on the amplitude of the correlation function (Fig. 2.5D). This is one major advantage of fFCS: the proper amplitudes can be obtained even when scattered light contributes significantly.

We fitted the correlation functions as done in Section 2.2 because the only change is the concentration, that is, the fraction of scattered photons that has been filtered from the total signal. The results from the fits are compared in Table 2.5.

Similarly, data from Figs. 2.3 and 2.4 were scatter filtered showing an improvement in the average number of molecules indicating that even at nanomolar concentration scattered photons can affect the amplitude of the correlation function (last columns in Tables 2.3 and 2.4).

After filtering the data for the contribution from scattered photons and generating the color auto- and cross-correlations functions, we are still restricted to the same factors as in the models described in previous sections, except that we have correctly extracted the amplitudes.

By using the scatter filtering, we can conclude the following:

**i.** The reaction time $t_R$ can be easily extracted when $t_T << t_R$.

**ii.** The amplitudes of the kinetic reaction terms ${}^{G,G}AC'^{(DA)}(t_c)$, ${}^{R,R}AC'^{(DA)}(t_c)$, and ${}^{G,R}CC'^{(DA)}(t_c)$ depend on the brightnesses of each state and their concentrations as in Section 2.2.

**iii.** Global fit FRET–FCS, including donor-only, improves stability of variable parameters.

**iv.** Correct correlation amplitudes at picomolar or higher concentrations can be obtained.

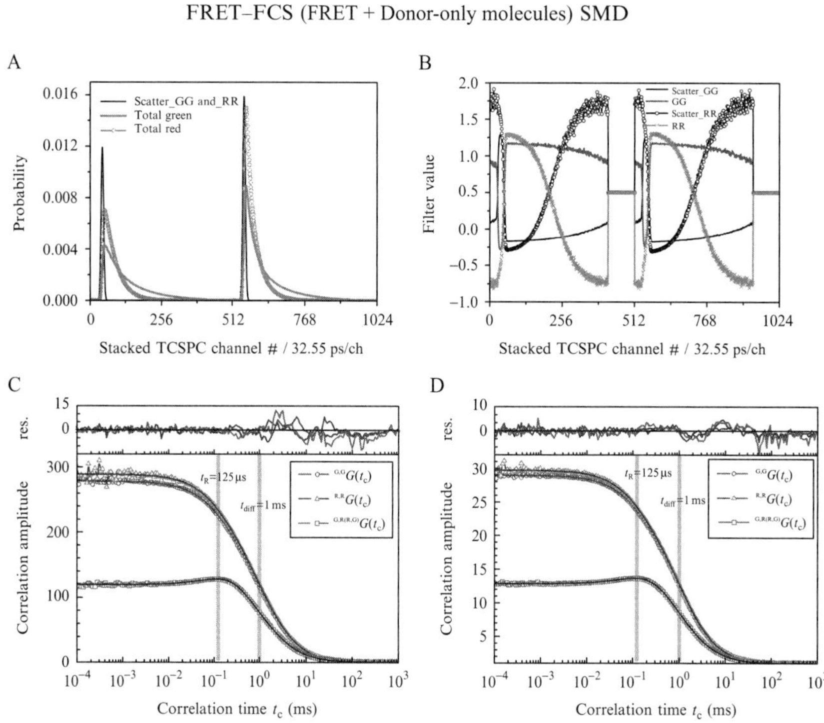


**Figure 2.5** FRET–FCS of DA molecules with donor-only (with scatter filter). Simulation of two-state dynamics in the presence of donor-only at single-molecule concentrations. (A) Stacked conditional probabilities (Eq. (2.28) with $d=2$) for the two selected species. One pattern was chosen to be the *scatter* and the rest was the mixture of DA molecules and donor-only. The stacked TCSPC channels (1–512, 513–1024) show the perpendicular and parallel polarized conditional probability distributions respectively. (B) Calculated filters according to Eq. (2.27) for the species selection. (C) Averaged color auto- and cross-correlation curves of simulated data for two-state dynamic molecules labeled with Donor and Acceptor (DA) mixed with 20% donor-only molecules at picomolar concentration. The average number of molecules in the detection volume $\langle N \rangle$ was 0.005. Overlaid fit curves correspond to Eqs. (2.23) and (2.24a). All other parameters where kept the same as done for Fig. 2.4. Fit results for a global fit using Eqs.(2.23) and (2.24a) are summarized in Table 2.5 and the residuals are shown in the top graph. (D) Averaged color auto- and cross-correlation curves of simulated data for the same two-state dynamic molecules as in (C) except that $\langle N \rangle$ was 0.05. Note that amplitudes are 10 times smaller corresponding to the factor of 10 larger total number of molecules. Residuals from the fit are shown in the top graph and the results are summarized in Table 2.5. (For color version of this figure, the reader is referred to the online version of this chapter.)

**Table 2.5** Results of fFCS analysis for simulations in Section 3.1 (scatter filtered) and Section 3.2 (species filtered)

| | Simulated values | Scatter filtered FCS (two species)[a,b] | Two-channel filtered FCS (four species)[a] | Four-channel filtered FCS (four species)[a] |
|---|---|---|---|---|
| $\langle N \rangle$ | 0.05 | 0.05 | 0.05 | 0.05 |
| $t_{\mathrm{diff}}$ (ms) | 1.0 | 0.992 | 0.985 | 0.986 |
| ${}^{\mathrm{G}}Q^{(1)}$ (kHz) | 75 | Fixed | | |
| ${}^{\mathrm{G}}Q^{(2)}$ (kHz) | 25 | 20.86 | | |
| ${}^{\mathrm{R}}Q^{(1)}$ (kHz) | 14.46 | Fixed | | |
| ${}^{\mathrm{R}}Q^{(2)}$ (kHz) | 40.38 | 40.17 | | |
| $x^{(\mathrm{D})}$ | 0.2 | 0.205 | 0.200 | 0.199 |
| $K$ | 0.6 | 0.596 | 0.592 | 0.599 |
| $t_{\mathrm{R}}$ (ms) | 0.125 | 0.127 | 0.125 | 0.126 |
| $E^{(1)}$ [c] | 0.25 | Fixed | | |
| $E^{(2)}$ [c] | 0.75 | 0.78 | | |
| $k_{1,2}$ ($\mathrm{ms}^{-1}$)[c] | 5 | 4.93 | 5.01[d] | 4.90[d] |
| $k_{2,1}$ ($\mathrm{ms}^{-1}$)[c] | 3 | 2.94 | 2.97[d] | 2.93[d] |

Experiment details are given in the captions of Figs. 2.5–2.7.
[a]Additional parameters were kept fixed during fitting: Offset = 1, $\omega_0 = 0.55$ μm, $z_0 = 2$ μm, ${}^{\mathrm{G}}Q^{(\mathrm{D})} =$ 100 kHz, cross talk $\alpha = 0.015$, $t_{\mathrm{diff}}{}^{(\mathrm{D})}/t_{\mathrm{diff}}{}^{(\mathrm{DA})} = 1$.
[b]Partially determined.
[c]Not a fit parameter: $k_{1,2} = \frac{1}{(1+K)t_{\mathrm{R}}}$; $k_{2,1} = K \cdot k_{1,2}$; $E^{(i)} = \frac{{}^{\mathrm{R}}Q^{(i)} - \alpha \cdot {}^{\mathrm{G}}Q^{(i)}}{{}^{\mathrm{R}}Q^{(i)} - \alpha \cdot {}^{\mathrm{G}}Q^{(i)} + {}^{\mathrm{G}}Q^{(i)} \frac{\Phi_{\mathrm{FA}}}{\Phi_{\mathrm{FD}(0)}} \frac{g_{\mathrm{R}}}{g_{\mathrm{G}}}}$.
[d]Calculated from Eqs. (2.18).

Finally, filters unequally weight the pseudo-species contribution of a mixed fluorescent species. Therefore, care must be taken with the amplitude of the auto- and cross-correlations, determined either from TAC gating or from scatter filtering. Furthermore, the temporal overlap with the scatter decay, in particular for those with very short lifetimes, can also affect the amplitudes of the correlation functions. This impact increases with the width of the scatter decay pattern. However, this is not the case if distinct differences in the decays are observed or if multidimensional filtering is applied (e.g., lifetime, polarization, and spectra). In addition, this artifact does not appear for correlations obtained from filters generated from individual species as discussed in the next section.

## 3.2. FRET and fFCS interconversion between states at SMD

Single-molecule FRET experiments using MFD are ideal for identifying heterogeneous solutions and dynamics (Kalinin, Valeri, Antonik, Felekyan, & Seidel, 2010; Widengren et al., 2006). A comprehensive review on FRET and MFD can be found in Sisamakis et al. (2010). In MFD, the emitted fluorescence is split by spectra and polarization (see Fig. 2.1). Bursts of photons from single molecular events are characterized by the FRET indicator $F_D/F_A$ (ratio of donor fluorescence over acceptor fluorescence) and the average fluorescence lifetime of donor in the presence of the acceptor, $\tau_{D(A)}$. The frequency of bursts is analyzed in 2D histograms, where $F_D/F_A$ is plotted against $\tau_{D(A)}$ for each burst. If FRET molecules exchange states at a rate faster than the diffusion time ($t_R < t_{diff}$), the width of distribution and the location of the 2D histogram represents, among other things, the dynamic equilibrium of the sample. For example, in a quasi-static equilibrium ($t_R >> t_{diff}$), a static FRET line in $F_D/F_A$ versus $\tau_{D(A)}$ plots defined as

$$\left(\frac{F_D}{F_A}\right)_{\text{static}} = \frac{\Phi_{FD(0)}}{\Phi_{FA}} \frac{\tau_{D(A)}}{\tau_{D(0)} - \tau_{D(A)}}. \qquad [2.29]$$

It represents all possible positions in the 2D histogram where each event is located. In Eq. (2.29), $\Phi_{FD(0)}$ and $\Phi_{FA}$ represent the quantum yields of the donor (without FRET) and acceptor fluorophores, respectively. The fluorescence lifetime of the donor without acceptor is $\tau_{D(0)}$ and $\tau_{D(A)}$ with an acceptor. When the bursts in the 2D histograms do not fall on this line, we can postulate the existence of faster dynamics (Kalinin et al., 2010). For the simplest case of a two-state system with single exponential decay times $\tau_{D(A)}{}^{(i)}$, a dynamic FRET line can be derived in the form

$$\left(\frac{F_D}{F_A}\right)_{\text{dyn}} = \frac{\Phi_{FD(0)}}{\Phi_{FA}} \frac{\left[\tau_{D(A)}{}^{(1)}\tau_{D(A)}{}^{(2)}\right]}{\tau_{D(0)}\left[\tau_{D(A)}{}^{(1)} + \tau_{D(A)}{}^{(2)} - \tau_{D(A)}\right] - \left[\tau_{D(A)}{}^{(1)}\tau_{D(A)}{}^{(2)}\right]}, \qquad [2.30]$$

where $\tau_{D(A)} = \langle\tau\rangle_f$ corresponds to the fluorescence weighted average lifetime determined by the maximum likelihood estimator (Maus et al., 2001). In this way, an estimation of the original states [1] and [2] can be obtained. The spread of the distribution over this dynamic FRET line also contains information on the timescale of the dynamics: the wider the spread, the slower the dynamics.

Figure 2.6 shows the $F_D/F_A - \tau_{D(A)}$ 2D histogram for the simulation presented in Section 3.1. The one-dimensional histograms are also shown as projections on the respective axes. For reference, we have added the static

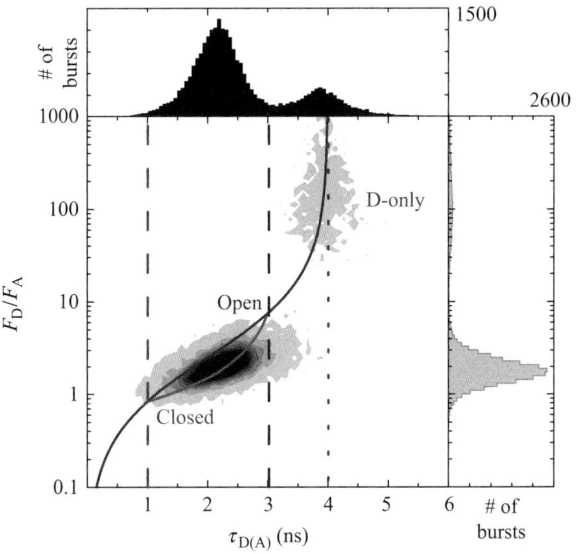


**Figure 2.6** 2D histogram $F_D/F_A$ versus $\tau_{D(A)}$ of simulated SMD data from Fig. 2.5C. Burstwise analysis of single-molecule simulations of the two-state dynamic molecules double-labeled with Donor and Acceptor (DA) and mixed with 20% fraction of donor-only (D-only) molecules. The simulation parameters are listed in Table 2.5 and are the same as for Fig 2.5C. The main FRET population shows a smeared distribution that is offset from the static FRET line given by Eq. (2.29) (wine). The dynamic FRET line (green) Eq. (2.30) shows the path taken by the conformational exchange. In this case, there are two conformational states, plus 20% fraction of donor-only labeled molecules. The open state was identified to have a lifetime $\tau_{D(A)}{}^{(open)} = 3.0$ns and rotational correlation time of $\rho = 1.0$ ns. For the closed state, $\tau_{D(A)}{}^{(closed)} = 1.0$ ns and the same rotational correlation time $\rho = 1.0$ ns. (For interpretation of the references to color in this figure legend, the reader is referred to the online version of this chapter.)

(wine) and dynamic (green) FRET lines corresponding to Eqs. (2.29) and (2.30). From here, one can immediately see that the sample is dynamic because the FRET population lies on the dynamic line. The end points of the dynamic line cross the static FRET line at the originating states $\tau_{D(A)}{}^{(open)} = 3.0$ and $\tau_{D(A)}{}^{(closed)} = 1.0$ ns. These lifetime values were determined by bi-exponential fitting of decay histograms generated from FRET bursts only (FRET sub-ensemble).

In Section 3.1, we filtered for scattered photons. However, one can build a species-specific correlation if also species-specific filters $f_j$ can be computed from experimentally determined or synthetically generated decay histograms $p_j$. We independently simulated a nanomolar concentration of each limiting state to generate proper decays and filters, which are shown in Fig. 2.7A and B,

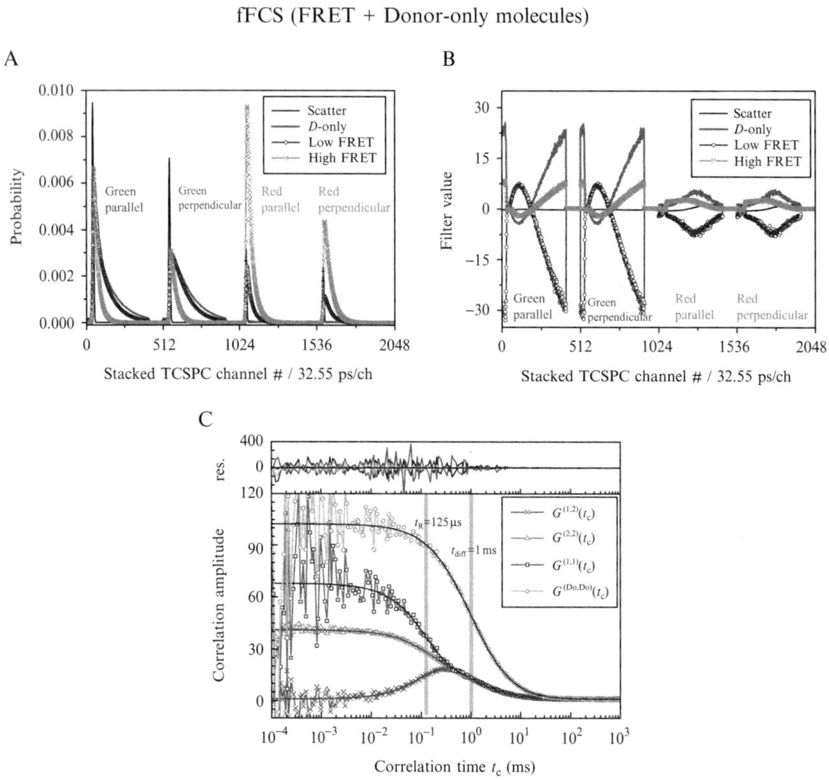


**Figure 2.7** fFCS for FRET molecules mixed with donor-only. Simulation of a molecule undergoing two-state dynamics in the presence of a donor-only species at single-molecule concentration. (A) Stacked conditional probabilities of the four detectors used, $d=4$ (Eq. 2.28), for the all selected species: *scatter* (black), donor-only (green), low FRET (blue), and high FRET (orange). (B) Calculated stacked filters according to Eq. (2.27) for the species selection shown in (A). Two green and two red channels were used to determine a species. (C) Averaged SCCF (green), SACFs of the high FRET and low FRET states (orange and blue respectively), and the SACF of the donor-only species (black) as defined by the four channel global filters. Global fits of these curves using Eqs. (2.31) and (2.32)are shown as solid lines. Residuals are shown in the top portion of the graph. The data correspond to the same simulation presented in Fig. 2.6. The results of the fits are given in Table 2.5. (For color version of this figure, the reader is referred to the online version of this chapter.)

respectively. The stacked histograms with dimensions $4L$ ($d=4$, two green and two red detection channels) and the global minimization algorithm proposed in Felekyan et al. (2012) were used for determining the best species separation.

From these four patterns (scatter, donor-only, low FRET or state [1], and high FRET or state [2]), the species-specific correlation curves for FRET molecules in state [1] $G^{(1,1)}(t_c)$, FRET molecules in state [2] $G^{(2,2)}(t_c)$, and the donor-only molecules $G^{(\text{Do,Do})}(t_c)$ can be calculated from the complete data trace. These are shown in Fig. 2.7C. Moreover, the species cross-correlation curve $G^{(1,2)}(t_c)$ between molecules in states [1] and [2] is also shown. The last two columns in Table 2.5 show the recovered fitting parameters of this simulation. We compare the cases where two channels ($d=2$) and four channels ($d=4$) were used to generate filters. Although there are no significant differences on the recovered parameters between these two methods, the amount of photons correlated increases when $d=4$. Therefore, the signal-to-noise ratio in the correlation curves also increases. Furthermore, species selectivity increases with the constraints posed by the additional detectors.

For a two-state dynamic system, assuming no triplet kinetics, the model of the species autocorrelation functions (SACFs) $G^{(i,i)}(t_c)$ and the species cross-correlation function (SCCF) $G^{(i,m)}(t_c)$ can be written as

$$\begin{aligned} G^{(1,1)}(t_c) &= 1+\frac{1}{N}G_{\text{diff}}(t_c)\left[1+\text{SAC}^{(1,1)}(t_c)\right], \\ G^{(2,2)}(t_c) &= 1+\frac{1}{N}G_{\text{diff}}(t_c)\left[1+\text{SAC}^{(2,2)}(t_c)\right], \\ G^{(1,2)}(t_c) &= 1+\frac{1}{N}G_{\text{diff}}(t_c)\left[1+\text{SCC}^{(1,2)}(t_c)\right], \end{aligned} \qquad [2.31]$$

where the corresponding kinetic terms are given by:

$$\begin{aligned} \text{SAC}^{(1,1)}(t_c) &= K\exp\left(-\frac{t_c}{t_R}\right), \\ \text{SAC}^{(2,2)}(t_c) &= K^{-1}\exp\left(-\frac{t_c}{t_R}\right), \\ \text{SCC}^{(1,2)}(t_c) &= -\exp\left(-\frac{t_c}{t_R}\right). \end{aligned} \qquad [2.32]$$

One can clearly notice the differences between this case and all previous model functions. The pre-exponential factors in Eq. (2.32) do not contain any complex information about the brightnesses of the states, and the contribution of donor-only sample is also removed. Moreover, the $\text{SCC}^{(i,m)}$ term shows maximum contrast with maximum negative correlation amplitude. In addition, the pre-exponentials on the $\text{SAC}^{(i,i)}$ terms depend only on the equilibrium constant $K$. The results from global fit to the data in Fig. 2.7 are given in Table 2.5.

By using species filtering, we can conclude the following:

**i.** The reaction time $t_R$ can be easily extracted when $t_T << t_R$.

**ii.** The amplitudes of the kinetic reaction terms $SAC^{(1,1)}(t_c)$ and $SAC^{(2,2)}(t_c)$ depend only on the equilibrium constant $K$.

**iii.** Maximum contrast on the $SCC^{(1,2)}(t_c)$ is obtained. It only depends on the reaction time $t_R$.

**iv.** Global fit FRET–fFCS improves stability of variable parameters.

**v.** Correct correlation amplitudes at picomolar or higher concentrations can be easily obtained.

**vi.** Donor-only contribution is removed.

In conclusion, the combination of MFD and FCS in the form of fFCS greatly simplifies analysis and the assignment of the involved species.

## 4. APPLICATIONS

One application of FRET–FCS is to study conformational transitions (Margittai, Widengren, et al., 2003; Price et al., 2011; Torres & Levitus, 2007). As example, we present a study done with Syntaxin 1 (Sx). Sx is one of the SNARE (soluble NSF attachment receptors) proteins that regulate synaptic vesicle release (Burkhardt, Hattendorf, Weis, & Fasshauer, 2008; Pobbati, Stein, & Fasshauer, 2006). The soluble domain of Sx is known to open when bound to SNAP25 allowing vesicle fusion and further neurotransmitter release. Munc-18, another Sx target, is believed to negative regulate exocytosis by locking Sx into a closed state (Burkhardt et al., 2008; Zhang et al., 2000). However, there is also evidence that Munc-18 could work as activator of exocytosis instead of inhibitor (Margittai, Fasshauer, Jahn, & Langen, 2003). What makes Sx a perfect example for FRET–FCS is that soluble domain of Sx in solution was shown to be in equilibrium between an "open" and "closed" conformations (Margittai, Widengren, et al., 2003).

We present the double mutant (G105C/S225C) randomly labeled with Alexa488 as the donor dye and Alexa594 as the acceptor dye (Felekyan et al., 2012; Margittai, Widengren, et al., 2003). Approximately, a 10-p$M$ solution of donor and acceptor labeled Sx in PBS buffer was measured in our MFD setup.

The single-molecule events were analyzed as described previously (Rothwell et al., 2003) and presented in Fig. 2.8 as a 2D histogram of $F_D/F_A$ versus $\tau_{D(A)}$ for qualitative analysis. For visual inspection, the static and dynamic FRET lines are shown as in Fig. 2.7. It is clear that the smeared population is off the static FRET line and it follows exactly the dynamic

FRET–FCS and fFCS for Syntaxin 1

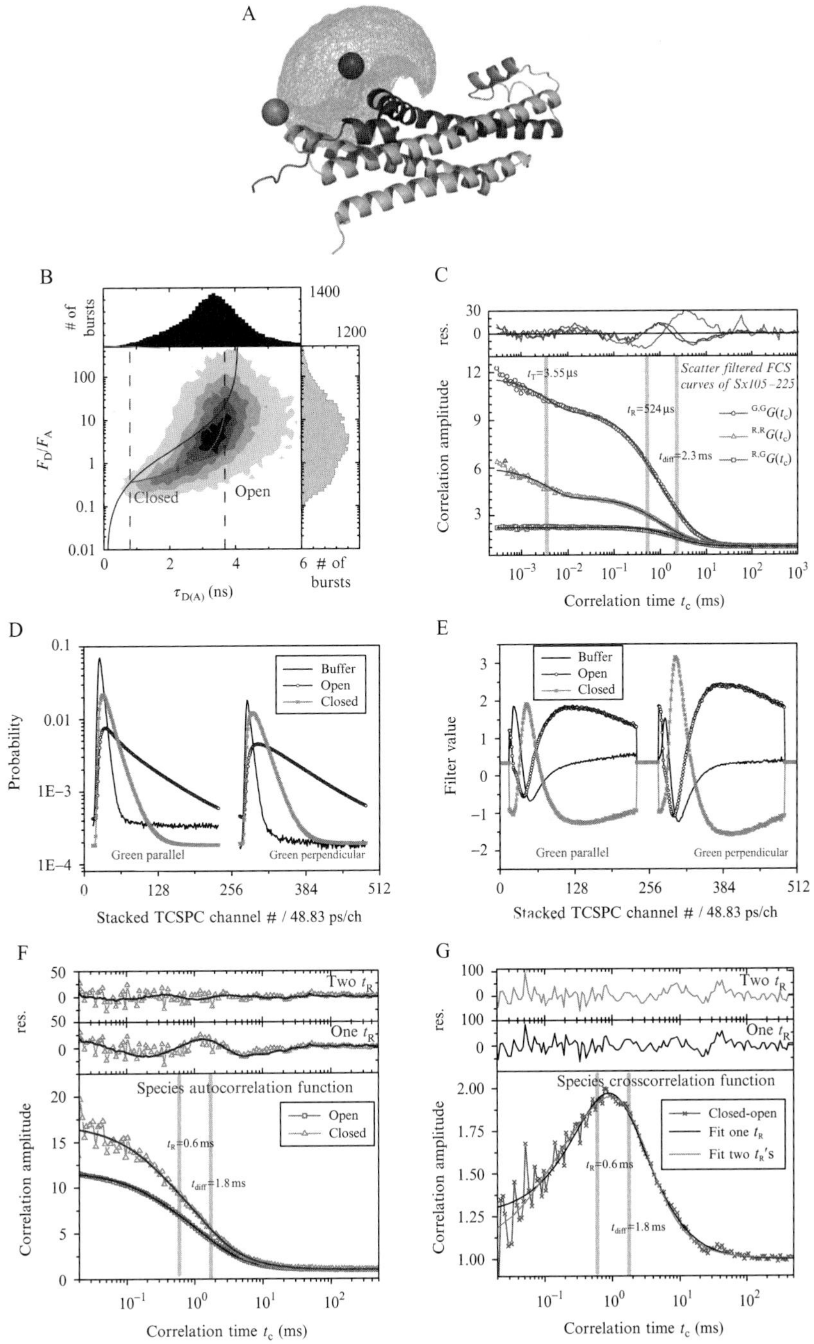

FRET line. The limiting lifetimes, that correspond to the "open" and "closed" states, are $\tau_{D(A)}^{(open)}=3.6$ and $\tau_{D(A)}^{(closed)}=0.8$ ns, respectively. These lifetimes were determined by bi exponential fitting the FRET subensemble TCSPC decay histograms.

The color auto- and cross-correlations were computed using the scatter filter described in Section 3.1. Species 1 consisted of the decay generated by the scattered photons (*scatter/buffer* measurement), and the rest came from the FRET sample. The ${}^{G,G}G(t_c)$, ${}^{R,R}G(t_c)$ autocorrelation functions, and the ${}^{R,G}G(t_c)$ cross-correlation function were fit using the global analysis (Eqs. 2.23 and 2.24a) with the addition of triplet kinetic terms ${}^{G,G}G_T(t_c)$, ${}^{R,R}G_T(t_c)$, and ${}^{G,R}G_T(t_c)$, respectively.

$$
\begin{aligned}
{}^{G,G}G(t_c) &= 1+\frac{1}{N}G_{\text{diff}}(t_c)\left[1+{}^{G,G}AC'^{(DA)}(t_c)\right]\left[1+{}^{G,G}G_T(t_c)\right],\\
{}^{R,R}G(t_c) &= 1+\frac{1}{N}G_{\text{diff}}(t_c)\left[1+{}^{R,R}AC'^{(DA)}(t_c)\right]\left[1+{}^{R,R}G_T(t_c)\right],\\
{}^{G,R}G(t_c) &= 1+\frac{1}{N}G_{\text{diff}}(t_c)\left[1+{}^{G,R}CC'^{(DA)}(t_c)\right]\left[1+{}^{G,R}G_T(t_c)\right].
\end{aligned}
\quad [2.33]
$$

**Figure 2.8** FRET–FCS and fFCS for Syntaxin 1. (A) Cartoon model of the soluble domain of Syntaxin 1. The Habc helices are shown in cyan and the H3 domain is shown in magenta. The mean position of the donor at position G105C (green) and acceptor at S225C (red) dyes are shown as solid spheres. The accessible volume calculation of the acceptor dye is shown as a mesh. (B) Burstwise analysis of the single-molecule events of randomly labeled Sx (G105C/S225C), with Alexa488 and Alexa594 as acceptor, are shown as 2D histogram $F_D/F_A$ versus $\tau_{D(A)}$. During the experiment, the background count rate in the green channel was $\langle B_G\rangle=4.4$ kHz, and $\langle B_R\rangle=0.39$ kHz for the red channel. A cross talk signal was estimated to be 4.2%. The calibrated green over red detection efficiency ($g_G/g_R$) was 0.8. The donor quantum yield $\Phi_{FD(0)}=0.8$, the acceptor quantum yield $\Phi_{FA}=0.4$, the donor-only lifetime $\tau_{D(0)}=4.0$ ns, and spectral *G*-factors 0.995 and 1.378 for the pair of green and red channels, respectively. The main FRET population follows the dynamic FRET line (green) Eq. (2.30). The open state was identified to have a fluorescence lifetime $\tau_{D(A)}{}^{(open)}=3.6$ ns and rotational correlation time of $\rho=1.5$ ns. For the closed state $\tau_{D(A)}{}^{(closed)}=0.8$ns with the same rotational correlation time. (C) FRET–FCS correlation curves overlaid by fit Eq. (2.15), including triplet contribution, with the residuals shown on the top graph. (D) Conditional probabilities used for generating the filters (buffer, open, and closed) shown in panel (E). (E) The filters used for calculating the fFCS curves. (F) SACF for open and closed states and (G) SCCF for the open to closed transition. Fits using Eqs. (2.31) and (2.32) are shown as solid lines and the residuals are shown on top graphs. One kinetic term yielded $t_R=0.6$ ms and an equilibrium constant of $K=1.5$. Fitting to two relaxation rates better describes all correlation functions and yields rates and fractions of $t_{R1}=1.1$ ms (84%) and $t_{R2}=0.08$ ms (16%). (For interpretation of the references to color in this figure legend, the reader is referred to the online version of this chapter.)

The visual inspection of the residuals indicated that the best fit fails to correctly describe all correlation functions in Fig. 2.8C. Assuming the simplest case of a two-state kinetic scheme, the mean relaxation time can be determined accurately, $t_R = 0.54$ ms. This relaxation time agrees to what has been reported previously for the double mutant (S91C/S225C), where the relaxation time was of $t_R = 0.8$ ms (Margittai, Widengren, et al., 2003).

Moreover, we apply fFCS using the two FRET states found by subensemble analysis and compute three different filters (scatter filter, "open" state, and "closed" state; Fig. 2.8D and E). The SACF and SCCF were computed according to Eq. (2.26) and shown in Fig. 2.8F and G. From the SCCF, one can extract the relaxation time with maximum amplitude contrast Eq. (2.32). This pattern can be fit with one component showing a relaxation rate $t_R = 0.6$ ms which agrees nicely with the 0.54 ms obtained by FRET–FCS. Residuals from global fitting SACF and SCCF with Eqs. (2.31) and (2.32) still show some deviations that are corrected with an additional kinetic term. Ignoring the additional kinetic term, we computed the kinetic rates $k_{o,c} = 0.62$ and $k_{c,o} = 0.94\ \mathrm{ms}^{-1}$ and the corresponding equilibrium constant $K_{o/c} = 1.5$ ($N^{(\mathrm{open})} = 0.09$, $N^{(\mathrm{closed})} = 0.06$) using Eq. (2.18). The numbers of molecules per species were obtained by fitting the species autocorrelation curves.

With two relaxation rates and corresponding amplitudes ($t_{R1} = 1.1$ ms (84%), $t_{R2} = 0.08$ ms (16%)), residuals show a better distribution. It is possible that the relaxation time of $t_R = 0.6$ ms obtained with one free parameter represents only the average behavior of a more complex kinetic scheme. At the moment, the meaning and source of this second time is still not fully understood and several factors that affect the SACF can be the source of the difference. Nevertheless, it is clear that the SCCF shows maximum contrasts and fit independently it can return an accurate relaxation time. However, the power of fFCS is also that it can help to distinguish more complex kinetics that with standard FRET–FCS is not possible.

## 5. DISCUSSION

For those implementing FRET–FCS, regardless of whether you are a first time user or an expert in the field, we consider it worthy to warn the reader about some of the complexities that so far have been only briefly mentioned or omitted for clarity. One example is triplet kinetics, which can alter the analytical form of the correlation function if it occurs on the same timescale as the kinetics reaction being studied. Also, some comments have to be made on the proper fundamentals when the brightnesses of the states are not known, or are wrongly determined. We can separate the interfering factors in two different categories: those that are time dependent and can affect the relaxation time and those that are intensity dependent and can affect the determination or uncertainty of the molecular brightnesses.

## 5.1. Relaxation time

### *5.1.1 Temporal boundaries for FRET–FCS*

It is important to highlight that for a two-state system, FRET–FCS is a reliable tool to obtain relaxation rates between the states. However, care needs to be taken when the triplet state reaction is not much faster than the relaxation kinetics. For example, if $t_T \approx t_R$ and the amplitude of the triplet kinetics is not negligible, the triplet kinetics and the kinetic reaction overlap. To illustrate this effect, we simulated a system shown in Fig. 2.9A, which considers dark states for donor and acceptor fluorophores. In the "regular" cross-correlation function ${}^{G,R}G(t_c)$ (Fig. 2.9B), dynamic and triplet terms overlap, and for shorter relaxation times (2.0 and 0.3 μs), the dynamic term becomes hardly distinguishable. However, in the SCCF (Fig. 2.9C), a clear anticorrelation term representing FRET dynamics is always present. This allows one to detect conformational dynamics and recover the correct value of $t_R$ irrespective of the presence of photophysical artifacts on a similar timescale.

### *5.1.2 Temporal boundaries for fFCS*

fFCS is a very powerful tool to resolve state interconversion. To test the resolution of this tool, we have made several simulations where the only changed parameter was the relaxation rate. Given the proper filters, a series of SCCF curves for relaxation times ranging from 2 ms to 0.3 μs can be easily recovered (Fig. 2.9C). This is close to four orders of magnitude and is possible because the donor triplet kinetics is not transferred into the SCCF when it is independent of the kinetics under study.

Afterpulsing is a characteristic of most detectors generating highly correlated events after each photon is detected. This artifact can propagate out to 10 μs (Bismuto, Gratton, & Lamb, 2001). In standard FRET–FCS, this can be removed by splitting the signal for each color with an additional 50/50 beamsplitter or with a beamsplitting polarizer. The later solution is implemented in a typical MFD setup. With this solution temporal resolution of below μs is easily accessible. However, if anisotropy information is required for species selection in fFCS, each of the four-detector channels in the MFD setup will need to be split again with 50/50 beamsplitters.

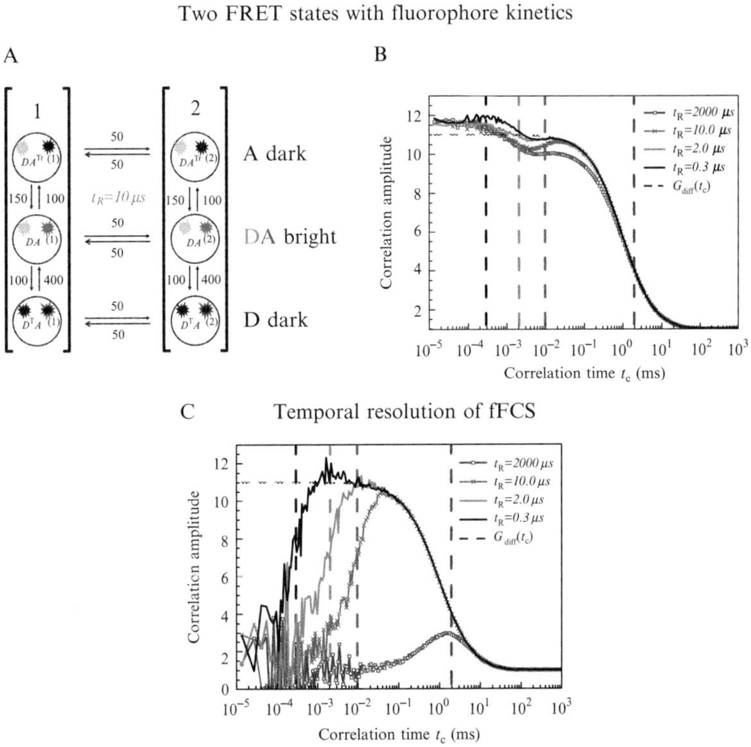


**Figure 2.9** (A) Simulated kinetic scheme consisting of six states: a bright low FRET ($DA^{(1)}$) and bright high FRET ($DA^{(2)}$) species; a dark low FRET ($D^TA^{(1)}$) and dark high FRET ($D^TA^{(2)}$) species with donor dye in the triplet state and species low FRET ($DA^{Tr(1)}$) and high FRET ($DA^{Tr(2)}$) where acceptor dye is in a dark state (representing, e.g., a *cis* state or triplet of Cy5). All kinetic rates are expressed in units of $ms^{-1}$; (B) $^{G,R}G(t_c)$ color cross-correlation functions for simulated various relaxation times (2 ms, 10, 2, and 0.3 μs); (C) SCCFs between FRET states (1) and (2) obtained by fFCS are nearly unaffected by photophysical artifacts always showing a pronounced anticorrelation term. (For color version of this figure, the reader is referred to the online version of this chapter.)

### 5.1.3 Alternative and complementary methods

We have shown that FCS recovers the correct relaxation times under various conditions. However, characterizing individual states with FCS alone can be challenging. Several approaches can complement FCS in complex cases. First, simple visual inspection of $F_D/F_A$ versus $\tau_{D(A)}$ 2D histograms allows one to identify dynamic populations, which show a "dynamic shift" toward longer fluorescence lifetimes (see also Sections 3.2 and 4) (Gopich & Szabo, 2012; Kalinin et al., 2010). Under favorable conditions, it is even possible to determine interconversion pathways (Gopich & Szabo, 2012; Kalinin et al., 2010) and to devise a kinetic model for FCS. Significant advances have been made also in understanding 1D histograms. Distributions of FRET efficiency can be predicted for any dynamic system (Gopich & Szabo, 2007, 2009) and fitted to experimental data using dynamic photon distribution analysis (dynamic PDA) (Kalinin et al., 2010; Santoso, Torella, & Kapanidis, 2010). PDA has a smaller dynamic range than FCS; however, it achieves maximum accuracy for relaxation times comparable to the diffusion time (Kalinin et al., 2010), that is, when the dynamic term is superimposed with the diffusion term in FCS. For this reason, and because of its ability to characterize individual states, dynamic PDA perfectly complements FRET–FCS. PDA can be additionally combined with the recently introduced burst variance analysis (Torella, Holden, Santoso, Hohlbein, & Kapanidis, 2011), which helps to identify dynamic populations.

## 5.2. Brightnesses uncertainty

Determination of at least one FRET efficiency and its corresponding brightness is required to have a meaningful fit in FRET–FCS. Although the absolute values are not needed, the ratios of the brightnesses with respect to a reference, preferably with respect to the donor brightness, suffice. To experimentally determine the molecular brightness, one can shift equilibrium to one of the states and use the preferred methodology (e.g., FCS, photon counting histogram (Chen, Müller, So, & Gratton, 1999), FIDA (Kask, Palo, Ullmann, & Gall, 1999), or TCSPC). Another alternative is to measure TCSPC of the mixture and carefully analyze the decay histograms. An attractive additional alternative is the use of two-dimensional fluorescence intensity distribution analysis (2D FIDA) (Kask et al., 2000) to obtain the brightnesses of each FRET state from the same measurement and use them for the FCS fit. We applied 2D FIDA to the data generated from the simulation of DA molecules in a mixture with donor-only (simulation presented in Fig. 2.4). In Table 2.6, we show the brightnesses obtained from 2D FIDA showing the maximum relative error close to 12%. These values are good estimates and can be used to fix the values in a global analyses defined by Eqs. (2.23) and (2.24a) or as good starting parameters when other input is known. We applied this methodology to the data shown in Fig. 2.4 and show results in Table 2.6. When treating $t_R$, and $x^{(D)}$ as free parameters, the maximum error in the kinetic rates ($k_{1,2}$ and $k_{2,1}$) is ~6%. Another option to consider is to have all FIDA parameters as initial values and fixing the donor-only brightnesses. In this case, the maximum error found for the kinetic rates was ~2%.

**Table 2.6** Comparison of brightnesses obtained from 2D FIDA and simulated ones

| Parameter | Simulated value | Dynamic FRET molecules and donor-only: 2D FIDA | | FRET–FCS[1)] | | | |
|---|---|---|---|---|---|---|---|
| | | Fit results | Relative error (%) | Fit (fixed $Q^{(i)}$) | Relative error (%) | Fit (free $Q^{(i)}$)[a] | Relative error (%) |
| ${}^{G}Q^{(1)}$ (kHz) | 75 | 70.8 | −5.6 | 70.8 | −5.6 | 74.56 | −0.6 |
| ${}^{G}Q^{(2)}$ (kHz) | 25 | 25.31 | 1.2 | 25.31 | 1.2 | 24.47 | −2.1 |
| ${}^{R}Q^{(1)}$ (kHz) | 14.46 | 14.43 | 8.1 | 14.43 | 8.1 | 13.79 | −4.6 |
| ${}^{R}Q^{(2)}$ (kHz) | 40.38 | 38.96 | −2.5 | 38.96 | −2.5 | 38.82 | −3.9 |
| ${}^{G}Q^{(D)}$ (kHz) | 100 | 99.22 | −0.8 | 99.22 | −0.8 | 99.22 | −0.8 |
| ${}^{R}Q^{(D)}$ (kHz) | 1.5 | 1.68 | 12 | 1.68 | 12 | 1.68 | 12 |
| $x^{(D)}$ | 0.2 | 0.22 | 10 | 0.202 | 2.5 | 0.198 | −1.0 |
| $K$ | 0.6 | 0.612 | 2.0 | 0.612 | 2.0 | 0.598 | −0.3 |
| $t_R$ (ms) | 0.125 | | | 0.132 | 5.6 | 0.127 | 1.6 |
| $E^{(1)}$ [b] | 0.25 | | | 0.26 | 4.0 | 0.24 | −4.0 |
| $E^{(2)}$ [b] | 0.75 | | | 0.74 | −1.3 | 0.75 | 0 |
| $k_{1,2}$ ($ms^{-1}$) [b] | 5 | | | 4.69 | −6.2 | 4.94 | −1.2 |
| $k_{2,1}$ ($ms^{-1}$) [b] | 3 | | | 2.87 | −4.3 | 2.95 | −1.7 |

[1)] Additional parameters were kept fixed during fitting: Offset = 1, $\omega_0 = 0.55\ \mu m$, $s_0 = 2\ \mu m$, $t_{diff}^{(D)} = t_{diff}^{(DA)} = 1\,ms$.
[a] Initial parameters were taken from 2D FIDA fitting results.
[b] Not a fit parameter: $k_{1,2} = \frac{1}{(1+K)t_R}$; $k_{2,1} = K \cdot k_{1,2}$; $E^{(i)} = \frac{{}^{R}Q^{(i)} - \alpha \cdot {}^{G}Q^{(i)}}{{}^{R}Q^{(i)} - \alpha \cdot {}^{G}Q^{(i)} + {}^{G}Q^{(i)} \frac{\Phi_{FA}}{\Phi_{FD(0)}} \frac{g_R}{g_G}}$.

## 5.3. Advantages of fFCS

If experimentally possible, fFCS possess several advantages over standard FRET–FCS. Pulsed excitation and TCSPC recording offer the possibility to carefully identify species from a mixture at a single-molecule level, via either standard $F_D/F_A - \tau_{D(A)}$ 2D histograms or subensemble analysis of decay histograms. In all, we can list them as follows:

- **i.** *fFCS fully recovers the amplitudes*: It was shown that by using the scatter filter proper amplitudes and therefore concentrations can be extracted even at picomolar concentrations. This can be applied as well when measurements are done in a highly scattering buffer.
- **ii.** *fFCS offers high degree of selectivity*: fFCS can be applied as a tool to compute species-specific correlation curves from a mixture of different species, provided the species can be distinguished based on lifetime or/ and anisotropy or/and emission spectrum.
- **iii.** *SACF and SCCF analytical functions are simplified*: Functional forms are brightness independent, although correct filters are required.
- **iv.** *Dynamic range is much broader*: Given that SCCF does not transfer donor triplet kinetic information allows fFCS to even find interconversion with sub-microsecond resolution.

## ACKNOWLEDGMENTS

This work was supported by the Deutsche Forschungsgemeinschaft (SE 1195/13-1) and EU (FP7-Health-2007-A-201837). H. S. acknowledges support from the Alexander von Humboldt foundation. We want to thank Enno Schweinberger for sample preparation and Jerker Widengren for many helpful discussions and introduction to FCS.

## REFERENCES

Al-Soufi, W., Reija, B., Novo, M., Felekyan, S., Kühnemuth, R., & Seidel, C. A. M. (2005). Fluorescence correlation spectroscopy, a tool to investigate supramolecular dynamics: Inclusion complexes of pyronines with cyclodextrin. *Journal of the American Chemical Society*, *127*, 8775–8784.

Bader, A. N., Hoetzl, S., Hofman, E. G., Voortman, J., Henegouwen, P., van Meer, G., et al. (2011). Homo-FRET imaging as a tool to quantify protein and lipid clustering. *ChemPhysChem*, *12*, 475–483.

Bismuto, E., Gratton, E., & Lamb, D. C. (2001). Dynamics of ANS binding to tuna apomyoglobin measured with fluorescence correlation spectroscopy. *Biophysical Journal*, *81*, 3510–3521.

Blanchard, S. C. (2009). Single-molecule observations of ribosome function. *Current Opinion in Structural Biology*, *19*, 103–109.

Böhmer, M., Wahl, M., Rahn, H. J., Erdmann, R., & Enderlein, J. (2002). Time-resolved fluorescence correlation spectroscopy. *Chemical Physics Letters*, *353*, 439–445.

Boukobza, E., Sonnenfeld, A., & Haran, G. (2001). Immobilization in surface-tethered lipid vesicles as a new tool for single biomolecule spectroscopy. *The Journal of Physical Chemistry B, 105*, 12165–12170.

Braslavsky, S. E., Fron, E., Rodriguez, H. B., Roman, E. S., Scholes, G. D., Schweitzer, G., et al. (2008). Pitfalls and limitations in the practical use of Förster's theory of resonance energy transfer. *Photochemical and Photobiological Sciences, 7*, 1444–1448.

Bruchez, M., Moronne, M., Gin, P., Weiss, S., & Alivisatos, A. P. (1998). Semiconductor nanocrystals as fluorescent biological labels. *Science, 281*, 2013–2016.

Burkhardt, P., Hattendorf, D. A., Weis, W. I., & Fasshauer, D. (2008). Munc18a controls SNARE assembly through its interaction with the syntaxin N-peptid. *The EMBO Journal, 27*, 923–933.

Chen, Y., & Müller, J. D. (2007). Determining the stoichiometry of protein heterocomplexes in living cells with fluorescence fluctuation spectroscopy. *Proceedings of the National Academy of Sciences of the United States of America, 104*, 3147–3152.

Chen, Y., Müller, J. D., So, P. T. C., & Gratton, E. (1999). The photon counting histogram in fluorescence fluctuation spectroscopy. *Biophysical Journal, 77*, 553–567.

Cisse, I., Okumus, B., Joo, C., & Ha, T. J. (2007). Fueling protein-DNA interactions inside porous nanocontainers. *Proceedings of the National Academy of Sciences of the United States of America, 104*, 12646–12650.

Clegg, R. M. (2009). Förster resonance energy transfer—FRET what is it why do it and how it's done. In: T. W. J. Gadella (Ed.), *Laboratory techniques in biochemistry and molecular biology (FRET and film techniques)*, Vol. 33, (pp. 1–57). Amsterdam: Elsevier.

Clegg, R. M., Murchie, A. I., Zechel, A., Carlberg, C., Diekmann, S., & Lilley, D. M. J. (1992). Fluorescence resonance energy transfer analysis of the structure of the four-way DNA junction. *Biochemistry, 31*, 4846–4856.

Curutchet, C., Franceschetti, A., Zunger, A., & Scholes, G. D. (2008). Examining Forster energy transfer for semiconductor nanocrystalline quantum dot donors and acceptors. *Journal of Physical Chemistry C, 112*, 13336–13341.

De Angelis, D. A. (1999). Why FRET over genomics? *Physiological Genomics, 1*, 93–99.

Diez, M., Zimmermann, B., Borsch, M., Konig, M., Schweinberger, E., Steigmiller, S., et al. (2004). Proton-powered subunit rotation in single membrane-bound F0F1-ATP synthase. *Nature Structural and Molecular Biology, 11*, 135–141.

Eggeling, C., Berger, S., Brand, L., Fries, J. R., Schaffer, J., Volkmer, A., et al. (2001). Data registration and selective single-molecule analysis using multi-parameter fluorescence detection. *Journal of Biotechnology, 86*, 163–180.

Eggeling, C., Kask, P., Winkler, D., & Jager, S. (2005). Rapid analysis of Forster resonance energy transfer by two-color global fluorescence correlation spectroscopy: Trypsin proteinase reaction. *Biophysical Journal, 89*, 605–618.

Ehrenberg, M., & Rigler, R. (1974). Rotational Brownian motion and fluorescence intensity fluctuations. *Chemical Physics, 4*, 390–401.

Elson, E. L., & Magde, D. (1974). Fluorescence correlation spectroscopy. I. Conceptual basis and theory. *Biopolymers, 13*, 1–27.

Enderlein, J., & Gregor, I. (2005). Using fluorescence lifetime for discriminating detector after pulsing in fluorescence-correlation spectroscopy. *The Review of Scientific Instruments, 76*, 033102.

Felekyan, S., Kalinin, S., Sanabria, H., Valeri, A., & Seidel, C. A. M. (2012). Filtered FCS: Species auto- and cross-correlation functions highlight binding and dynamics in biomolecules. *ChemPhysChem, 13*, 1036–1053.

Felekyan, S., Kalinin, S., Valeri, A., & Seidel, C. A. M. (2009). Filtered FCS and species cross correlation function. In P. Ammasi & T. C. S. Peter (Eds.), *Proceedings of SPIE multiphoton microscopy in the biomedical sciences IX* (Vol. 7183). *Proceedings of SPIE—The International Society for Optical Engineering*, pp. 71830D-1–71830D-14.

Felekyan, S., Kühnemuth, R., Kudryavtsev, V., Sandhagen, C., Becker, W., & Seidel, C. A. M. (2005). Full correlation from picoseconds to seconds by time-resolved and time-correlated single photon detection. *The Review of Scientific Instruments, 76*, 083104.

Förster, T. (1948). Zwischenmolekulare Energiewanderung und Fluoreszenz. *Annals of Physics, 437*, 55–75.

Fries, J. R., Brand, L., Eggeling, C., Köllner, M., & Seidel, C. A. M. (1998). Quantitative identification of different single-molecules by selective time-resolved confocal fluorescence spectroscopy. *The Journal of Physical Chemistry A, 102*, 6601–6613.

Gansen, A., Valeri, A., Hauger, F., Felekyan, S., Kalinin, S., Toth, K., et al. (2009). Nucleosome disassembly intermediates characterized by single-molecule FRET. *Proceedings of the National Academy of Sciences of the United States of America, 106*, 15308–15313.

Giepmans, B. N. G., Adams, S. R., Ellisman, M. H., & Tsien, R. Y. (2006). The fluorescent toolbox for assessing protein location and function. *Science, 312*, 217–224.

Gonçalves, M. S. T. (2009). Fluorescent labeling of biomolecules with organic probes. *Chemical Reviews, 109*, 190–212.

Gopich, I. V., & Szabo, A. (2007). Single-molecule FRET with diffusion and conformational dynamics. *The Journal of Physical Chemistry B, 111*, 12925–12932.

Gopich, I. V., & Szabo, A. (2009). Decoding the pattern of photon colors in single-molecule FRET. *The Journal of Physical Chemistry B, 113*, 10965–10973.

Gopich, I. V., & Szabo, A. (2012). Theory of the energy transfer efficiency and fluorescence lifetime distribution in single-molecule FRET. *Proceedings of the National Academy of Sciences of the United States of America, 109*, 7747–7752.

Gregor, I., & Enderlein, J. (2007). Time-resolved methods in biophysics. 3. Fluorescence lifetime correlation spectroscopy. *Photochemical and Photobiological Sciences, 6*, 13–18.

Gurunathan, K., & Levitus, M. (2009). FRET fluctuation spectroscopy of diffusing biopolymers: Contributions of conformational dynamics and translational diffusion. *The Journal of Physical Chemistry B, 114*, 980–986.

Gurunathan, K., & Levitus, M. (2010). FRET fluctuation spectroscopy of diffusing biopolymers: Contributions of conformational dynamics and translational diffusion. *The Journal of Physical Chemistry B, 114*, 980–986.

Ha, T., Enderle, T., Ogletree, D. F., Chemla, D. S., Selvin, P. R., & Weiss, S. (1996). Probing the interaction between two single molecules: Fluorescence resonance energy transfer between a single donor and a single acceptor. *Proceedings of the National Academy of Sciences of the United States of America, 93*, 6264–6268.

Hanson, J. A., & Yang, H. (2008). Quantitative evaluation of cross correlation between two finite-length time series with applications to single-molecule FRET. *The Journal of Physical Chemistry B, 112*, 13962–13970.

Henzler-Wildman, K., & Kern, D. (2007). Dynamic personalities of proteins. *Nature, 450*, 964–972.

Henzler-Wildman, K. A., Thai, V., Lei, M., Ott, M., Wolf-Watz, M., Fenn, T., et al. (2007). Intrinsic motions along an enzymatic reaction trajectory. *Nature, 450*, 838–844.

Joo, C., McKinney, S. A., Nakamura, M., Rasnik, I., Myong, S., & Ha, T. (2006). Real-time observation of RecA filament dynamics with single monomer resolution. *Cell, 126*, 515–527.

Kalinin, S., Valeri, A., Antonik, M., Felekyan, S., & Seidel, C. A. M. (2010). Detection of structural dynamics by FRET: A photon distribution and fluorescence lifetime analysis of systems with multiple states. *The Journal of Physical Chemistry B, 114*, 7983–7995.

Kalinin, S., Kuhnemuth, R., Vardanyan, H., & Seidel, C. A. M. (2012). A 4 ns hardware photon correlator based on a general-purpose field-programmable gate array development board implemented in a compact setup for fluorescence correlation spectroscopy. *The Review of Scientific Instruments, 83*, 096105.

Kapusta, P., Wahl, M., Benda, A., Hof, M., & Enderlein, J. (2007). Fluorescence lifetime correlation spectroscopy. *Journal of Fluorescence, 17*, 43–48.
Kask, P., Palo, K., Fay, N., Brand, L., Mets, Ü., Ullmann, D., et al. (2000). Two-dimensional fluorescence intensity distribution analysis: Theory and application. *Biophysical Journal, 78*, 1703–1713.
Kask, P., Palo, K., Ullmann, D., & Gall, K. (1999). Fluorescence-intensity distribution analysis and its application in biomolecular detection technology. *Proceedings of the National Academy of Sciences of the United States of America, 96*, 13756–13761.
Kim, S. A., Heinze, K. G., Bacia, K., Waxham, M. N., & Schwille, P. (2005). Two-photon cross-correlation analysis of intracellular reactions with variable stoichiometry. *Biophysical Journal, 88*, 4319–4336.
Knight, J. L., Mekler, V., Mukhopadhyay, J., Ebright, R. H., & Levy, R. M. (2005). Distance-restrained docking of rifampicin and rifamycin SV to RNA polymerase using systematic FRET measurements: Developing benchmarks of model quality and reliability. *Biophysical Journal, 88*, 925–938.
Kudryavtsev, V., Sikor, M., Kalinin, S., Mokranjac, D., Seidel, C. A. M., & Lamb, D. C. (2012). Combining MFD and PIE for accurate single-pair Forster resonance energy transfer measurements. *ChemPhysChem, 13*, 1060–1078.
Kühnemuth, R., & Seidel, C. A. M. (2001). Principles of single molecule multiparameter fluorescence spectroscopy. *Single Molecules, 2*, 251–254.
Lakowicz, J. R. (2006). *Principles of fluorescence spectroscopy*. New York: Springer.
Lamb, D. C., Schenk, A., Rocker, C., Scalfi-Happ, C., & Nienhaus, G. U. (2000). Sensitivity enhancement in fluorescence correlation spectroscopy of multiple species using time-gated detection. *Biophysical Journal, 79*, 1129–1138.
Levitus, M. (2010). Relaxation kinetics by fluorescence correlation spectroscopy: Determination of kinetic parameters in the presence of fluorescent impurities. *Journal of Physical Chemistry Letters, 1*, 1346–1350.
Lippincott-Schwartz, J., Snapp, E., & Kenworthy, A. (2001). Studying protein dynamics in living cells. *Nature Reviews. Molecular Cell Biology, 2*, 444–456.
Magde, D., Elson, E. L., & Webb, W. W. (1972). Thermodynamic fluctuations in a reacting system—Measurement by fluorescence correlation spectroscopy. *Physical Review Letters, 29*, 705–708.
Magde, D., Elson, E. L., & Webb, W. W. (1974). Fluorescence correlation spectroscopy. II. An experimental realization. *Biopolymers, 13*, 29–61.
Majumdar, D. S., Smirnova, I., Kasho, V., Nir, E., Kong, X. X., Weiss, S., et al. (2007). Single-molecule FRET reveals sugar-induced conformational dynamics in LacY. *Proceedings of the National Academy of Sciences of the United States of America, 104*, 12640–12645.
Margittai, M., Fasshauer, D., Jahn, R., & Langen, R. (2003). The Habc domain and the SNARE core complex are connected by a highly flexible linker. *Biochemistry, 42*, 4009–4014.
Margittai, M., Widengren, J., Schweinberger, E., Schröder, G. F., Felekyan, S., Haustein, E., et al. (2003). Single-molecule fluorescence resonance energy transfer reveals a dynamic equilibrium between closed and open conformations of syntaxin 1. *Proceedings of the National Academy of Sciences of the United States of America, 100*, 15516–15521.
Maus, M., Cotlet, M., Hofkens, J., Gensch, T., De Schryver, F. C., Schaffer, J., et al. (2001). An experimental comparison of the maximum likelihood estimation and nonlinear least-squares fluorescence lifetime analysis of single molecules. *Analytical Chemistry, 73*, 2078–2086.
Mekler, V., Kortkhonjia, E., Mukhopadhyay, J., Knight, J., Revyakin, A., Kapanidis, A. N., et al. (2002). Structural organization of bacterial RNA polymerase holoenzyme and the RNA polymerase-promoter open complex. *Cell, 108*, 599–614.
Michalet, X., Weiss, S., & Jager, M. (2006). Single-molecule fluorescence studies of protein folding and conformational dynamics. *Chemical Reviews, 106*, 1785–1813.

Mickler, M., Hessling, M., Ratzke, C., Buchner, J., & Hugel, T. (2009). The large conformational changes of Hsp90 are only weakly coupled to ATP hydrolysis. *Nature Structural and Molecular Biology, 16*, 281–286.

Mocsár, G., Kreith, B., Buchholz, J., Krieger, J. W., Langowski, J., & Vamosi, G. (2012). Multiplexed multiple-tau auto- and cross-correlators on a single field programmable gate array. *The Review of Scientific Instruments, 83*, 046101.

Müller, B. K., Zaychikov, E., Bräuchle, C., & Lamb, D. C. (2005). Pulsed interleaved excitation. *Biophysical Journal, 89*, 3508–3522.

Muschielok, A., Andrecka, J., Jawhari, A., Bruckner, F., Cramer, P., & Michaelis, J. (2008). A nano-positioning system for macromolecular structural analysis. *Nature Methods, 5*, 965–971.

Nettels, D., Gopich, I. V., Hoffmann, A., & Schuler, B. (2007). Ultrafast dynamics of protein collapse from single-molecule photon statistics. *Proceedings of the National Academy of Sciences of the United States of America, 104*, 2655–2660.

Pandey, M., Syed, S., Donmez, I., Patel, G., Ha, T., & Patel, S. S. (2009). Coordinating DNA replication by means of priming loop and differential synthesis rate. *Nature, 462*, 940–943.

Pobbati, A. V., Stein, A., & Fasshauer, D. (2006). N- to C-terminal SNARE complex assembly promotes rapid membrane fusion. *Science, 313*, 673–676.

Price, E. S., Aleksiejew, M., & Johnson, C. K. (2011). FRET-FCS detection of intralobe dynamics in calmodulin. *The Journal of Physical Chemistry B, 115*, 9320–9326.

Price, E. S., DeVore, M. S., & Johnson, C. K. (2010). Detecting intramolecular dynamics and multiple Forster resonance energy transfer states by fluorescence correlation spectroscopy. *The Journal of Physical Chemistry B, 114*, 5895–5902.

Rigler, R., & Mets, Ü. (1992). Diffusion of single molecules through a Gaussian laser beam. *SPIE, 1921*, 239–248.

Rothwell, P. J., Berger, S., Kensch, O., Felekyan, S., Antonik, M., Wöhrl, B. M., et al. (2003). Multiparameter single-molecule fluorescence spectroscopy reveals heterogeneity of HIV-1 reverse transcriptase:primer/template complexes. *Proceedings of the National Academy of Sciences of the United States of America, 100*, 1655–1660.

Roy, R., Hohng, S., & Ha, T. (2008). A practical guide to single-molecule FRET. *Nature Methods, 5*, 507–516.

Sahoo, H., & Schwille, P. (2011). FRET and FCS—Friends or foes? *ChemPhysChem, 12*, 532–541.

Santoso, Y., Torella, J. P., & Kapanidis, A. N. (2010). Characterizing single-molecule FRET dynamics with probability distribution analysis. *ChemPhysChem, 11*, 2209–2219.

Schaefer, D. W. (1973). Dynamics of number fluctuations: Motile microorganisms. *Science, 180*, 1293–1295.

Schuler, B., & Eaton, W. A. (2008). Protein folding studied by single-molecule FRET. *Current Opinion in Structural Biology, 18*, 16–26.

Schwille, P. (2001). Fluorescence correlation spectroscopy. Theory and applications. In R. R. E. L. Elson (Ed.), *Vol. 65* (pp. 360–378). Berlin: Springer.

Sindbert, S., Kalinin, S., Nguyen, H., Kienzler, A., Clima, L., Bannwarth, W., et al. (2011). Accurate distance determination of nucleic acids via Förster resonance energy transfer: Implications of dye linker length and rigidity. *Journal of the American Chemical Society, 133*, 2463–2480.

Sisamakis, E., Valeri, A., Kalinin, S., Rothwell, P. J., & Seidel, C. A. M. (2010). Accurate single-molecule FRET studies using multiparameter fluorescence detection. *Methods in Enzymology, 475*, 455–514.

Slaughter, B. D., Allen, M. W., Unruh, J. R., Bieber Urbauer, R. J., & Johnson, C. K. (2004). Single-molecule resonance energy transfer and fluorescence correlation spectroscopy of calmodulin in solution. *The Journal of Physical Chemistry B, 108*, 10388–10397.

Sun, Y. S., Day, R. N., & Periasamy, A. (2011). Investigating protein-protein interactions in living cells using fluorescence lifetime imaging microscopy. *Nature Protocols*, *6*, 1324–1340.

Sun, Y. S., Wallrabe, H., Seo, S. A., & Periasamy, A. (2011). FRET microscopy in 2010: The legacy of Theodor Forster on the 100th anniversary of his birth. *ChemPhysChem*, *12*, 462–474.

Szollosi, J., Damjanovich, S., & Matyus, L. (1998). Application of fluorescence resonance energy transfer in the clinical laboratory: Routine and research. *Cytometry*, *34*, 159–179.

Tan, Y. W., & Yang, H. (2011). Seeing the forest for the trees: Fluorescence studies of single enzymes in the context of ensemble experiments. *Physical Chemistry Chemical Physics*, *13*, 1709–1721.

Thompson, N. L. (1991). Fluorescence correlation spectroscopy. In: J. R. Lakowicz (Ed.), *Topics in fluorescence spectroscopy*, Vol. 1, (pp. 337–378). New York: Plenum Press.

Torella, J. P., Holden, S. J., Santoso, Y., Hohlbein, J., & Kapanidis, A. N. (2011). Identifying molecular dynamics in single-molecule FRET experiments with burst variance analysis. *Biophysical Journal*, *100*, 1568–1577.

Torres, T., & Levitus, M. (2007). Measuring conformational dynamics: A new FCS-FRET approach. *The Journal of Physical Chemistry B*, *111*, 7392–7400.

Ulbrich, M. H., & Isacoff, E. Y. (2007). Subunit counting in membrane-bound proteins. *Nature Methods*, *4*, 319–321.

Vogel, S. S., Thaler, C., & Koushik, S. V. (2006). Fanciful FRET. *Science's STKE*, *2006*, re2.

Wang, Y. F., Guo, L., Golding, I., Cox, E. C., & Ong, N. P. (2009). Quantitative transcription factor binding kinetics at the single-molecule level. *Biophysical Journal*, *96*, 609–620.

Weidtkamp-Peters, S., Felekyan, S., Bleckmann, A., Simon, R., Becker, W., Kuhnemuth, R., et al. (2009). Multiparameter fluorescence image spectroscopy to study molecular interactions. *Photochemical and Photobiological Sciences*, *8*, 470–480.

Widengren, J., Kudryavtsev, V., Antonik, M., Berger, S., Gerken, M., & Seidel, C. A. M. (2006). Single-molecule detection and identification of multiple species by multiparameter fluorescence detection. *Analytical Chemistry*, *78*, 2039–2050.

Widengren, J., Mets, Ü., & Rigler, R. (1995). Fluorescence correlation spectroscopy of triplet states in solution: A theoretical and experimental study. *The Journal of Physical Chemistry*, *99*, 13368–13379.

Widengren, J., Rigler, R., & Elson, E. L. (2001). Photophysical aspects of FCS measurements. In F. P. Schäfer, J. P. Toennies & W. Zinth (Eds.), *Fluorescence correlation spectroscopy* (pp. 276–301). Berlin, Heidelberg, New York: Springer-Verlag.

Widengren, J., Rigler, R., & Mets, Ü. (1994). Triplet-state monitoring by fluorescence correlation spectroscopy. *Journal of Fluorescence*, *4*, 255–258.

Woźniak, A. K., Schröder, G., Grubmüller, H., Seidel, C. A. M., & Oesterhelt, F. (2008). Single molecule FRET measures bends and kinks in DNA. *Proceedings of the National Academy of Sciences of the United States of America*, *105*, 18337–18342.

Wu, P. G., & Brand, L. (1994). Resonance energy-transfer—Methods and applications. *Analytical Biochemistry*, *218*, 1–13.

Wu, X., & Schultz, P. G. (2009). Synthesis at the interface of chemistry and biology. *Journal of the American Chemical Society*, *131*, 12497–12515.

Zhang, W., Efanov, A., Yang, S. N., Fried, G., Kolare, S., Brown, H., et al. (2000). Munc-18 associates with syntaxin and serves as a negative regulator of exocytosis in the pancreatic beta-cell. *The Journal of Biological Chemistry*, *275*, 41521–41527.

CHAPTER THREE

# Fluorescence Fluctuation Spectroscopy Approaches to the Study of Receptors in Live Cells

**David M. Jameson**[*,1], **Nicholas G. James**[*], **Joseph P. Albanesi**[†]
[*]Department of Cell and Molecular Biology, John A. Burns School of Medicine, University of Hawaii, Honolulu, Hawaii, USA
[†]Department of Pharmacology, University of Texas Southwestern Medical Center, Dallas, Texas, USA
[1]Corresponding author: e-mail address: djameson@hawaii.edu

## Contents



## Abstract

Communication between cells and their environment, including other cells, is often mediated by cell surface receptors. Fluorescence methodologies are among the most important techniques used to study receptors and their interactions, and in the past decade, fluorescence fluctuation spectroscopy (FFS) approaches have been increasingly utilized. In this overview, we illustrate how diverse FFS approaches have been used to elucidate important aspects of receptor systems, including interactions of receptors

*Methods in Enzymology*, Volume 519
ISSN 0076-6879
http://dx.doi.org/10.1016/B978-0-12-405539-1.00003-8

with their ligands and receptor oligomerization and clustering. We also describe the most popular methods used to introduce fluorescent moieties into the biological systems. Finally, specific attention will be given to cell maintenance and transfection strategies especially as related to microscopy studies.

## 1. INTRODUCTION

In order for multicellular organisms to exist and to function, they must be able to communicate with their surroundings, that is, to exchange chemicals and information. Communication between cells and their environment, including other cells, is often mediated by cell surface receptors, which are integral plasma membrane proteins. Signal transduction pathways are usually initiated by binding of extracellular ligands (such as growth factors, cytokines, hormones, neurotransmitters, etc.) to these plasma membrane receptors. Hundreds of different types of cell surface receptors have been identified to date—by far the largest category of these are the seven-transmembrane-spanning type (7TM) receptors.

Interestingly, nearly all of the 7TM receptors found to date are G-protein-coupled receptors (GPCRs). In fact, it is estimated that 3–4% of the human genome encodes for more than 800 different putative GPCRs and about half of currently used drugs target GPCRs (Lundstrom, 2009). Different GPCRs have selectivity for different G proteins, which in turn will regulate different signaling pathways. G proteins are heterotrimers consisting of α, β, and γ subunits. Mammals express approximately 20 αs, 5 βs, and 12 γs. In the basal state, α(GDP)-βγ is inactive. The GPCR catalyzes the exchange of GTP for GDP on the α subunit and the dissociation of α(GTP) from the βγ dimer (Fig. 3.1). Both α(GTP) and βγ regulate downstream effectors. In addition to the GPCRs, other plasma membrane receptors include the receptor tyrosine kinases (such as the insulin receptor), nutrient receptors (such as the transferrin receptor), and ligand-gated ion channels (such as the nicotinic acetylcholine receptor). Of course, only a small subset of receptor genes are turned on in any given cell type, and hence, a typical cell may express only 50–100 different types of receptors. A smaller subset of receptors, including the nuclear hormone receptors, are soluble cytoplasmic proteins that translocate to the nucleus upon binding ligands, such as steroids or vitamins. Activated nuclear receptors (NRs) may also be considered as ligands of their own receptor, DNA, as they often serve as transcription factors or regulators.

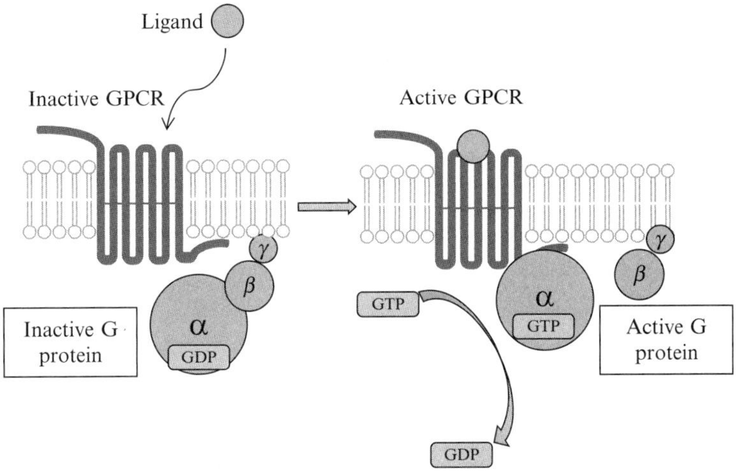


**Figure 3.1** Schematic illustrating the activation of a GPCR by a ligand and the subsequent activation of a G-protein via GDP/GTP exchange. (For color version of this figure, the reader is referred to the online version of this chapter.)

Most nutrient uptake and signaling receptors bind to their ligands on the plasma membrane where they undergo lateral movement to promote their oligomerization and their association with signaling and trafficking elements; the insulin receptor and EGF receptor (EGFR) are among the most studied systems. Most, though not all, receptors are rapidly internalized for downregulation, for subsequent recycling to the plasma membrane, and/or for assembly into intracellular signaling complexes (e.g., Bethani, Skånland, Dikic, & Acker-Palmer, 2010). In contrast, as already mentioned, NRs often reside in the cytoplasm under resting conditions and traffic to the nucleus upon ligand binding. Hence, an understanding of receptor function must include knowledge of its oligomerization state and a dynamic picture of its life cycle, which means studying their movement both in the cell membrane and in the cell interior.

Fluorescence methodologies have been successfully used to study properties of membrane proteins for several decades. One of the first such applications to elucidate the movement of membrane proteins was the seminal study by Frye and Edidin (1970) who used fluorescent antibodies to visualize the mobility of the major histocompatibility complex in the cell membrane. Soon after it was developed, the technique of fluorescence recovery after photobleaching (FRAP) was applied to study the movement of receptors in cell membranes. For example, wheat germ agglutinin receptor complexes on the surface of human embryo fibroblasts were studied using FRAP in 1976 (Jacobson, Derzko, Wu, Hou, & Poste, 1976), and these studies revealed that most of these receptors (>75%) were mobile with diffusion coefficients in the range of $2\times10^{-11}$ to $2\times10^{-10}$ cm$^2$/s. On the other hand, concanavalin A receptors were largely immobile on the surface of 3T3 fibroblasts. For several decades, FRAP remained one of the principal fluorescence methods to study receptor movements in the plasma membranes of cells, and it is still popular today. In the early 1980s, Axelrod developed the total internal reflection fluorescence (TIRF) approach, which became an important tool in receptor studies (Axelrod, 1981). With TIRF optics, the evanescent field associated with the exciting light only penetrates a short distance into the cell ($\sim$100–200 nm) and, therefore, is ideal for illuminating fluorophores associated with the cell membrane. In the early 1970s, Magde, Elson, and Webb published seminal papers on the theory and application of fluorescence fluctuation analysis (FFS) (Elson & Magde, 1974; Magde, Elson, & Webb, 1972, 1974). Widespread applications of FFS had to wait though until the 1990s when the advent of confocal and two-photon microscopy greatly reduced the observation volume and thus significantly improved the sensitivity of the method, even extending it to single-molecule levels (Denk, Strickler, & Webb, 1990; Eigen & Rigler, 1994; Maiti, Haupts, & Webb, 1997). In recent years, the coupling of FFS with TIRF has significantly enhanced the scope of both methods.

In the past decade, FFS approaches have been increasingly used to study receptors. Since there are many articles in this, and the previous, volume describing in detail the basics of FFS as well as the latest developments in instrumentation and analysis, we shall not dwell on the methodology. Rather, in the next section, we shall point out the main FFS methodologies that have been used to date in receptor studies. These include traditional FCS studies of diffusional rates of receptors and their ligands, photon counting histogram (PCH) and number and brightness (N&B) approaches toward measuring receptor stoichiometry, cross-correlation and two-color approaches to characterize receptor complexes, and image correlation spectroscopy (ICS) approaches to monitor receptor mobility (see Jameson, Ross, & Albanesi, 2009 for an introduction to these methods).

## 2. SELECTED FFS STUDIES

In this section, we give examples of studies of receptor systems that utilize FFS. We note that this area was also reviewed in 2007 by Briddon and Hill (2007). This section is organized from the perspective of the cell biologist, namely, we have delineated specific aspects of receptor interactions and illustrated how FFS methods have been applied to address these issues. Although this report focuses on proteins, we should note that FFS approaches are also highly useful for investigations of lipid bilayers, as reviewed by Chiantia, Ries, and Schwille (2009).

## 2.1. Determination of receptor densities

For many years, radioligand binding was the method of choice for measuring the number and affinity of a particular receptor on a cell surface. This approach, often carried out at 4 °C to inhibit internalization of the ligand–receptor complex, typically involves separation of free and bound ligand by centrifugation or filtration. The time required to carry out these procedures may compromise the binding analysis, particularly if interactions are weak and the half-lives of ligand–receptor complexes are short. The development of fluorescently labeled ligands and antibodies encouraged the use of spectrofluorimetric methods, for example, fluorescence flow cytometry, to quantify cell surface receptor number without the need to separate bound from free ligand (Sklar, 1987). More recently, receptor number densities have been measured in live cells using a variety of image correlation methods (Kolin & Wiseman, 2007). Although FFS analyses of receptors have focused on their mobility and oligomerization (see below), several groups have used the FCS approach to measure expression levels of endogenous receptors on the cell surface. A critical step in any fluorescence-based characterization of endogenous receptors is the development of fluorescent ligand analogues with pharmacological properties reasonably similar to those of the parent molecules. For example, Hegener et al. (2004) generated an Alexa 532-labeled form of norepinephrine to characterize β2-adrenergic receptors on the surface of primary hippocampal neurons. Measurements carried out at 20 °C indicated that these neurons had a receptor density of $\sim 4.5\ \mu m^{-2}$, equivalent to $\sim$2700 receptors/cell. Moreover, the autocorrelation analyses obtained from these binding experiments revealed the presence of two distinctly migrating populations of ligand–receptor complexes, having diffusion constants of $\sim 5 \times 10^{-8}$ and $\sim 6 \times 10^{-10}\ cm^2/s$. More recently, Chen et al. (2009) used an FITC-labeled aptamer to determine that HeLa and leukemia CCRF-CEM cells express $\sim$550 and $\sim$1300 copies per $\mu m^2$ of the membrane receptor PTK7 (protein tyrosine kinase 7), a regulator of cell polarity. These authors suggest that fluorophore-labeled aptamers, which can provide high-binding affinity and specificity, are potentially important tools in future FFS analyses of ligand–receptor interactions.

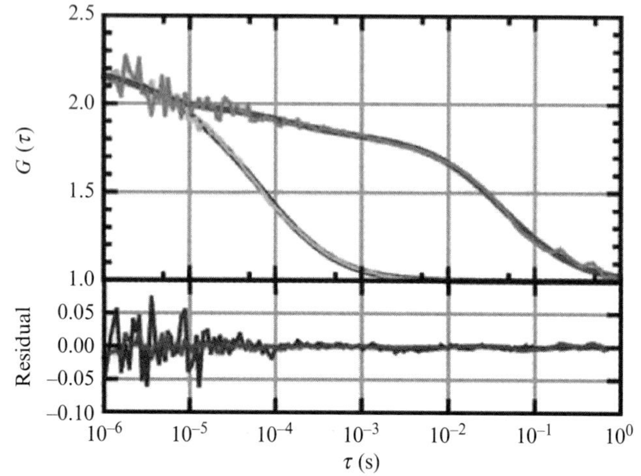


**Figure 3.2** FCS results of Miagi and Maruyama (2010). Autocorrelation curves of bound Rh-EGF to CHO cells expressing EGFR–EGFP (red, right curve), and free Rh-EGF (green, left curve) are shown with their fits. (For interpretation of the references to color in this figure legend, the reader is referred to the online version of this chapter.)

FCS has also been used to measure the number of expressed, fluorescently labeled receptors on the plasma membrane. Although of less physiological significance than the estimation of endogenous receptor number, quantification of expressed receptors is an important step in the rigorous characterization of any ligand–receptor system. As an example of such an analysis, Miagi and Maruyama (2010) used rhodamine-labeled EGF (Rh-EGF) and an EGFP-receptor construct (EGFR–EGFP) to study EGFRs in Chinese hamster ovary (CHO) cells, which do not express endogenous EGFR. The results are shown in Fig. 3.2. The green line corresponds to the autocorrelation curve for Rh-EGF free in solution, which had a diffusion coefficient of $94.3 \pm 4.7\ \mu m^2/s$, while the red line corresponds to Rh-EGF bound to the cell surface with a diffusion coefficient of $0.13 \pm 0.03\ \mu m^2/s$. The FCS data allowed them to estimate the number of expressed EGFR–EGFP receptors on the CHO cell surface to be $1.01 \pm 0.76 \times 10^5$ (number of cells = 15). By way of comparison, Scatchard analysis of the binding of Rh-EGF to these cells gave $1.70 \pm 0.11 \times 10^5$ EGFR–EFGP receptors.

## 2.2. Measurement of binding affinities

As indicated in the previous chapters, FCS is able to determine the concentration and the diffusional properties of fluorescent molecules in solutions and in cells. Perhaps the first use of FCS to study a receptor–ligand interaction was the report from Rigler's laboratory (Rauer, Neumann, Widengren, & Rigler, 1996) wherein the detergent-solubilized acetylcholine receptor of *Torpedo californica* was isolated as mixed micelle complexes and the ligand α-bungarotoxin was labeled with tetramethylrhodamine. The main intent of this study was to elucidate the association and dissociation rate constants for the ligand binding. A more recent use of FCS to study ligand–receptor interactions, wherein ligands were labeled with fluorophores, was the study by Winter, McPhee, Van Orden, Roess, and Barisas (2011). These researchers utilized insulin labeled at the N-terminus with FITC and found two classes of insulin-binding sites with dissociation constants ($K_d$) of 0.11 and 75 n*M*. Swift, Burger, Massotte, Dahms, and Cramb (2007) used two-photon excitation fluorescence cross-correlation spectroscopy (TPE-FCCS) to study the interaction of both agonists and antagonists with the human μ-opioid receptor. Recombinant receptors were expressed containing a C-terminal hexahistidine tag that could then be targeted with fluorescently labeled antibodies (Alexa 488 was used to label the antibodies). In one of their studies, biotinylated enkephalin peptides were labeled with streptavidin-conjugated quantum dots (QDs) and equilibrium-binding constants of the ligand–receptor complexes were determined. TPE-FCCS was also used by Savatier, Jalaguier, Ferguson, Cavaillès, and Royer (2010) to measure the $K_d$ for human estrogen receptor (ER), α and β, with the coactivator partner TIF2 in COS-7 cells. ERα and ERβ were fused with the blue fluorescent protein (FP) cerulean, while TIF2 was fused to the mCherry FP such that cross talk among the FPs was minimized. $K_d$ values were determined for receptor–TIF2 complexes in the presence of agonist or antagonist, as well as for the unliganded system.

## 2.3. Receptor oligomerization state and clustering

It is known that many membrane-bound receptors are active as dimers, but not as monomers, and that ligand binding often induces dimerization, as in the case of the EGFR (Fig. 3.3). Hence, determination of the oligomerization state of a receptor, in the presence and absence of ligand, is an active research topic. Several FFS methods have been developed in the past decade to elucidate the stoichiometry of protein complexes, that is, the number of protein subunits in an oligomeric complex. These methods, described in impressive detail in other chapters in this, and the previous, volume, include the PCH method (Chen, Müller, So, & Gratton, 1999), the related fluorescence intensity distribution analysis (FIDA) method (Kask, Palo,

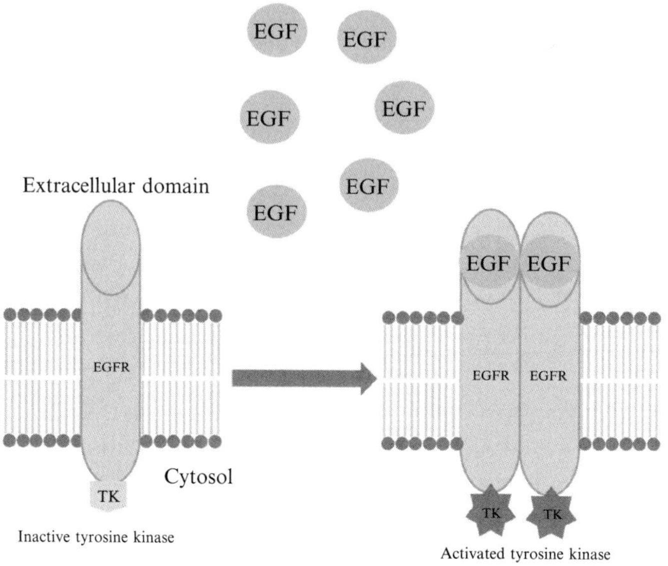


**Figure 3.3** Schematic illustrating dimerization and activation of the EGFR in response to binding of the EGF ligand. (For color version of this figure, the reader is referred to the online version of this chapter.)

Ullmann, & Gall, 1999), and the time-integrated fluorescence cumulant analysis approach (Wu & Müller, 2005). The basic idea underlying these methods is that, in addition to providing information on fluorophore diffusion, one can determine the inherent "brightness" of the molecule under observation. True molecular brightness, for example, counts per second per molecule, is a function of photophysical parameters such as the molecule's extinction coefficient, or two-photon cross section, its quantum yield, as well as instrument parameters such as laser power, detector efficiencies and optical settings. If one has access to a molecule of standard "brightness," for example, a monomeric EGFP, then one can use that standard to calibrate the system and ascertain the "brightness" of the target molecule. For example, Chen, Wei, and Müller (2003) used monomeric EGFP and a dimeric tandem EGFP as standards to study the oligomerization of NRs, specifically the testicular receptor 4, the retinoic acid receptor, and the retinoid X receptor. Homocomplexes of receptors were observed as a function of either protein concentration or ligand binding. One interesting feature of the analysis outlined in this work is that it can be applied up to relatively high protein concentrations (for FFS), for example, ~10 μM. Saffarian, Li, Elson, and Pike (2007) used FFS and the FIDA analysis to investigate clustering of the

EGFRs in response to cholesterol concentrations. In this chapter, EGFP was fused to the C-terminus of the EGFR. The data were analyzed using a modification of the FIDA method, specifically for use in a two-dimensional system using "quantal brightness," an analytical approach that provides explicit information on the distribution of the fluorescent molecules among clusters of increasing size. The results indicated that the EGFR existed as an equilibrium between single receptors and clusters of receptors and that depletion and augmentation of cholesterol led, respectively, to increased and decreased clustering of the receptor. Very recently, Herrick-Davis, Grinde, Lindsley, Cowan, and Marzurkiewicz (2012) used FCS and the PCH method to demonstrate that the serotonin 5-HT$_{2C}$ receptors within the plasma membranes of HEK293 and rat hippocampal neurons were dimeric.

Another FFS approach to elucidate protein oligomerization is the N&B analysis method developed in Enrico Gratton's laboratory (see, e.g., Digman, Dalal, Horwitz, & Gratton, 2008). This approach allows a pixel by pixel analysis of the "brightness" of a signal, which in turn can provide information on the oligomerization state of the labeled proteins. This approach was utilized by Ross, Digman, Gratton, Albanesi, and Jameson (2011) who used TIRF and FFS to study the oligomerization state of the large GTPase dynamin in the cell membrane. As shown in Fig. 3.4, dynamin-EGFP is predominantly tetrameric on the plasma membrane although some larger aggregates are also present. A comprehensive and elegant FFS study using N&B analysis was recently carried out by Hellriegel, Caiolfa, Corti, Sidenius, and Zamai (2011) on an EGFP construct of GPI-anchored urokinase plasminogen activator receptor, which demonstrated that binding of the amino-terminal fragment of urokinase plasminogen activator is sufficient to induce dimerization of the receptor. That work discusses many of the technical issues involved in this type of study and the N&B analysis.

## 2.4. Analysis of nuclear receptors

FFS approaches, often in conjunction with FRAP and FRET (Förster resonance energy transfer), are being increasingly employed to define the molecular mechanisms of gene regulation by NRs and other transcriptional regulators. Properties of NRs examined by FFS include homo- and heterodimerization, transport into and out of the nucleus, and association with DNA and the transcription machinery. As with cell surface receptors, the majority of FFS investigations of NRs have utilized FCS analysis to

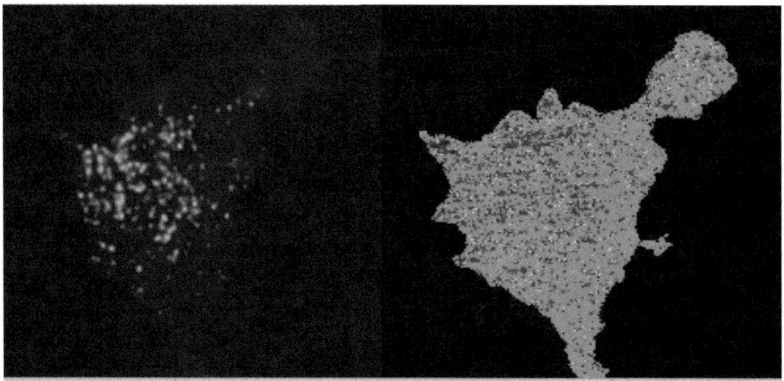

| Region | Brightness (B) | True molecular brightness ($\varepsilon$) | Ratio to mEGFP |
|---|---|---|---|
| Calibration EGFP | 1.068 | 1.36 cpsm | 1× |
| Small aggregates | 1.262 | 5.24 cpsm | 4× |
| Large aggregates | 1.690 | 13.8 cpsm | 10× |

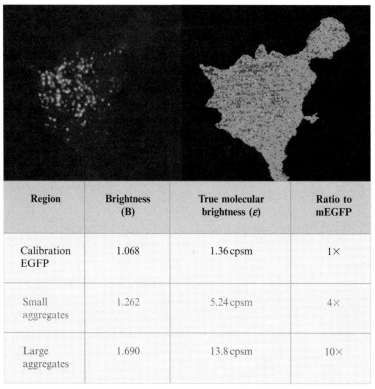

**Figure 3.4** Results of Ross et al. (2011) illustrating N&B analysis of TIRF data on dynamin2-EGFP oligomerization state in the plasma membrane of a mouse embryo fibroblast. Specifically, the analysis indicates that the majority of the dynamin2-EGFP is present as a tetramer. The image in the upper left corresponds to the TIRF intensity, while the image in the upper right corresponds to the N&B analysis, which shows pixels (green) with brightness levels four times those of the monomer standard. (See Color Insert.)

monitor ligand-dependent changes in diffusion constants (Mazza, Stasevich, Karpova, & McNally, 2012). Once in the nucleus, NRs face a daunting variety of potential diffusional barriers, including subnuclear compartments (e.g., PML bodies, Cajal bodies, and nuclear speckles), nonspecific DNA-binding sites, and perhaps even a nuclear cytoskeleton. However, both FCS and FRAP analyses have indicated that NRs are highly mobile when unligated (Lionnet, Wu, Grünwald, Singer, & Larson, 2010). Wu, Corbett, and Berland (2009) used TPE-FCS to study the intracellular mobility of the nuclear import receptors karyopherin-α and karyopherin-β, both free and associated with their cargoes, namely, EGFP linked to nuclear localization signals. Based on FRAP measurements, ligand binding has negligible effects on the mobility of some NRs (e.g., peroxisome proliferator-activated receptor (PPAR), retinoic acid receptor, and thyroid hormone receptor) but significantly slows or immobilizes others (e.g., glucocorticoid receptor, ER). However, Gelman et al. (2006) suggested that these differences merely reflected the relatively low temporal and spatial resolution afforded by FRAP and showed using FCS that in fact, the mobility of PPAR is significantly reduced upon ligand binding. A compilation and analysis of FCS data on the mobility of NRs is presented by Lionnet et al. (2010).

The majority of NRs function as homo- or heterodimers, and the nature of their dimerization partners can determine whether they act as transcriptional activators or repressors. FCS analysis is generally not useful to monitor these dimerization events, as the diffusion constant for a globular protein is inversely proportional to the cube root of its mass and, hence, would only decrease by $\sim$20% upon dimerization. In fact, as shown by Meseth, Wohland, Rigler, and Vogel (1999), to resolve two components by FCS, their diffusion times must differ by at least a factor of 1.6. To overcome this limitation, PCH analysis has been used to measure NR homodimerization (Chen et al., 2003) and dual-color FFS to measure heterodimerization (Chen & Müller, 2007).

# 3. CHOICE OF FLUOROPHORES: GENERAL CONSIDERATIONS

Obviously, before one can contemplate carrying out FFS, one must introduce a fluorophore into the system. Various approaches are currently used and several of the most popular shall be discussed. It is important to note that the fluorophore utilized must have a high intrinsic brightness, by which we mean a high extinction coefficient and a good quantum yield. The reason for this requisite sensitivity is that the FFS method, inherently, can usually be implemented only at low fluorophore concentrations, typically less than 0.01 μ*M* (although exceptions exist). This concentration restriction is seldom a problem since receptors are usually present in low numbers, although these numbers vary from system to system.

## 3.1. Fluorescent antibodies

One of the oldest methods utilized for introducing a fluorophore into a biological system is to label an antibody with a covalently attached probe. In this approach, a fluorescent moiety is covalently attached to one of the antibody's amino acid residues, most commonly lysine or cysteine. Perhaps the most comprehensive and thorough source for information on the chemistries of probe-protein labeling is "The Molecular Probes® Handbook—A Guide to Fluorescent Probes and Labeling Technologies," specifically the 11th edition. Texts which discuss the chemistry of protein reactive groups and diverse labeling reagents used for protein conjugation include Hermanson (1996) and Wong and Jameson (2012). Most researchers buy antibodies already labeled with a probe, often fluorescein- or rhodamine-based, though now antibodies labeled with the Alexa series of probes are also available. Regardless of whether the labeled antibody is obtained commercially or prepared "in-house," characterization of the functional properties of the labeled antibody is important. Usually, though not always, the fluorescent labels are on secondary antibodies (i.e., an antibody that recognizes the primary antibody).

Assuming suitable antibodies are available, immunofluorescence (IF) analysis of endogenous, untagged receptors should be performed to confirm that the tagged proteins localize correctly in the cell. To ascertain the specificity of the antibodies, it is desirable to carry out both IF and Western blotting in untreated cells and in cells depleted of the target proteins by RNAi. The specificity of the labeled antibody to its target should also be determined. This specificity is usually accomplished using Western blotting. We have noted a tendency of many researchers to use antibody labeled with FITC, a probe with notoriously poor photostability properties in the microscope. Although most (though not all) studies we have seen explicitly monitor and discuss the photostability issue, we note that use of more photostable probes, for example, from the Alexa (Life Technologies, Grand Isle, NY, USA), Cy (GE Healthcare, Piscataway, NJ, USA), DyLight (Thermo Fisher Scientific, Pittsburgh, PA, USA), or ATTO (ATTO-TEC GmbH, Siegen, Germany) series, would mitigate this problem significantly.

## 3.2. Fluorescent proteins

The use of FPs, that is, GFP-like proteins, to label cellular target proteins such as receptors has increased dramatically in recent years. Needless to say, there are now a great many members of the FP family with diverse excitation and emission properties. New FPs are being reported almost every month; a very useful review of this topic has been written by Stepanenko et al. (2011). A very recent article by Jones, Ehrhardt, and Frommer (2012) has a title which neatly summarizes the field, namely, "A never ending race for new and improved fluorescent proteins." It is manifestly impractical for us to try to summarize this huge field, but we can point out some of the most salient aspects of choosing FPs for use with FFS, which may give those new to the field some guidelines to consider. A couple of considerations arise immediately from the biological point of view. For example, one may ask how long it takes for the FP to generate the fluorophore in the cell; this issue is usually only important when time-critical events are being monitored. A recent review on novel FPs by Wu, Piatkovich, Lionnet, Singer, and Verkhusha (2011) discusses, among other aspects, some of the so-called "fluorescent timers," which are FPs with variable maturation times. Another concern, which in fact is rarely addressed, is whether the target protein linked to the FP still possesses its inherent biological activity. It is often difficult to isolate the target-FP/fusion protein complex from eukaryotic cells, and one is usually relegated to showing that the cell appears normal or that it still can perform some particular function, such as endocytosis. In some cases, the target-FP can be isolated. For example, in our own studies on dynamin2-EGFP constructs, we isolated the recombinant protein from Sf9 cells and were gratified to determine that its GTPase activity matched that of wild-type dynamin (Fig. 3.5).

Another consideration when choosing an FP is the instrumentation available, for example, one or two-photon sources, since the excitation wavelengths available will certainly limit which FPs can be utilized. One of the most important properties of FPs, for FFS studies, is their inherent brightness, that is, how many counts per second per molecule can they deliver? This consideration is clearly important since it will determine the number of FPs that can be detected in the context of the inherent autofluorescence of the cell. An excellent review of the two-photon properties of various orange and red FPs, including two-photon absorption cross sections and two-photon brightness, is given by Drobizhev, Tillo, Makarov, Hughes, and Rebane (2009). Photostability of the FP is also a concern as is the type of "blinking" exhibited by the FP. This "blinking" is caused by dissociation and binding of protons to various amino acid residues, for example, Tyr-66 (Haupts, Maiti, Schwille, & Webb, 1998). Clearly, blinking on the right timescale may cause a fluctuation in the fluorescence signal, which can be measured by FCS and which should not be confused with other sources of signal fluctuation. We shall not go into more detail on these points since numerous articles in the primary literature address these issues (see, e.g., Liu, Kim, & Heikal, 2006; Ringemann et al., 2008; Ward, 2006; Wong, Banks, Abu-Arish, & Fradin, 2007). As stated, new and interesting FPs appear almost every month, for example, Shcherbakova, Hink, Joosen, Gadella, and Verkhusha (2012) very recently reported a new orange FP called

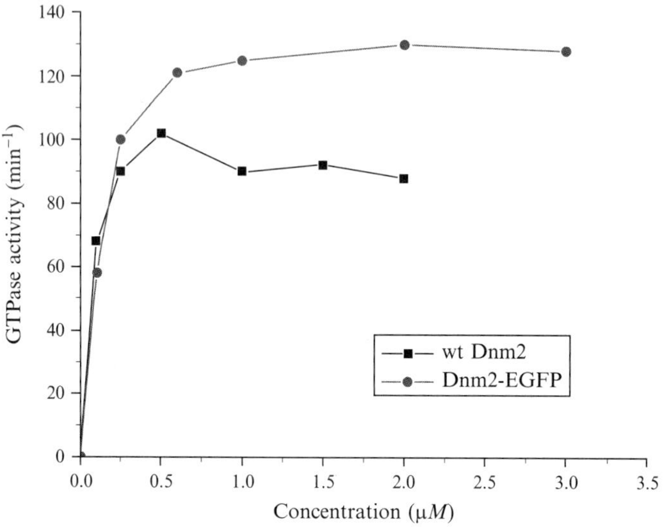


**Figure 3.5** GTPase activities of C-terminally $His_6$-tagged (black) and EGFP-tagged (red) dynamin 2 as a function of dynamin concentration. We had previously shown that $His_6$-tagged dynamin behaves identically to endogenous, untagged dynamin purified from bovine brain. The specific activities of dynamins are strongly dependent on their states of self-association; hence, these activities increase as dynamins polymerize in a concentration-dependent manner. As evident from this figure, the GTPase activities of EGFP-tagged and untagged dynamins are essentially identical. (For interpretation of the references to color in this figure legend, the reader is referred to the online version of this chapter.)

LSSmOrange (LSS stands for large-Stokes shift), which has very favorable optical properties for use with FFS or FRET measurements. Their report describes the spectroscopic properties of this FP along with several other new FPs and nicely illustrates the type of spectroscopic information required to evaluate a new FP. Finally, we should note that the positioning of FP tags should be chosen to minimize disruption of folding, activity, or interactions. Most plasma membrane receptors have extracellular N-termini and cytoplasmic C-termini. N-terminal tagging of these receptors may be problematic, as the tag must be placed after the signal sequence or it will interfere with or be removed during processing at the endoplasmic reticulum. In addition, an extracellular FP tag may impair ligand binding. For these reasons, FPs are most often fused to the C-termini of receptors, affording the additional advantage that maturation of the FP provides evidence that the entire receptor has been translated. However, even with C-terminal tags, it must be ascertained that the FP does not interfere with critical cytoplasmic interactions, for example, with downstream signaling molecules, intracellular trafficking machinery, or posttranslational modifications, such as phosphorylation or ubiquitylation. These considerations do not apply to NRs, which are soluble proteins and therefore have both N- and C- termini available for tagging.

## 3.3. Biomolecular fluorescence complementation

A variant on genetically encoding FPs onto target proteins is the biomolecular fluorescence complementation (BiFC) method, which allows the study of protein–protein interaction (Ghosh, Hamilton, & Regan, 2000; Nagai, Sawano, Park, & Miyawaki, 2001). The BiFC method involves the attachment of nonfluorescent N- and C-termini of an FP, typically split between β-sheets 7 and 8, to the target proteins thought to interact. Upon association, if the three-dimensional properties of the complex permit, the FP halves can come together, which promotes refolding of the FP and maturation of the fluorescent moiety, producing a fluorescent marker of association (Fig. 3.6). Formation of the fluorescent signal after association is not instantaneous. The refolding of FP, in BiFC assays, has a half-time of seconds to minutes, and the maturation of the chromophore requires several minutes (Hu, Chinenov, & Kerppola, 2002). This process,

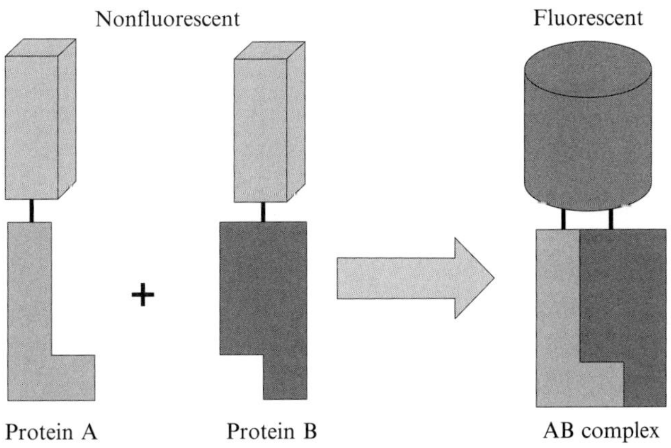


**Figure 3.6** Schematic illustrating the principle underlying the biomolecular fluorescence complementation (BiFC) method. Specifically, one of the target proteins (A) is fused with part of a fluorescent protein, while the other target protein (B) is fused with the other part of the fluorescent protein. If the two target proteins interact and form a complex (AB), then it is possible for the fluorescent protein parts to fold together and form a fluorescent adduct. (For color version of this figure, the reader is referred to the online version of this chapter.)

formation of a mature tertiary protein attached to two target proteins, is generally considered to be irreversible, thus having the potential to capture transient interactions. However, under some conditions, BiFC has been shown to be reversible (Anderie & Schmid, 2007). This method of protein–protein detection has found great popularity among the GPCR aficionados as a means of detecting agonist-dependent protein-interacting partners, as it offers several unique advantages over FRET (Rose, Briddon, & Holliday, 2010). Although FRET measurements can be acquired in real time and are fully reversible, the analysis is often more difficult than with BiFC because appropriate controls are needed to determine energy transfer efficiency in the presence of potential complicating factors such as signal bleed-through or homo-FRET. Kilpatrick, Briddon, and Holliday (2012) combined FFS and PCH with BiFC to study the effect of β-arrestin adaptors and endocytic mechanisms on the diffusion and particle brightness of GFP-tagged neuropeptide Y receptors in the plasma membrane of HEK293 cells. In this report, Kilpatrick et al. noted that the complex photophysics of the yellow fluorescent protein (YFP), typically used in the BiFC approach, creates some difficulties with the use of FFS with BiFC. To overcome this problem, they developed a novel BiFC system based on a version of the superfolder (sf) GFP. As mentioned earlier, Herrick-Davis et al. (2012) used FFS combined with BiFC to provide evidence for the dimeric state of the 5-$HT_{2C}$ receptor.

## 3.4. HaloTags/SNAP/FlAsH

*HaloTags*: In the HaloTag approach, one's protein of interest is fused recombinantly to a mutated form of bacterial haloalkane dehalogenase. The dehalogenase normally catalyzes the removal of halides from halogenated aliphatic hydrocarbons through nucleophilic displacement mechanisms; however, the mutant form loses the ability to release the covalent ester bond formed with an aspartic acid residue during catalysis (Los & Wood, 2006; Los et al., 2005, 2008). Hence, reaction of this mutant dehalogenase with an appropriate halogenated aliphatic substrate, covalently attached to a fluorophore, will result in the covalent linkage of the fluorescent probe to the halogenase and thus the protein of interest (Fig. 3.7). This method allows for the introduction of different fluorophores onto the target protein without having to modify the construct. Since the dehalogenase is a bacterial enzyme, the labeling is specific and cross-reaction with mammalian proteins is eliminated. This method is suitable for cell surface

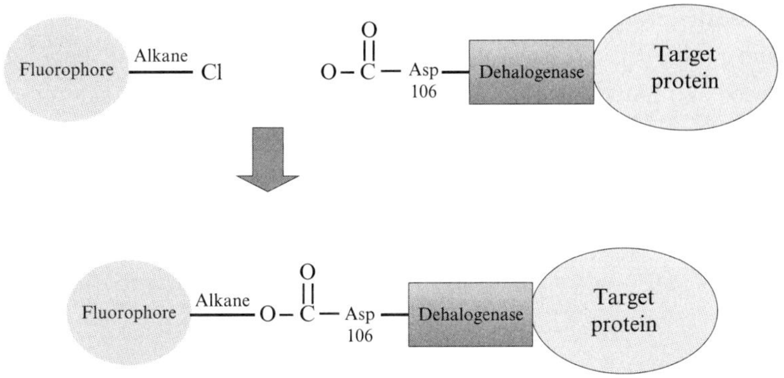


**Figure 3.7** Schematic illustrating the principle underlying the HaloTag method. A fluorescent probe is linked to a haloalkane, and the bacterial dehalogenase enzyme is recombinantly fused to the target protein. The fluorescent haloalkane can form a stable covalent linkage with the dehalogenase. (For color version of this figure, the reader is referred to the online version of this chapter.)

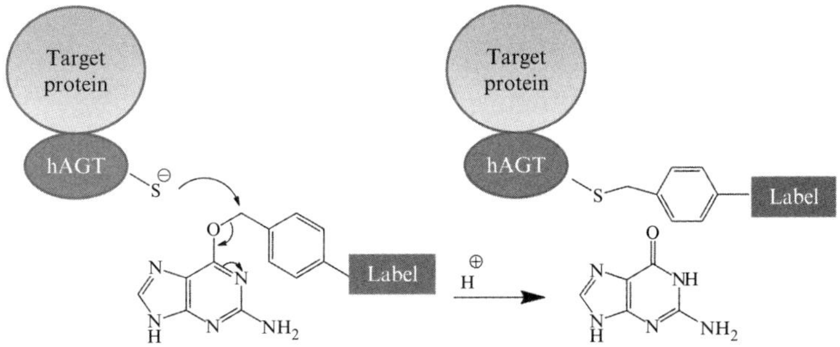


**Figure 3.8** Mechanism of covalent labeling of hAGT, which is fused to a target protein, with a fluorescent label linked onto an $O^6$-benzylguanine derivative. *Figure modified from Keppler et al. (2003).* (For color version of this figure, the reader is referred to the online version of this chapter.)

proteins, such as receptors, which are accessible to the fluorescent substrate. Of course, the same considerations regarding the fusion of the dehalogenase and target protein genes mentioned in the section on FP constructs, also apply for the HaloTag approach.

*SNAP Tags*: Another approach that allows for labeling *in vivo* is the SNAP-tag method (Keppler et al., 2003). Covalent labeling of a target protein is accomplished by construction of a fusion protein with $O^6$-alkylguanine-DNA alkyltransferase (hAGT). hAGT irreversibly transfers the alkyl group from its substrate to a cysteine residue (Fig. 3.8). Substrate specificity is relatively low as hAGT can use $O^6$-benzylguanine (BG) or $O^6$-benzylguanosine with substituted benzyl rings as substrates

(Damoiseaux, Keppler, & Johnsson, 2001). Optimization of the specificity was shown by Xu-Welliver, Leitao, Kanugula, Meehan, and Pegg (1999) who used a mutated form of hGAT, namely, G160W hGAT, which has approximately five times the specific activity for BG *in vitro* compared to hGAT. The high activity of hGAT to a substrate that is typically inert makes this system ideal for targeted labeling of a protein of interest.

*FlAsH*: The Roger Tsien lab developed a uniquely specific probe, termed FlAsH, by modifying fluorescein to contain two arsenoxides (Griffin, Adams, & Tsien, 1998). Arsenoxides have a high affinity for closely paired cysteines and the two on FlAsH confer a strong affinity for a CCXGCC sequence, which is uncommon in naturally occurring proteins (Fig. 3.9). To implement this method, the target sequence is introduced genetically into the protein of interest and the FlAsH reagent is added to the cell's medium, whereupon it can be taken up by the cell (the 1,2-ethanedithiol (EDT) moieties on the fluorophore render it sufficiently hydrophobic to pass through the plasma membrane) and react with the cysteine sequence. Toxicity of the arsenoxides is minimized by labeling the cells in the presence of EDT. Testing of the specificity of FlAsH for different sequences led to the discovery that FlAsH has a preference for the CCPGCC motif, which adopts a hairpin rather than helical structure (Adams et al., 2002). At the same time, experiments to adjust the emission, by using rhodamine or 3,6-dihydroxyxanthone instead of fluorescein, were conducted, thus making the probe more attractive for microscopy. These experiments led to the creation of the red-emitting ReAsH as well as a blue-emitting CHOxAsH, expanding the spectroscopic diversity. A major advantage of the FlAsH tag is its small size combined with its selectivity for the CCPGCC motif. Hoffmann et al. (2005), for example, demonstrated that a GPCR labeled

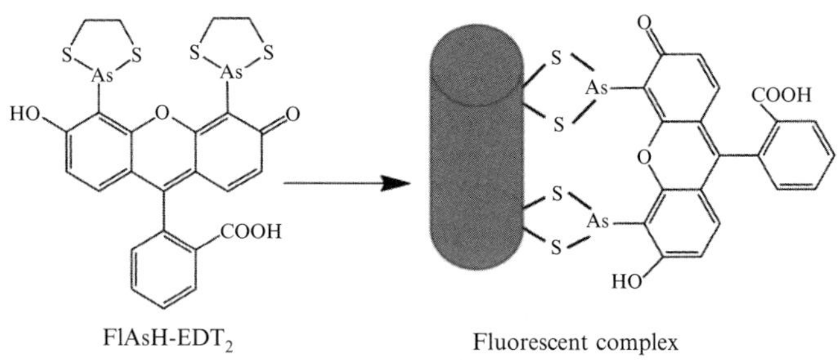


**Figure 3.9** FlAsH reagent forming a fluorescent complex with an α-helix of a target protein that contains the CCXGCC FlAsH motif. (For color version of this figure, the reader is referred to the online version of this chapter.)

with a CFP/YFP FRET pair could not support coupling to adenylyl cyclase, whereas a CFP/FlAsH pair maintained this biological function. On the other hand, a potential limitation of the FlAsH tag approach is the possibility of significant fluorescent backgrounds.

### 3.5. Quantum dots

QDs are semiconductor nanocrystals, typically a few hundred to a few thousand atoms of cadmium sulfide, cadmium selenide, or cadmium telluride, which have a characteristic emission, due to exciton confinement, based on the size of the particle (first reported by Rossetti, Nakahara, & Brus, 1983). The core emission is typically weak and always unstable and so a "shell" is added to insulate the core. The shell material, for example, zinc sulfide, is typically almost entirely unreactive and completely insulates the core. A layer of organic ligands can be covalently attached to the surface of the shell, which provides a surface for conjugation to biological (antibodies, streptavidin, lectins, and nucleic acids) and nonbiological species and which makes the QDs water-soluble. QDs exhibit absorption over a wide range, extending into the visible region. Emissions can be tuned from the visible to the infrared by varying the size of the QD. In recent years, QDs have found applications in biological imaging, due to their inherent brightness and photostability. This popularity has been driven, in part, by innovations in QD surface modifications, which has facilitated the process of linking them to biological targets; a recent review of this topic has appeared (Bruchez, 2011). Heuff, Swift, and Cramb (2007) explicitly discussed the challenges involved in the use of QDs with FCS owing to their blinking characteristics. Several groups have used QDs and FFS to study receptors. For example, QDs were used by Boyle et al. (2011) to study the clustering of the T-cell receptor (TCR) with *k*-space ICS. Interestingly, the blinking statistics differed between small clusters of QDs compared to single QDs or large clusters, which allowed this function to serve as a reporter of nanometer-scale changes in QD-labeled TCR organization after T-cell activation by antigen.

## 4. CELLS: GENERAL CONSIDERATIONS

An important initial consideration for those embarking on FFS analysis of receptors is the choice of cell type. For example, to study the properties of recombinant, GFP-tagged EGFRs with minimal interfering contributions from endogenous unlabeled receptors, one may wish to utilize

NIH 3T3 cells, which express ~650,000 PDGF receptors (Liapi, Raynaud, Anderson, & Evain-Brion, 1990) but fewer than 10,000 EGFRs per cell (Velu et al., 1987). On the other hand, A431 cells, which express 2–4 million EGFRs per cell, may be chosen to study the trafficking of the fluorescently tagged ligand, EGF, bound to endogenous unlabeled receptors. HeLa cells, with approximately 50,000 EGFRs per cell (Berkers, van Bergen en Henegouwen, & Boonstra, 1991), could serve as a compromise. Once suitable cells for a given experiment are chosen, the success of any live-cell imaging analysis depends on the ability to maintain these cells in a healthy, functional state throughout their visualization. Critical factors that must be generally controlled, including nutrient supply, pH, and phototoxicity have been defined in numerous reviews (e.g., Frigault, Lacoste, Swift, & Brown, 2009) and are discussed only briefly here.

## 4.1. Cell growth and transfection

To increase the surface density of receptors and ensure that they are poised to respond to their specific ligands, cells are grown for various times, often overnight, in serum-free or serum-depleted (e.g., 0.05–0.5%) medium. For live-cell imaging experiments, cells are typically plated in dishes with coverslip bottoms, such as MatTek 35 mm #1.5 dishes having an average coverslip thickness of 0.17 mm. Slides may be uncoated or, if necessary for cell attachment, coated with collagen, polylysine, or fibronectin, for example.

Transient transfection of fluorescently tagged constructs is routinely performed using a commercially available gene transfer reagent, selected by experience for each cell type. A comparison of the efficiency and cytotoxicity of six transfection reagents in nine commonly used cell lines was recently reported (Yamato, Dai, & Moursi, 2010). It is important to establish the suitability of a particular transfection system for each cell type prior to functional experiments, although FuGENE HD (Promega) was found to be the preferred reagent for many cell types in the Yamato et al. (2010) study.

## 4.2. Maintenance of cell viability

Mammalian cells are most often grown in medium containing 5–10% fetal bovine serum buffered with a $CO_2$/bicarbonate system to maintain pH within appropriate physiological range, typically 7.2–7.4. Phenol red, a pH indicator that turns orange at pH 7.0, then yellow below pH 6.8, is routinely included in culture medium at a concentration of ~15 mg/L.

Acidification of the medium may be reflective of bacterial contamination or to the simple release of metabolic and other waste products from the cultured cells themselves. An orange color serves as a warning that the medium should be changed. Because Phenol red interferes with fluorescence imaging due to its high extinction coefficient, most conventional culture media are currently available in a Phenol red-free form. When cells are transferred from the $CO_2$ incubator to the often uncontrolled atmospheric conditions of the microscope stage for analysis, reduction of $CO_2$ levels can lead to alkalinization of the medium in less than an hour, accompanied by a change in Phenol red color to pink and then purple. To reduce the need for $CO_2$, bicarbonate buffer is often supplemented with or replaced by 10–25 m*M* HEPES buffer, which maintains pH even without $CO_2$ equilibration, and allows both survival and proliferation of cells for ~10 h provided they are kept in complete medium (Frigault et al., 2009). However, even when using HEPES, the medium can alkalinize to unhealthy levels with time and should be monitored. Moreover, HEPES has been reported to form cytotoxic products, such as hydrogen peroxide, upon exposure to illumination (Zigler, Lepe-Zuniga, Vistica, & Gery, 1985). The Leibovitz L-15 culture medium has been developed to circumvent problems arising from the use of HEPES buffer or the absence of bicarbonate/$CO_2$. A common cause of cell stress is expression of abnormally high levels of recombinant proteins, which may also result in the generation of physiologically irrelevant data. Whenever possible, fluorescently tagged proteins should be expressed in cells depleted of their endogenous counterparts (e.g., by RNAi), using inducible promoters to control expression levels. Characteristics of dead or dying cells include detachment and rounding, formation of surface blebs and large vacuoles, and deformation of mitochondria. Numerous indicators of cell viability are available. For example, alamarBlue (Life Technologies) can be used to monitor cell metabolism and MitoTracker stain (Life Technologies) to visualize changes in mitochondrial morphology.

## 4.3. Photobleaching and phototoxicity

A critical consideration in live-cell imaging is to achieve a balance between maximal signal and minimal photobleaching and phototoxicity. Although photobleaching is embraced in FRAP approaches, it should be avoided or certainly minimized in most other fluorescence measurements (we note that photobleaching of the acceptor in FRET pairs is sometimes utilized). Considerable effort has been expended during the past few decades to develop fluorescent probes with enhanced photostability. Fluorescein is still probably one of the most popular probes in use, but it is notorious for its tendency to photobleach in the microscope. For these reasons, researchers turned early on to rhodamine, a much more photostable probe, and then to newer families of probes such as the Alexa or ATTO series. We should note that not all photobleaching processes are irreversible, for example, an excited fluorophore may be transiently trapped in a triplet state, which can eventually decay to the ground state and hence resurrect the functional fluorophore (Periasamy, Bicknese, & Verkman, 1996). Phototoxicity may be coupled to fluorescence excitation, which can result in the production of toxic products, often reactive oxygen species, or to fluorescent-independent processes such as light-induced cell heating (Tinevez et al., 2012). Photodamage can be minimized by reducing illumination to a level that maintains a tolerable signal-to-noise ratio. An important contribution to photobleaching of GFPs is their ability, in the excited state, to donate electrons to cellular components such as FAD, FMN, and $NAD^+$, a process which has been termed "oxidative reddening" (Bogdanov et al., 2009). Oxidative reddening is influenced by the culture medium, which in turn controls the levels of biological electron acceptors. For example, the photostability of EGFP (but not the red FPs) is increased by approximately ninefold by growing cells in vitamin-depleted DMEM (Bogdanov et al., 2009). Phototoxicity can also be reduced by adding antioxidants, such as ascorbic acid or Trolox, to the culture medium.

## 4.4. Autofluorescence

Background fluorescence originates both from cellular components and from the cell culture medium. As stated, Phenol red is a major source of autofluorescence and should be removed from imaging medium. However, even in medium without Phenol red, background intensities from cell-free regions should be measured. Cellular sources of autofluorescence include, but are not limited to, flavins and flavoproteins (at 500–600 nm), reduced pyridine nucleotides (at 400–500 nm), and aromatic amino acids. Dead, damaged, crowded, or otherwise stressed cells increase autofluorescence, again highlighting the importance of maintaining a healthy culture environment. An interesting study of the lifetimes associated with the autofluorescence of stem cells in different metabolic stages has recently appeared (Stringari, Sierra, Donovan, & Gratton, 2012).

## 5. SUMMARY

In this overview, we have endeavored to elucidate the nature of cell surface receptors and the type of information that FFS methods can provide. We have also noted the principal methodologies presently used to introduce fluorophores into biological systems. Finally, we discussed key issues regarding cell maintenance and health, since no matter how sophisticated the spectroscopic technique may be, if the target cell is not functioning properly, the information attained may be suspect. With appropriate regard to all of the caveats that accompany instrumentation, choice of fluorophore, data analysis, cell transfection, and cell maintenance, however, FFS methodologies truly do open a window in the life of the cell that promises to illuminate an important undiscovered biological landscape.

## ACKNOWLEDGMENTS

This work was supported by grants RO1GM076665 (D. M. J.), R21NS072754 (D. M. J.), and RO1GM075401 (J. P. A.) from the National Institutes of Health and a grant from Allergan, Inc. We thank Barbara Barylko for carrying out the GTPase assays shown in Fig. 3.5 and Lei Wang for generation of the dynamin2-EGFP construct.

## REFERENCES

Adams, S. R., Campbell, R. E., Gross, L. A., Martin, B. R., Walkup, G. K., Yao, Y., et al. (2002). New bisarsenical ligands and tetracysteine motifs for protein labeling in vitro and in vivo: Synthesis and biological applications. *Journal of the American Chemical Society*, *124*, 6063–6076.

Anderie, I., & Schmid, A. (2007). In vivo visualization of actin dynamics and actin interactions by BiFC. *Cell Biology International*, *31*, 1131–1135.

Axelrod, D. (1981). Cell-substrate contacts illuminated by total internal reflection fluorescence. *The Journal of Cell Biology*, *89*, 141–145.

Berkers, J. A., van Bergen en Henegouwen, P. M., & Boonstra, J. (1991). Three classes of epidermal growth factor receptors on HeLa cells. *The Journal of Biological Chemistry*, *266*, 922–927.

Bethani, I., Skånland, S. S., Dikic, I., & Acker-Palmer, A. (2010). Spatial organization of transmembrane receptor signalling. *The EMBO Journal*, *29*, 2677–2688.

Bogdanov, A. M., Bogdanova, E. A., Chudakov, D. M., Gorodnicheva, T. V., Lukyanov, S., & Lukyanov, K. A. (2009). Cell culture medium affects GFP photostability: A solution. *Nature Methods*, *6*, 859–860.

Bogdanov, A. M., Mishin, A. S., Yampolsky, I. V., Belousov, V. V., Chudakov, D. M., Subach, F. V., et al. (2009). Green fluorescent proteins are light-induced electron donors. *Nature Chemical Biology*, *5*, 459–461.

Boyle, S., Kolin, D. L., Bieler, J. G., Schneck, J. P., Wiseman, P. W., & Edidin, M. (2011). Quantum dot fluorescence characterizes the nanoscale organization of T cell receptors for antigens. *Biophysical Journal*, *101*, L57–L59.

Briddon, S. J., & Hill, S. J. (2007). Pharmacology under the microscope: The use of fluorescence correlation spectroscopy to determine the properties of ligand-receptor complexes. *Trends in Pharmacological Sciences, 28*, 637–645.

Bruchez, M. P. (2011). Quantum dots find their stride in single molecule tracking. *Current Opinion in Chemical Biology, 15*, 775–780.

Chen, Y., & Müller, J. D. (2007). Determining the stoichiometry of protein heterocomplexes in living cells with fluorescence fluctuation spectroscopy. *Proceedings of the National Academy of Sciences of the United States of America, 104*, 3147–3152.

Chen, Y., Müller, J. D., So, P. T., & Gratton, E. (1999). The photon counting histogram in fluorescence fluctuation spectroscopy. *Biophysical Journal, 77*, 553–567.

Chen, Y., Munteanu, A. C., Huang, Y. F., Phillips, J., Zhu, Z., Mavros, M., et al. (2009). Mapping receptor density on live cells by using fluorescence correlation spectroscopy. *Chemistry, 15*, 5327–5336.

Chen, Y., Wei, L.-N., & Müller, J. D. (2003). Probing protein oligomerization in living cells with fluorescence fluctuation spectroscopy. *Proceedings of the National Academy of Sciences of the United States of America, 100*, 15492–15497.

Chiantia, S., Ries, J., & Schwille, P. (2009). Fluorescence correlation spectroscopy in membrane structure elucidation. *Biochimica et Biophysica Acta, 1788*, 225–233.

Damoiseaux, R., Keppler, A., & Johnsson, K. (2001). Synthesis and application of chemical probes for human $O^6$-alkylguanine-DNA alkyltransferases. *ChemBioChem, 2*, 83–100.

Denk, W., Strickler, J. H., & Webb, W. W. (1990). Two-photon laser scanning fluorescence microscopy. *Science, 248*, 73–76.

Digman, M. A., Dalal, R., Horwitz, A. F., & Gratton, E. (2008). Mapping the number of molecules and brightness in the laser scanning microscope. *Biophysical Journal, 94*, 2320–2332.

Drobizhev, M., Tillo, S., Makarov, N. S., Hughes, T. E., & Rebane, A. (2009). Absolute two-photon absorption spectra and two-photon brightness of orange and red fluorescent proteins. *The Journal of Physical Chemistry B, 113*, 855–859.

Eigen, M., & Rigler, R. (1994). Sorting single molecules: Application to diagnostics and evolutionary biotechnology. *Proceedings of the National Academy of Sciences of the United States of America, 91*, 5740–5747.

Elson, E. L., & Magde, D. (1974). Fluorescence correlation spectroscopy. I. Conceptual basis and theory. *Biopolymers, 13*, 1–27.

Frigault, M. M., Lacoste, J., Swift, J. L., & Brown, C. M. (2009). Live-cell microscopy—Tips and tools. *Journal of Cell Science, 122*, 753–767.

Frye, L. D., & Edidin, M. (1970). The rapid intermixing of cell surface antigens after formation of mouse-human heterokaryons. *Journal of Cell Science, 7*, 319–335.

Gelman, L., Feige, J. N., Tudor, C., Engelborghs, Y., Wahli, W., & Desvergne, B. (2006). Integrating nuclear receptor mobility in models of gene regulation. *Nuclear Receptor Signaling, 4*, 1–4.

Ghosh, I., Hamilton, A. D., & Regan, L. (2000). Antiparallel leucine zipper-directed protein reassembly: Application to the green fluorescent protein. *Journal of the American Chemical Society, 122*, 5658–5659.

Griffin, B. A., Adams, S. R., & Tsien, R. Y. (1998). Specific covalent labeling of recombinant protein molecules inside live cells. *Science, 281*, 269–272.

Haupts, U., Maiti, S., Schwille, P., & Webb, W. W. (1998). Dynamics of fluorescence fluctuations in green fluorescent protein observed by fluorescence correlation spectroscopy. *Proceedings of the National Academy of Sciences of the United States of America, 95*, 13573–13578.

Hegener, O., Prenner, L., Runkel, F., Baader, S. L., Kappler, J., & Häberlein, H. (2004). Dynamics of beta2-adrenergic receptor-ligand complexes on living cells. *Biochemistry, 43*, 6190–6199.

Hellriegel, C., Caiolfa, V. A., Corti, V., Sidenius, N., & Zamai, M. (2011). Number and brightness image analysis reveals ATF-induced dimerization kinetics of uPAR in the cell membrane. *The FASEB Journal, 25*, 2883–2897.

Hermanson, G. T. (1996). *Bioconjugate techniques*. San Diego, CA: Academic Press.

Herrick-Davis, K., Grinde, E., Lindsley, T., Cowan, A., & Marzurkiewicz, J. E. (2012). Oligomer size of the serotonin 5-HT2C receptor revealed by fluorescence correlation spectroscopy with photon counting histogram analyses: Evidence for homodimers without monomers or tetramers. *The Journal of Biological Chemistry, 287*, 23604–23614.

Heuff, R. F., Swift, J. L., & Cramb, D. T. (2007). Fluorescence correlation spectroscopy using quantum dots: Advances, challenges and opportunities. *Physical Chemistry Chemical Physics, 28*, 1870–1880.

Hoffmann, C., Gaietta, G., Bünemann, M., Adams, S. R., Oberdorff-Maas, S., Behr, B., et al. (2005). FlAsH-based FRET approach to determine G-protein coupled receptor activation in living cells. *Nature Methods, 2*, 171–176.

Hu, C. D., Chinenov, Y., & Kerppola, T. K. (2002). Visualization of interactions among bZIP and Rel family proteins in living cells using biomolecular fluorescence complementation. *Molecular Cell, 9*, 789–798.

Jacobson, K., Derzko, Z., Wu, E. S., Hou, Y., & Poste, G. (1976). Measurement of the lateral mobility of cell surface components in single, living cells by fluorescence recovery after photobleaching. *Journal of Supramolecular Structure, 5*, 565–576.

Jameson, D. M., Ross, J. A., & Albanesi, J. P. (2009). Fluorescence fluctuation spectroscopy: Ushering in a new age of enlightenment in cellular dynamics. *Biophysical Reviews, 1*, 105–118.

Jones, A. M., Ehrhardt, D. W., & Frommer, W. D. (2012). A never ending race for new and improved fluorescent proteins. *BMC Biology, 10*, 39–41.

Kask, P., Palo, K., Ullmann, D., & Gall, K. (1999). Fluorescence-intensity distribution analysis and its application in biomolecular detection technology. *Proceedings of the National Academy of Sciences of the United States of America, 96*, 13756–13761.

Keppler, A., Gendreizig, S., Gronemeyer, T., Pick, H., Vogel, H., & Johnsson, K. (2003). A general method for the covalent labeling of fusion proteins with small molecules in vivo. *Nature Biotechnology, 21*, 86–89.

Kilpatrick, L. E., Briddon, S. J., & Holliday, N. D. (2012). Fluorescence correlation spectroscopy, combined with biomolecular complementation, reveals the effects of β-arrestin complexes and endocytic targeting on the membrane mobility of neuropeptide Y receptors. *Biochimica et Biophysica Acta, 1823*, 1068–1081.

Kolin, D. L., & Wiseman, P. W. (2007). Advances in image correlation spectroscopy: Measuring number densities, aggregation states, and dynamics of fluorescently labeled macromolecules in cells. *Cell Biochemistry and Biophysics, 49*, 141–164.

Liapi, C., Raynaud, F., Anderson, W. B., & Evain-Brion, D. (1990). High chemotactic response to platelet-derived growth factor of a teratocarcinoma differentiated mesodermal cell line. *In Vitro Cellular and Developmental Biology, 26*, 388–392.

Lionnet, T., Wu, B., Grünwald, D., Singer, R. H., & Larson, D. R. (2010). Nuclear physics: Quantitative single-cell approaches to nuclear organization and gene expression. *Cold Spring Harbor Symposia on Quantitative Biology, 75*, 113–126.

Liu, Y., Kim, H.-R., & Heikal, A. A. (2006). Structural basis of fluorescence fluctuation dynamics of green fluorescent proteins in acidic environments. *The Journal of Physical Chemistry B, 110*, 24138–24146.

Los, G. V., Darzins, A., Karassina, N., Zimprich, C., Learish, R., McDougall, M. G., et al. (2005). HaloTag interchangeable labeling technology for cell imaging and protein capture. *Cell Notes, 11*, 2–6.

Los, G. V., Encell, L. P., McDougall, M. G., Hartzell, D. D., Karassina, N., Zimprich, C., et al. (2008). HaloTag: A novel protein labeling technology for cell imaging and protein analysis. *Chemistry and Biology, 3*, 373–382.

Los, G. V., & Wood, K. V. (2006). The HaloTag: A novel technique for cell imaging and protein analysis. *Methods in Molecular Biology, 356*, 195–208.

Lundstrom, K. (2009). An overview on GPCRs and drug discovery: Structure-based drug design and structural biology on GPCRs. *Methods in Molecular Biology, 552*, 51–66.

Magde, D., Elson, E. L., & Webb, W. W. (1972). Thermodynamics fluctuations in a reacting system: Measurement by fluorescence correlation spectroscopy. *Physical Review Letters, 29*, 705–708.

Magde, D., Elson, E. L., & Webb, W. W. (1974). Fluorescence correlation spectroscopy. II. An experimental realization. *Biopolymers, 13*, 29–61.

Maiti, S., Haupts, U., & Webb, W. W. (1997). Fluorescence correlation spectroscopy: Diagnostics for sparse molecules. *Proceedings of the National Academy of Sciences of the United States of America, 94*, 11753–11757.

Mazza, D., Stasevich, T. J., Karpova, T. S., & McNally, J. G. (2012). Monitoring dynamic binding of chromatin proteins in vivo by fluorescence correlation spectroscopy and temporal image correlation spectroscopy. *Methods in Molecular Biology, 833*, 177–200.

Meseth, U., Wohland, T., Rigler, R., & Vogel, H. (1999). Resolution of fluorescence correlation measurements. *Biophysical Journal, 76*, 1619–1631.

Miagi, H., & Maruyama, I. N. (2010). Analysis of ligand-receptor interaction the surface of living cells by fluorescence correlation spectroscopy. *The Open Spectroscopy Journal, 4*, 28–31.

Nagai, T., Sawano, A., Park, E. S., & Miyawaki, A. (2001). Circularly permuted green fluorescent proteins engineered to sense $Ca^{2+}$. *Proceedings of the National Academy of Sciences of the United States of America, 98*, 3197–3202.

Periasamy, N., Bicknese, S., & Verkman, A. S. (1996). Reversible photobleaching of fluorescein conjugates in air-saturated viscous solutions: Singlet and triplet state quenching by tryptophan. *Photochemistry and Photobiology, 63*, 265–271.

Rauer, B., Neumann, E., Widengren, J., & Rigler, R. (1996). Fluorescence correlation spectrometry of the interaction kinetics of tetramethylrhodamine alpha-bungarotoxin with Torpedo californica acetylcholine receptor. *Biophysical Chemistry, 58*, 3–12.

Ringemann, C., Schönle, A., Giske, A., von Middendorff, C., Hell, S. W., & Eggeling, C. (2008). Enhancing fluorescence brightness: Effect of reverse intersystem crossing studied by fluorescence fluctuation spectroscopy. *ChemPhysChem, 9*, 612–624.

Rose, R. H., Briddon, S. J., & Holliday, N. D. (2010). Bimolecular fluorescence complementation: Lighting up seven transmembrane domain receptor signalling networks. *British Journal of Pharmacology, 159*, 738–750.

Ross, J. A., Digman, M. A., Gratton, E., Albanesi, J. P., & Jameson, D. M. (2011). Oligomerization state of dynamin 2 in cell membranes using TIRF and number and brightness analysis. *Biophysical Journal, 100*, L15–L17.

Rossetti, R., Nakahara, S., & Brus, L. E. (1983). Quantum size effects in the redox potentials, resonance Raman spectra and electronic spectra of CdS crystallites in aqueous solutions. *The Journal of Chemical Physics, 79*, 1086–1088.

Saffarian, S., Li, Y., Elson, E. L., & Pike, L. J. (2007). Oligomerization of the EGF receptor investigated by live cell fluorescence intensity distribution analysis. *Biophysical Journal, 93*, 1021–1031.

Savatier, J., Jalaguier, S., Ferguson, M. L., Cavaillès, V., & Royer, C. A. (2010). Estrogen receptor interactions and dynamics monitored in live cells by fluorescence cross-correlation spectroscopy. *Biochemistry, 49*, 772–781.

Shcherbakova, D. M., Hink, M. A., Joosen, L., Gadella, T. W., & Verkhusha, V. V. (2012). An orange fluorescent protein with a large Stokes shift for single-excitation multicolor FCCS and FRET imaging. *Journal of the American Chemical Society, 134*, 7913–7923.

Sklar, L. A. (1987). Real-time spectroscopic analysis of ligand-receptor dynamics. *Annual Review of Biophysics and Biophysical Chemistry, 16*, 479–506.

Stepanenko, O. V., Stepanenko, O. V., Shcherbakova, D. M., Kuznetsova, I. M., Turoverov, K. K., & Verkhusha, V. V. (2011). Modern fluorescent proteins: From chromophore formation to novel intracellular applications. *BioTechniques*, *51*, 313–327.

Stringari, C., Sierra, R., Donovan, P. J., & Gratton, E. (2012). Label-free separation of human embryonic stem cells and their differentiating progenies by phasor fluorescence lifetime microscopy. *Journal of Biomedical Optics*, *4*, 046012.

Swift, J. L., Burger, M. C., Massotte, D., Dahms, T. E. S., & Cramb, D. T. (2007). Two-photon excitation fluorescence cross-correlation assay for ligand-receptor binding: Cell membrane nanopatches containing the μ-opioid receptor. *Analytical Chemistry*, *79*, 6783–6791.

Tinevez, J. Y., Dragavon, J., Baba-Aissa, L., Roux, P., Perret, E., Canivet, A., et al. (2012). A quantitative method for measuring phototoxicity of a live cell imaging microscope. *Methods in Enzymology*, *506*, 291–309.

Velu, T. J., Beguinot, L., Vass, W. C., Willingham, M. C., Merlino, G. T., Pastan, I., et al. (1987). Epidermal-growth-factor-dependent transformation by a human EGF receptor proto-oncogene. *Science*, *238*, 1408–1410.

Ward, W. W. (2006). Biochemical and physical properties of green fluorescent protein. *Methods of Biochemical Analysis*, *47*, 39–65.

Winter, P. W., McPhee, J. T., Van Orden, A. K., Roess, D. A., & Barisas, B. G. (2011). Fluorescence correlation spectroscopic examination of insulin and insulin-like growth factor 1 binding in live cells. *Biophysical Chemistry*, *159*, 303–310.

Wong, F. H., Banks, D. S., Abu-Arish, A., & Fradin, C. (2007). A molecular thermometer based on fluorescent protein blinking. *Journal of the American Chemical Society*, *129*, 10302–10303.

Wong, S. S., & Jameson, D. M. (2012). *Chemistry of protein and nucleic acid cross-linking and conjugation*. New York, NY: Tayor and Francis.

Wu, J., Corbett, A. H., & Berland, K. M. (2009). The intracellular mobility of nuclear import receptors and NLS cargoes. *Biophysical Journal*, *96*, 3840–3849.

Wu, B., & Müller, J. D. (2005). Time-integrated fluorescence cumulant analysis in fluorescence fluctuation spectroscopy. *Biophysical Journal*, *89*, 2721–2735.

Wu, B., Piatkovich, K. D., Lionnet, T., Singer, R. H., & Verkhusha, V. V. (2011). Modern fluorescent proteins and imaging to study gene expression, nuclear localization, and dynamics. *Current Opinion in Cell Biology*, *23*, 310–317.

Xu-Welliver, M., Leitao, J., Kanugula, S., Meehan, W. J., & Pegg, A. E. (1999). Role of codon 160 in the sensitivity of human $O^6$-alkylguanine-DNA alkyltransferase to $O^6$-benzylguanine. *Biochemical Pharmacology*, *58*, 1279–1285.

Yamato, S., Dai, J., & Moursi, A. M. (2010). Comparison of transfection efficiency of non-viral gene transfer reagents. *Molecular Biotechnology*, *46*, 287–300.

Zigler, J. S., Jr., Lepe-Zuniga, J. L., Vistica, B., & Gery, I. (1985). Analysis of the cytotoxic effects of light-exposed HEPES-containing culture medium. *In Vitro Cellular and Developmental Biology*, *21*, 282–287.

CHAPTER FOUR

# Studying the Protein Corona on Nanoparticles by FCS

**G. Ulrich Nienhaus**[*,†,1], **Pauline Maffre**[*], **Karin Nienhaus**[*]

[*]Institute of Applied Physics and Center for Functional Nanostructures (CFN), Karlsruhe Institute of Technology (KIT), Wolfgang-Gaede-Straße 1, Karlsruhe, Germany

[†]Department of Physics, University of Illinois at Urbana-Champaign, Urbana, Illinois, USA

[1]Corresponding author: e-mail address: uli@illinois.edu

## Contents



## Abstract

Engineered nanoparticles (NPs) have found widespread application in technology and medicine. Whenever they come in contact with a living organism, interactions take place between the surfaces of the NPs and biomatter, in particular proteins, which are currently not well understood. We have introduced fluorescence correlation spectroscopy (FCS) and dual-focus FCS (2fFCS) to measure protein adsorption onto small NPs (~10–30 nm diameter). FCS allows us to measure, with subnanometer precision and as a function of protein concentration, the increase in hydrodynamic radius of the NPs due to protein adsorption. Investigations of the adsorption of a number of important serum

*Methods in Enzymology*, Volume 519
ISSN 0076-6879
http://dx.doi.org/10.1016/B978-0-12-405539-1.00004-X

proteins onto negatively charged, carboxyl-functionalized NPs revealed a stepwise increase of the NP size due to protein binding, clearly indicating that a protein monolayer enshrouds the NP. Structure-based calculations of the protein surface potentials reveal positively charged patches through which the proteins interact electrostatically with the negatively charged NP surfaces; the observed protein layer thickness is correlated with the molecular dimensions of the proteins binding in suitable orientations.

## 1. INTRODUCTION

Nanoparticles (NPs), that is, nanoscale objects with all three spatial dimensions in the range of 1–100 nm (International Organization for Standardization, 2008), may originate from natural events, for example, forest fires and volcano eruptions, or may be "human-made" as unintended byproducts of industrial activity, for example, exhaust fumes. Moreover, recent advances of nanotechnology have led to the design and production of engineered NPs with well-controlled size, shape, charge, chemical composition, and solubility. They can be tailor-made for specific medical and technological applications, and some of them have already found widespread use in consumer products. Lately, concerns have arisen that unintended exposure to engineered NPs may pose risks to human health if their incorporation and distribution within the body cannot be tightly controlled (Colvin, 2003; Nel, Xia, Mädler, & Li, 2006). These concerns call for detailed investigations into the interactions of NPs with biological systems and the subsequent bioresponses.

Owing to their nanoscale dimensions, NPs are capable of invading all parts of the human body including tissues, cells, and even subcellular compartments. For cell intrusion, they can exploit the cellular endocytosis machinery (Jiang, Röcker, et al., 2010; Lunov, Syrovets, et al., 2011; Lunov, Zablotskii, et al., 2011), much like many biological NPs, that is, viruses. Therefore, they hold great promise as tools for biomedical applications such as targeted drug delivery or gene therapy. However, NPs may also pose severe biological hazards if incorporated unintentionally. NPs may exhibit properties that are entirely new and not expected from the bulk material. Importantly, an enhanced chemical reactivity may result from their large surface-to-volume ratios. Consequently, nanotoxicology has become a very active research field in recent years.

NPs present in the environment enter the body spontaneously through the lung, gut, or skin. For biomedical applications, NP exposure is mainly through intravenous injection, for example, as contrast agents in magnetic resonance imaging (Lunov, Syrovets, Buchele, et al., 2010; Lunov, Syrovets, Röcker, et al., 2010). Upon incorporation, NPs come in contact with biological fluids such as blood plasma or lung epithelial lining fluid, where they encounter a large variety of dissolved biomolecules including lipids and proteins. For kinetic reasons, one expects that the NP surface initially becomes coated with highly mobile and most abundant proteins. Subsequently, these proteins should be replaced by proteins with a higher binding affinity but are less prevalent. A protein adhesion layer, also known as "protein corona," forms and may completely enshroud the NP (Cedervall, Lynch, Lindman, et al., 2007). Therefore, the initial encounter between a NP and a cell, which triggers the endocytosis machinery by activating specific receptors, is largely governed by the properties of the protein layer rather than those of the bare NP surface. To be able to control the biological effects of NPs including uptake prevention or targeted delivery to specific cells or tissues, it is of utmost importance to understand the structural and dynamic properties of the protein corona at the molecular level (Treuel & Nienhaus, 2012).

A wide variety of biophysical techniques have been employed to characterize protein adsorption onto NPs, including circular dichroism (Shang, Wang, Jiang, & Dong, 2007; Treuel, Malissek, Gebauer, & Zellner, 2010), Fourier transform infrared spectroscopy (Wang, Bai, Jiang, & Nienhaus, 2012; Zhang & Yan, 2005), Raman spectroscopy and surface-enhanced Raman spectroscopy (Shao, Lu, Wang, Luo, & Duo Duo Ma, 2009) as well as fluorescence spectroscopy (Shang et al., 2012), size-exclusion chromatography and isothermal titration calorimetry (Cedervall, Lynch, Foy, et al., 2007), dynamic light scattering (Jans, Liu, Austin, Maes, & Huo, 2009), and surface plasmon resonance (Cedervall, Lynch, Foy, et al., 2007; Cheng, Wang, Borghs, & Chen, 2011). Moreover, one- and two-dimensional polyacrylamide gel electrophoresis (1D-/2D-PAGE) have been utilized (Aggarwal, Hall, McLeland, Dobrovolskaia, & McNeil, 2009) to identify corona proteins by comparing 2D protein gels to a 2D master map of the plasma (Kim et al., 2007; Lemarchand et al., 2006). In most experiments, NPs are exposed to biological liquids *in vitro* or *in vivo* and, for the analysis of protein adsorption, subsequently separated from the solutions by centrifugation, dialysis, gel filtration, and—if applicable—magnetic separation (Aggarwal et al., 2009; Jansch, Stumpf, Graf, Rühl, & Müller, 2012; Monopoli et al., 2011). However, separation and washing steps may severely affect protein binding and unbinding and, therefore, modify the protein layer on the NPs.

Here, we describe a fluorescence correlation spectroscopy (FCS)-based experimental approach that we have developed to study protein adsorption onto small fluorescent NPs with diameters below ~20–30 nm (Jiang, Weise, et al., 2010; Maffre, Nienhaus, Amin, Parak, & Nienhaus, 2011; Röcker, Pötzl, Zhang, Parak, & Nienhaus, 2009). FCS enables us to monitor minute changes in NP diffusivity due to protein adsorption from a solution with a well-defined protein concentration. The *in situ* measurement avoids additional separation steps to remove unbound protein from the solution. By using the Stokes–Einstein relation, the diffusion coefficient can be converted into an (average) hydrodynamic radius, which FCS allows us to determine with subnanometer precision.

## 2. SAMPLE PREPARATION

### 2.1. Proteins

NPs are administered intravenously in the majority of biomedical applications, for example, as contrast agents in MRI or as vehicles for targeted drug delivery. Consequently, we have studied protein corona formation by using a number of important blood serum proteins as examples. Human serum albumin (HSA) is the most abundant protein in human blood plasma and a carrier of a number of poorly water-soluble molecules including steroid hormones. Transferrin is another important blood plasma protein involved in iron transport and delivery. The apolipoproteins apoA-I and apoE4 function as transporters for lipid molecules in the blood by binding a large number of lipid and cholesterol molecules to form water-soluble lipoproteins, which are then transported to their destinations in the body. Stock solutions of apoA-I and apoE4 and freeze-dried preparations of HSA and apotransferrin were purchased from Sigma–Aldrich (St. Louis, MO). To study the adsorption of specific proteins onto NPs by FCS, pure protein preparations are required, with minimal amounts of denatured or aggregated proteins. To prepare stock solutions from lyophilized preparations, we slowly add protein powder to phosphate-buffered saline (PBS, PAALabs, Cölbe, Germany) solution. To dissolve the protein, the solution is gently agitated by hand. Stock solutions are aliquoted (50 μl) in LoBind tubes (Eppendorf, Hamburg, Germany) and stored for a few days at 4 °C. For some proteins, stock solutions may also be shock-frozen in liquid nitrogen and stored at −20 °C for later use.

## 2.2. Nanoparticles

We monitor protein adsorption onto fluorescent NPs by using FCS so that the added volume due to the protein corona can be measured by the changed NP diffusivity. Considering that typical protein dimensions are in the range of 3–10 nm, the relative increase in size due to protein corona formation will be rather small for larger NPs (diameter > 30 nm), so the method is most sensitive for small NPs. Because the presence of aggregates is detrimental to FCS, key to success is an excellent colloidal stability of the NPs so that they have a negligible tendency to aggregate. Moreover, they should have a well-defined diameter, that is, small size dispersion. Many synthesis routes have been devised to produce NPs with well-defined shapes and properties from a diverse set of materials, including polymers, silica, metals, metal chalcogenides, and semiconductors (e.g., CdSe, CdTe). Frequently, inorganic cores are made water soluble by encasing them with a ligand or polymer shell. In our laboratory, many FCS studies of protein corona formation were carried out with FePt nanocrystals embedded in an amphiphilic polymer synthesized from dodecylamine and poly(isobutylene-alt-maleic anhydride) (Pellegrino et al., 2004). The polymer coating presents carboxyl groups on the NP surface so that the NPs are negatively charged and, due to their electrostatic repulsion, colloidally stable. Organic fluorophores (DY-636) are covalently bound within the polymer shell of the FePt NPs to render them fluorescent. CdSe/ZnS quantum dots are also excellent NPs with a narrow size distribution. They are brightly autofluorescent and so do not need additional fluorescence labeling. To make them water soluble and colloidally stable, small thiol-containing molecules such as D-penicillamine or lipoic acid can be attached to their surfaces (Breus, Heyes, & Nienhaus, 2007; Breus, Heyes, Tron, & Nienhaus, 2009).

## 2.3. Preparation of a protein concentration series by serial dilution

FCS experiments allow us to measure binding curves by exposing NPs in nanomolar dilutions to a wide range of protein concentrations and, thereby, yield information on the tendency of the protein to adsorb. To obtain equally spaced points on a logarithmic protein concentration scale, we use the so-called serial dilution procedure. A pipette is adjusted to a fixed volume of typically 10 μl so as to avoid errors due to repeated pipette volume change. In a first Eppendorf tube, 10 μl of protein stock solution is mixed with 10 μl of PBS (pH 7.4) to obtain a sample with the protein concentration reduced by a factor of 2. Subsequently, 10 μl of the 1:1 mixture is again mixed with 10 μl of PBS. This dilution procedure is repeated until the range of concentrations is considered sufficient for the experiments. Afterward, the protein solutions are equilibrated for another hour for complete reconstitution. All protein solutions are stored at 4 °C until used.

### 2.4. Sample cell and sample loading

For FCS experiments, we use a simple sandwich sample cell consisting of two standard cover slips (20 × 20 and 32 × 24 mm$^2$) separated by two strips of double-sided adhesive tape (thickness 0.2 μm), leaving an 1-mm-wide channel for the sample solution in the middle. Before assembling the sample cell, the cover glasses are flamed with a propane torch to remove fluorescent impurities. All procedures are carried out in a flow box to avoid air-borne contaminants.

Ten minutes prior to the FCS measurement, 10 μl of a ~1 n*M* NP solution is added to 10 μl of protein solution. The NP–protein solution is allowed to equilibrate for ~8 min at room temperature before loading it into the sandwich. The same sandwich cell is used for measuring all points of a binding curve, starting with the lowest protein concentration. Although 4 μl of the solution suffices to completely fill the channel of the sandwich, we flow essentially the entire 20 μl of sample solution through the channel to completely replace the preceding solution. The sequential refilling of the sample cell minimizes effects of protein adsorption onto its surfaces. Especially when working with highly dilute protein solutions, great care needs to be taken to avoid errors due to protein adsorption.

## 3. EXPERIMENTAL PROCEDURES

### 3.1. Experimental setups

In our laboratory, conventional FCS experiments are performed on a home-built confocal microscope similar to those published earlier (Anikin et al., 2005; Schenk, Ivanchenko, Röcker, Wiedenmann, & Nienhaus, 2004). The instrument design is based on an inverted epifluorescence microscope (Axiovert 200, Carl Zeiss, Göttingen, Germany). 532- or 635-nm excitation light from solid-state laser sources is routed to the back port of the microscope by a single-mode optical fiber (QSMJ, OZ Optics, Ottawa, Canada). The emitted light is collected by a water immersion objective (UPLAPO 60×/1.2 W, Olympus, Hamburg,

Germany) and passes through a dichroic mirror (z532/633xr, AHF, Tübingen, Germany) and suitable band pass filters. The light is focused onto a 62.5-μm diameter gradient index fiber (Thorlabs, Newton, NJ, USA) replacing the confocal pinhole. Photons are registered by an avalanche photodiode (SPCM-CD3017, Perkin Elmer, Fremont, CA) and processed by a digital correlator (ALV-5000/E, ALV, Langen, Germany). Measurements are performed with an excitation power of typically 6 μW.

Alternatively, we use dual-focus FCS (2fFCS) (Dertinger et al., 2007), which is discussed in detail by Dertinger and Enderlein in Chapter 8. Conventional FCS is very sensitive to refractive index mismatch, variations in cover slide thickness, and optical saturation effects and, therefore, requires continuous, careful instrument calibration. By contrast, 2fFCS is much more robust due to its internal length standard. The setup is based on a time-resolved confocal microscopy system (Microtime 200, PicoQuant, Berlin, Germany) shown schematically in Fig. 4.1. Instead of using a single excitation laser, the light from two orthogonally polarized pulsed 640-nm diode lasers (LDH-P-C-640B, Picoquant, Berlin, Germany) is combined by a polarizing beam splitter. The lasers are pulsed in an alternate fashion with an interval of 25 ns between successive pulses, which is much longer than the fluorescence lifetime of typical dye molecules used in these experiments (~0.5 ns in the case of DY-636), which ensures that the fluorescence decays

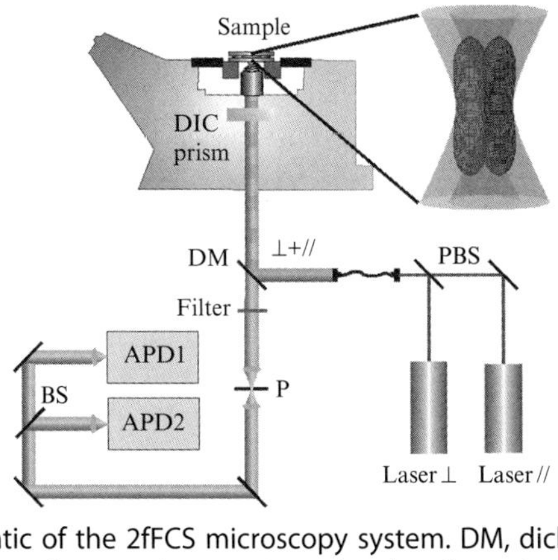


**Figure 4.1** Schematic of the 2fFCS microscopy system. DM, dichroic mirror; BS, beam splitter; PBS, polarizing beam splitter; APD, avalanche photo diode; P, pinhole. *Adapted from Maffre et al. (2011).* (For color version of this figure, the reader is referred to the online version of this chapter.)

completely between successive pulses. Both laser outputs are coupled into a polarization maintaining optical fiber so that the light emerging from the fiber can be collimated into a beam consisting of a train of laser pulses with alternating orthogonal polarizations. The beam is passed through a dichroic mirror (470/635 nm, AHF) and a Nomarski DIC prism (U-DICTHC, Olympus, Hamburg, Germany), which deflects the different polarization components in two different directions into the objective (UPLSAPO 60XW, Olympus) of an inverted microscope (IX71, Olympus). In the sample volume, two partially overlapping excitation foci are generated with a lateral separation of 404 nm (for our prism). Typical total power of each laser is 3 μW. The fluorescence light is collected by the same objective and transmitted through the prism and the dichroic mirror. Behind the pinhole (150 μm), the light is collimated, split, and focused onto two avalanche photodiodes (APD, SPCM-AQR-13, Perkin Elmer, Rodgau, Germany). A single photon counting card (HydraHarp 400 picosecond event timer and TCSPC module, PicoQuant) records the arrival times of detected photons with picosecond time resolution so that the photons can be assigned unambiguously to one or the other of the two foci. Autocorrelation functions for each detection volume as well as cross-correlation functions between the two detection volumes are calculated from the photon arrival time traces using SymphoTime software (PicoQuant).

## 3.2. Microscope calibration

Prior to the FCS measurement, the objective lens is examined and cleaned with ethanol if necessary. The position of the correction collar on the objective is inspected to ensure that it is appropriate for the thickness of the cover slips used for assembling the sample sandwich. Experimentally, the optimal position of the correction collar is found by analyzing fluorescence emission time traces of Atto 655 carboxylate in buffer. The correction collar is adjusted such that the highest detected fluorescence brightness correlates with the expected dimensions of the confocal volume (Fig. 4.2). This adjustment is maintained for all measurements.

Particular attention is paid to the fringe patterns of the laser reflection from the cover slide surfaces. For each measurement, we aim for a position in the channel of the sample sandwich with symmetric laser reflections. Small distortions of the concentric rings indicate the presence of dust particles or air bubbles in the sample solution which cause a distortion of the point spread function or affect the reproducibility of the results.

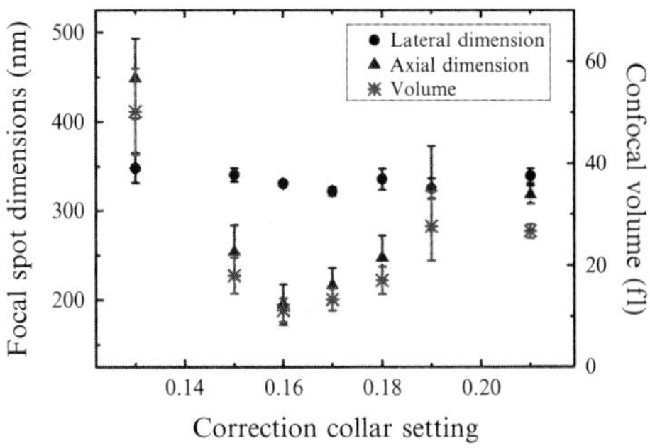


**Figure 4.2** Dependence of the confocal spot dimensions (Gauss-Lorentz model of the confocal volume, Dertinger et al., 2007) on the correction collar setting. (For color version of this figure, the reader is referred to the online version of this chapter.)

## 3.3. Data collection

Samples are illuminated continuously for typically ~5–10 min so that, for NP concentrations of 1 n*M*, ~10,000 single molecule bursts are analyzed. During the experiment, humidity is controlled near the sample to avoid evaporation of the sample solution. The laboratory is air conditioned to 22 °C, and the sample temperature, *T*, is measured continuously during the experiments. Usually, at least three independent sets of measurements are taken and averaged.

Conventional FCS experiments are typically performed by using a digital correlator (see Section 3.1), which computes the (normalized) intensity autocorrelation function directly from the intensity fluctuations of the emission time traces. While this procedure is convenient for the experimenter because the large amount of time-dependent fluorescence emission data is reduced to an autocorrelation function "on the fly," it can be disadvantageous that the raw data are not available. We usually acquire ~10 independent autocorrelations over typically 1 min, and discard those that show artifacts due to impurities and average the good ones. In 2fFCS, two time traces related to the fluorescence emission emanating from the two foci are acquired and are thus available for subsequent editing. FCS experiments are notoriously sensitive to the presence of bright aggregates, and therefore, those parts of the time traces that show excessively high intensities are eliminated prior to computation of the correlation curves. Otherwise, the autocorrelation curve will be distorted at longer times.

# 4. DATA ANALYSIS

## 4.1. Correlation function analysis

In this section, we briefly sketch the analysis of the pair-correlation functions for 2fFCS (and FCS as a limiting case). The normalized cross-correlation function is defined as

$$G(\tau)=\frac{\langle\delta F_1(t+\tau)\delta F_2(t)\rangle}{\langle F_1(t)\rangle\langle F_2(t)\rangle}=\frac{\langle F_1(t+\tau)F_2(t)\rangle}{\langle F_1(t)\rangle\langle F_2(t)\rangle}-1, \qquad [4.1]$$

where $F_1(\delta F_1)$ and $F_2(\delta F_2)$ are the intensities and intensity variations around the mean detected from focus 1 and focus 2, respectively, at time $t$; the lag time is denoted by $\tau$. The experimentally determined $G(\tau)$ is subsequently fitted with a model function that represents three-dimensional diffusion of a single fluorescent species within the confocal detection volume from which the diffusional correlation time, $\tau_D$, or the diffusion coefficient, $D=r_0{}^2/4\tau_D$, is extracted. Here, $r_0$ denotes the radial extension of the focal volume. An additional exponential reaction term captures dye blinking and is required to achieve a good fit but is not of further relevance here. The diffusional pair-correlation function is given by

$$G(\tau)=\frac{1}{\langle C\rangle(4\pi D\tau)^{3/2}}\frac{\iint\limits_{V_1,V_2} W(\mathbf{r}_1)\cdot W(\mathbf{r}_2)\cdot\exp\left[-\frac{(\mathbf{r}_1-\mathbf{r}_2-\delta\hat{\mathbf{x}})^2}{4D\tau}\right]\mathrm{d}\mathbf{r}_1\mathrm{d}\mathbf{r}_2}{\left(\int\limits_V W(r)\mathrm{d}\mathbf{r}\right)^2}, \qquad [4.2]$$

with the average fluorophore concentration, $\langle C\rangle$, and the distance between the two foci, $\delta$. $W$ is the normalized molecular detection function (MDF), that is, the spatial probability density distribution to detect fluorophores within the confocal volume.

In conventional FCS analysis, the MDF is usually approximated by a three-dimensional Gaussian with radial and axial extensions $r_0$ and $z_0$,

$$W(r)=\exp\left[-2\frac{x^2+y^2}{r_0^2}\right]\exp\left[-2\frac{z^2}{z_0^2}\right], \qquad [4.3]$$

and there is only one detection volume; so the distance between the two foci is $\delta = 0$. Defining the effective observation volume as

$$V_{\text{eff}} = \frac{\left(\int_V W(r)\mathrm{d}\mathbf{r}\right)^2}{\int_V W^2(r)\mathrm{d}\mathbf{r}} = \pi^{3/2} r_0^2 z_0, \quad [4.4]$$

we obtain the well-known expression for the normalized diffusion autocorrelation function (Haustein & Schwille, 2007),

$$G(\tau) = \frac{1}{\langle C \rangle V_{\text{eff}}} \left(1 + \frac{\tau}{\tau_{\text{D}}}\right)^{-1} \left(1 + \left(\frac{r_0}{z_0}\right)^2 \frac{\tau}{\tau_{\text{D}}}\right)^{-1/2}. \quad [4.5]$$

Taking $\delta \neq 0$ yields an expression that contains an additional exponential term,

$$G(\tau) = \frac{1}{\langle C \rangle V_{\text{eff}}} \left(1 + \frac{\tau}{\tau_{\text{D}}}\right)^{-1} \left(1 + \left(\frac{r_0}{z_0}\right)^2 \frac{\tau}{\tau_{\text{D}}}\right)^{-1/2} \exp\left[-\frac{\delta^2}{r_0^2} \frac{1}{1 + \tau/\tau_{\text{D}}}\right]. \quad [4.6]$$

To obtain a description of the MDF that is more closer to reality than a three-dimensional Gaussian, Dertinger and coworkers (Dertinger et al., 2007) introduced a modified Gauss–Lorentz shape, for which a closed-form expression cannot be derived, so that a numerical integration along the $z$-axis is required to compute the correlation function. With this procedure, we fit our two 2fFCS autocorrelations and the cross-correlation simultaneously to determine the free parameters using a Matlab routine (PicoQuant).

To include the effect of dye blinking due to intersystem crossing in the fit, the diffusional correlation function is multiplied by

$$G_{\text{T}}(\tau) = 1 + A\exp[-k\tau]. \quad [4.7]$$

The parameters, $A$ and $k$, are not of interest here but are technically required to obtain a proper fit of the autocorrelation function.

## 4.2. Computation of hydrodynamic radii

The diffusion coefficients that are determined by fits are converted into hydrodynamic radii by using the Stokes–Einstein relation,

$$R_{\mathrm{H}} = \frac{k_{\mathrm{B}} T}{6\pi\eta D}, \quad [4.8]$$

with the Boltzmann constant, $k_{\mathrm{B}}$, and the solution viscosity, $\eta$. This conversion requires the sample temperature to be measured. Moreover, for protein concentrations exceeding $\sim 100\ \mu M$, the protein effect on the solution viscosity becomes significant and needs to be corrected for. To this end, we employ a linear approximation for the contribution of the solute to the solution viscosity based on the intrinsic viscosity of the proteins, $[\eta]$, so that the viscosity of the protein solution, $\eta$, is given by

$$\eta = \left([\eta] c_{\mathrm{Protein}} + 1\right)\eta_0. \quad [4.9]$$

Here, the viscosity of the pure solvent is denoted by $\eta_0$ and the protein concentration by $c$ (in g/cm$^3$). For example, the intrinsic viscosities of HSA and apoA-I are 4.2 cm$^3$/g (Sigma–Aldrich) and 9.2 cm$^3$/g (Edelstein & Scanu, 1980).

# 5. PROTEIN CORONA FORMATION MEASURED BY FCS

Below we present some typical experimental results on protein adhesion onto polymer-embedded, carboxyl-functionalized FePt NPs with hydrodynamic radii of $\sim$6 nm. We have obtained excellent data by using conventional FCS on HSA and transferrin in the past (Jiang, Weise, et al., 2010; Röcker et al., 2009). However, the 2fFCS method is more robust and reliable, and so we prefer to use it for single-color FCS applications. We note that our current 2fFCS setup provides only 640-nm excitation, so we are limited to single-color experiments with suitable red-emitting fluorophores.

## 5.1. Concentration dependence of protein adhesion

The tendency of serum proteins to bind to NPs can be investigated by 2fFCS on fluorescent NPs diffusing freely in solutions that contain the proteins in concentrations ranging over several orders of magnitude. NP concentrations have to be adjusted to $\sim$1 n$M$ to make sure that intensity fluctuations, on which the FCS method is based, are large. Obviously, the protein concentration range should be chosen suitably for observing the transition

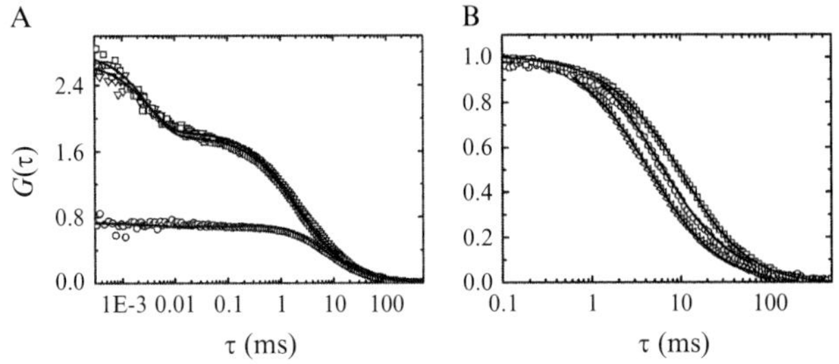


**Figure 4.3** Fluorescence intensity fluctuation correlation curves of NPs dissolved in buffer solutions of apoE4. (A) Measured (symbols) and fitted (lines) 2fFCS autocorrelation (triangles and squares) and cross-correlation (circles) functions of polymer-coated FePt NPs in the presence of 7.2 μ*M* apoE4. (B) Cross-correlation curves plotted for three selected protein concentrations (0 n*M*, 14 n*M*, 7.2 μ*M* aopE4, from left to right), normalized to 1 at $\tau = 0.1$ ms for comparison.

from uncoated to protein-coated NPs, with protein concentrations equally spaced on a logarithmic scale, as described in Section 2.3.

Typical examples of measured 2fFCS correlation curves are shown in Fig. 4.3 for 6-nm radius, carboxyl-functionalized FePt NPs exposed to apoE4. In Fig. 4.3A, the autocorrelation curves of the two foci and the cross-correlation curve are plotted for 7.2 μ*M* apoE4. The autocorrelations display two pronounced decay processes. The step in the millisecond time range is due to NP diffusion and, therefore, reveals the particle size via the Stokes–Einstein relation, whereas the step on the microsecond time scale arises from dye blinking due to intersystem crossing to the triplet state; dye molecules in the triplet state do not fluoresce until they have returned to the ground singlet state and are reexcited. This process is strongly suppressed in the cross-correlation function owing to the small overlap of the two detection volumes and the short time scale of this process.

In Fig. 4.3B, cross-correlation curves are plotted for three selected protein concentrations (0 n*M*, 14 n*M*, 7.2 μ*M* from left to right), normalized to 1 at $\tau = 0.1$ ms for comparison. The curves shift with increasing protein concentration toward longer times, indicating that the effective size of the NPs grows due to protein adsorption. The effect is small, however; so precise data are needed for a quantitative analysis of protein binding.

From fitting the correlation functions, the diffusion coefficient is extracted and converted into a hydrodynamic radius, $R_H$. Figure 4.4 shows the concentration dependence of $R_H$ for four serum proteins. We note that the error bars represent deviations from the mean based on at least three independent data sets and not statistical errors from an individual measurement. A stepwise increase is always observed, with a midpoint

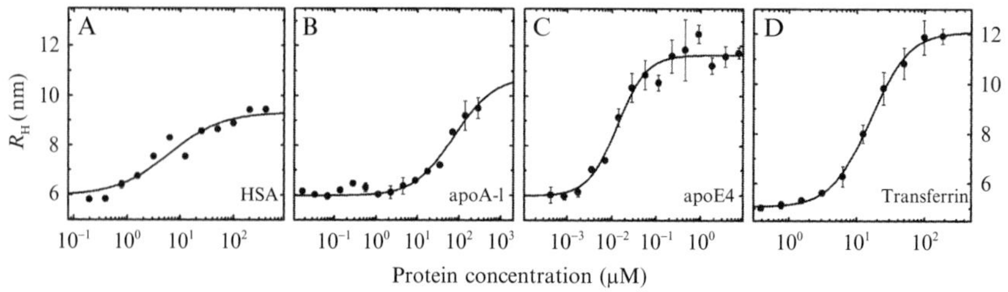


**Figure 4.4** Hydrodynamic radius, $R_H$, of the FePt NPs, plotted as a function of the concentration of (A) HSA, (B) apoA-I, (C) apoE4, and (D) apotransferrin. The solid lines are fit results using Eqs. (4.10) and (4.11). Data in (A–D) were measured by using 2fFCS and standard FCS, respectively.

**Table 4.1** Parameters of protein adsorption onto FePt NPs

| Protein | $R_H(N_{max})$ (nm) | $\Delta R_H$ (nm) | $K$ (μM) | $n$ | $N_{max}$ |
|---|---|---|---|---|---|
| HSA[a] | 9.3 ± 0.3 | 3.3 ± 0.3 | 9.9 ± 4.7 | 0.9 ± 0.2 | 27 ± 3 |
| ApoA-I[b] | 10.8 ± 1.5 | 4.8 ± 1.4 | 140 ± 60 | 1.0 ± 0.3 | 52 ± 10 |
| ApoE4[b] | 11.7 ± 0.3 | 5.7 ± 0.2 | 0.021 ± 0.003 | 1.4 ± 0.2 | 65 ± 3 |
| Transferrin[a] | 13.0 ± 0.5 | 7.0 ± 0.4 | 26 ± 6 | 1.7 ± 0.2 | 22 ± 3 |

[a]Jiang, Weise, et al. (2010), Maffre et al. (2011), Röcker et al. (2009).
[b]Maffre et al. (2011).

concentration $K$ that varies by almost four orders of magnitude between the proteins (Table 4.1). Moreover, the increase in $R_H$ depends on the type of protein adhering to the NP. A quantitative analysis can be based on a simple model that attributes the increase of $R_H$ to the binding of a certain number, $N$, of proteins with molecular volume, $V_P$, to the NP of volume $V_0$,

$$R_H(N) = \sqrt[3]{\frac{3}{4\pi}(V_0 + NV_P)} = R_H(0)\sqrt[3]{1 + cN}. \qquad [4.10]$$

Here, $R_H(0)$ is the hydrodynamic radius of the bare NP, and $c$ is the volume ratio $V_P/V_0$. In view of the data in Fig. 4.4, which resemble a binding isotherm, we model the dependence of $N$ on the protein concentration by using the Hill equation (Röcker et al., 2009),

$$N = N_{max}\frac{1}{1 + (K/[\mathrm{P}])^n}. \qquad [4.11]$$

Here, $N_{max}$ is the number of proteins bound in saturation and $K$ is the midpoint concentration, or "apparent" dissociation coefficient

(Table 4.1). The "Hill coefficient" $n$ controls the steepness of the binding curve. We emphasize here that protein adsorption to NPs, although it apparently can be perfectly modeled with the Hill equation, is not an equilibrium binding process because it is at least partially irreversible (data not shown).

## 5.2. Structure of the protein corona

Figure 4.4 clearly reveals that the thickness of the protein corona, $\Delta R_H$, depends on the particular protein species adsorbed. Interestingly, the thickness of the protein corona is correlated with the molecular dimensions of the proteins as revealed by structural studies including protein crystallography. In Fig. 4.5, we compare simplified cartoon structures of the four proteins with the thicknesses of their protein coronae.

HSA is a protein with a molecular mass of 67 kDa (He & Carter, 1992) and is approximately shaped like an equilateral triangular prism, with sides of ~8 nm and a height of ~3 nm. The ~3.3-nm radius increase upon adsorption of HSA observed with 2fFCS implies that HSA molecules adsorb via their triangular surfaces onto the NPs.

Lipid-free human apoA-I, the principal component of high-density lipoprotein, has a molecular mass of 28 kDa. It consists of two domains, an N-terminal domain forming a four-helix bundle and a structurally less well organized C-terminal domain (Ajees, Anantharamaiah, Mishra, Hussain, & Murthy, 2006; Saito et al., 2003; Silva, Hilliard, Fang, Macha, & Davidson, 2005). There is evidence that apoA-I may be more flexible in solution than in the crystal (Davidson & Thompson, 2007; Thomas, Bhat, & Sorci-Thomas, 2008). Its overall shape has been described by a prolate ellipsoid with an axial ratio of 5.5:1 on the basis of analytical ultracentrifugation, viscometric, and fluorescence data (Barbeau, Jonas, Teng, & Scanu, 1979; Davidson et al., 1999). Förster resonance energy transfer (FRET) studies have indicated that the interdomain distance in solution is even smaller than in the crystal structure (Koyama et al., 2009). An estimated 4.5 nm thickness correlates well with the $\Delta R_H$ increase of 4.8 nm measured by 2fFCS. Importantly, also for this protein, a single-protein layer is compatible with the experimental data.

Human apoE4 is another soluble apolipoprotein with a molecular mass of 34 kDa (Hatters, Peters-Libeu, & Weisgraber, 2006). Like apoA-I, apoE4 has two structural domains (Fig. 4.5), an N-terminal, elongated four-helix

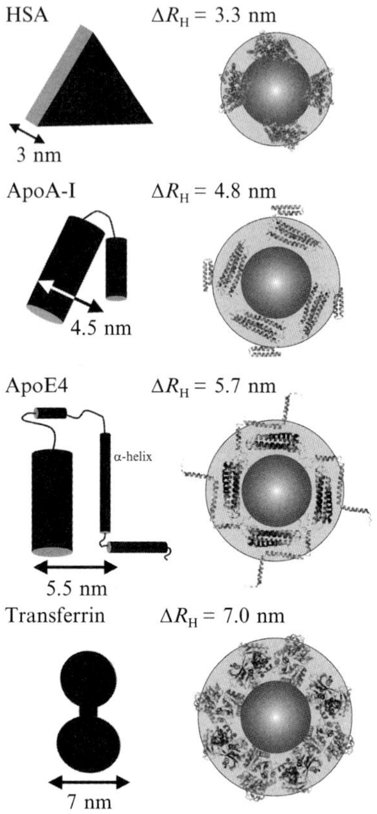


**Figure 4.5** The thickness of the protein corona correlates with the molecular dimensions of the adsorbed proteins. Left column: simplified representations of the proteins including approximate dimensions. Right column: schematic depiction of protein monolayers formed around the NPs. (For color version of this figure, the reader is referred to the online version of this chapter.)

bundle, and a C-terminal, highly α-helical domain the structure of which is not yet known (Wilson, Wardell, Weisgraber, Mahley, & Agard, 1991). Recently, it was reported that apoE4 is, similar to apoA-I, ellipsoidal, with an axial ratio of ~7:1 (Garai & Frieden, 2010). According to an electron paramagnetic resonance study, the C-terminal domain forms a long α-helix that runs parallel to the helix bundle at a distance of ~2 nm (Hatters, Budamagunta, Voss, & Weisgraber, 2005). If we assume that the four-helix bundle of apoE4 lies flat on the NP surface and if we add the typical diameter of a single α-helix separated by 2 nm, we obtain an overall thickness of 5–6 nm for the protein corona, which nicely matches the observed $\Delta R_H$ (Fig. 4.5).

Transferrin is an important and abundant human plasma (glyco-)protein with a molecular mass of 80 kDa. Its 679 amino acid polypeptide chain folds into two globular lobes having the shape of a prolate ellipsoid of $4.2 \times 5 \times 7$ nm$^3$ connected by a three-turn helix (Wally et al., 2006). The observed increase $\Delta R_H = 7$ nm at saturation as measured by FCS is again compatible with a monolayer of protein molecules formed around the NPs (Fig. 4.5) so that the $4.2 \times 10$ nm$^2$ face attaches to the NP. The size increase is also in accordance with the hydrodynamic diameter of 7.4 nm (Armstrong, Wenby, Meiselman, & Fisher, 2004).

## 5.3. Protein electrostatics and adsorption tendency

From all examples in the previous section, the thickness of the protein corona was compatible with the assumption that proteins adsorb in certain orientations as a monolayer. It is presently still an open question as to how far the protein structure is altered by the protein–NP interactions. Electrostatic forces are expected to play a significant role in protein binding to our negatively charged, carboxylated NPs, but only over distances shorter than the screening (Debye) length, which is below 1 nm at the ionic strengths of our solutions. Therefore, the charge distributions on the protein surfaces should be relevant for binding rather than the net charges. In Fig. 4.6, we compare cartoon representations of the proteins (left column) with space-filling models colored according to their electrostatic surface potentials (right column).

On the surface of apoA-I, there are two rather extended negatively charged patches. In close vicinity to the larger patch, a small area of positive electrostatic potential is visible that may mediate binding to our negatively charged NPs (inside the dashed ellipse). By attaching with this patch to the NP surface, the apoA-I molecules are expected to form a layer of 4–5 nm thickness, which is in good agreement with our experimental findings (Fig. 4.5). The high midpoint concentration of 140 $\mu M$ is consistent with the fairly weakly charged spot. Transferrin features two regions of positive surface potential on the side of the two lobes. The measured $\Delta R_H$ is compatible with a protein orientation that is required for binding with these two spots to the NP surface. HSA, while being overall negatively charged at pH 7.4, exhibits a pronounced positive patch on one of the triangular protein surfaces, which is likely to promote the interaction with the negatively

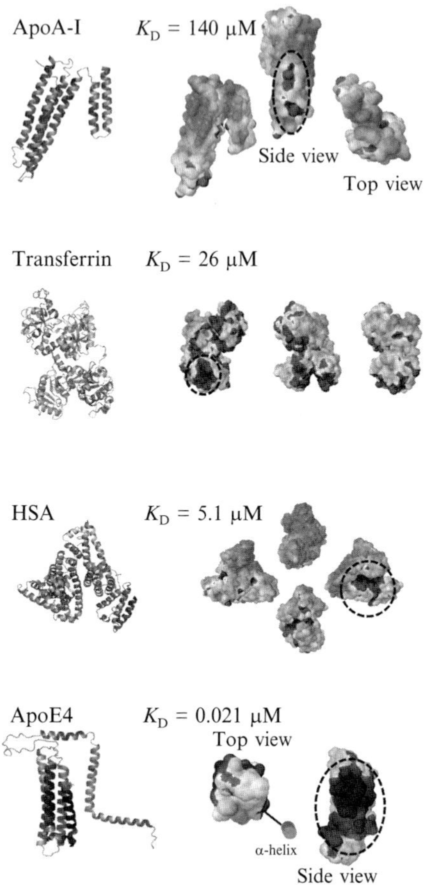


**Figure 4.6** Electrostatic interactions govern protein adsorption. Left column: cartoon representations of apoA-I (protein data bank accession (pdb) code 2A01), apo transferrin (pdb code 2HAU), HSA (pdb code 1UOR), and apoE4 (pdb code 1GS9). For apoE4, only the structure of the 22-K domain (4-helix bundle) has been solved. Right column: space-filling models colored to indicate their surface electrostatics at pH 7.4 (blue/light gray: negative potential, red/dark gray: positive potential; range: −5 to +5 kT/e; calculated online at http://kryptonite.nbcr.net/pdb2pqr/ (Dolinsky, Nielsen, McCammon, & Baker, 2004)). The positively charged patches are marked by the dashed ellipses. (See Color Insert.)

charged carboxyl groups on the NP surfaces (Fig. 4.6). Finally, apoE4, with a midpoint concentration of 21 n*M*, is the protein with the highest tendency to adsorb onto the carboxylated NPs. Indeed, this protein features a rather large patch of positive charge (Fig. 4.6), which most likely is responsible for the high tendency of apoE4 to bind to the negatively charged NPs.

## 6. CONCLUSIONS

Here, we have shown that FCS and 2fFCS, in particular, are superb methods to monitor protein adsorption onto small NPs with subnanometer resolution. Upon exposure to a wide range of protein concentrations, a step-wise increase of the overall NP size was observed for all proteins studied so far using this single-protein exposure assay. The concentration dependence can be modeled by a binding isotherm (Eq. 4.11) despite the fact that protein adhesion is clearly not an equilibrium process. The quantitative analysis yields the increase in hydrodynamic radius, $\Delta R_H$, that is, the average thickness of the protein corona, the midpoint concentration (apparent binding affinity), and the Hill coefficient describing the steepness of the transition. From $\Delta R_H$, we can compute the number of proteins bound in saturation. Based on the known molecular structures, the thickness of the protein corona can be related to a particular orientation of the protein. Widely different midpoint concentrations of proteins can be explained by the presence of positively charged surface patches on the proteins (for binding to negatively charged NPs).

Proteins are flexible biomacromolecules that can fluctuate among a vast number of conformational substates at room temperature (Frauenfelder, Nienhaus, & Johnson, 1991; Nienhaus, Müller, McMahon, & Frauenfelder, 1997; Parak, Heidemeier, & Nienhaus, 1988). Their adsorption onto NPs constitutes a major energetic perturbation that is expected to induce conformational changes. However, from our FCS data, it appears that the overall protein structure is essentially conserved because the thickness of the adsorption layer correlates with the molecular dimensions. Moreover, the binding tendency can be explained by the surface charge distribution of the native protein. Nevertheless, further work will be necessary to characterize the structure of the adsorbed proteins, for example, by single-pair FRET (Nienhaus, 2006; Schuler, 2007), which may yield more detailed insights into the structural properties of the protein corona surrounding NPs.

## ACKNOWLEDGMENTS

We thank Prof. W. Parak, Dr. Carlheinz Röcker, and Mr. Stefan Brandholt for their collaboration. This research was supported by the Deutsche Forschungsgemeinschaft (DFG) through the Center for Functional Nanostructures (CFN) and Schwerpunktprogramm (SPP) 1313, Grants NI291/7 and NI291/8.

## REFERENCES


Aggarwal, P., Hall, J. B., McLeland, C. B., Dobrovolskaia, M. A., & McNeil, S. E. (2009). Nanoparticle interaction with plasma proteins as it relates to particle biodistribution, biocompatibility and therapeutic efficacy. *Advanced Drug Delivery Reviews*, *61*, 428–437.

Ajees, A. A., Anantharamaiah, G. M., Mishra, V. K., Hussain, M. M., & Murthy, H. M. (2006). Crystal structure of human apolipoprotein A-I: Insights into its protective effect against cardiovascular diseases. *Proceedings of the National Academy of Sciences of the United States of America*, *103*, 2126–2131.

Anikin, K., Röcker, C., Wittemann, A., Wiedenmann, J., Ballauff, M., & Nienhaus, G. U. (2005). Polyelectrolyte-mediated protein adsorption: Fluorescent protein binding to individual polyelectrolyte nanospheres. *The Journal of Physical Chemistry B*, *109*, 5418–5420.

Armstrong, J. K., Wenby, R. B., Meiselman, H. J., & Fisher, T. C. (2004). The hydrodynamic radii of macromolecules and their effect on red blood cell aggregation. *Biophysical Journal*, *87*, 4259–4270.

Barbeau, D. L., Jonas, A., Teng, T., & Scanu, A. M. (1979). Asymmetry of apolipoprotein A-I in solution as assessed from ultracentrifugal, viscometric, and fluorescence polarization studies. *Biochemistry*, *18*, 362–369.

Breus, V. V., Heyes, C. D., & Nienhaus, G. U. (2007). Quenching of CdSe-ZnS core-shell quantum dot luminescence by water-soluble thiolated ligands. *Journal of Physical Chemistry C*, *111*, 18589–18594.

Breus, V. V., Heyes, C. D., Tron, K., & Nienhaus, G. U. (2009). Zwitterionic biocompatible quantum dots for wide pH stability and weak nonspecific binding to cells. *ACS Nano*, *3*, 2573–2580.

Cedervall, T., Lynch, I., Foy, M., Berggard, T., Donnelly, S. C., Cagney, G., et al. (2007). Detailed identification of plasma proteins adsorbed on copolymer nanoparticles. *Angewandte Chemie (International Ed. in English)*, *46*, 5754–5756.

Cedervall, T., Lynch, I., Lindman, S., Berggard, T., Thulin, E., Nilsson, H., et al. (2007). Understanding the nanoparticle-protein corona using methods to quantify exchange rates and affinities of proteins for nanoparticles. *Proceedings of the National Academy of Sciences of the United States of America*, *104*, 2050–2055.

Cheng, Y., Wang, M., Borghs, G., & Chen, H. (2011). Gold nanoparticle dimers for plasmon sensing. *Langmuir*, *27*, 7884–7891.

Colvin, V. L. (2003). The potential environmental impact of engineered nanomaterials. *Nature Biotechnology*, *21*, 1166–1170.

Davidson, W. S., Arnvig-McGuire, K., Kennedy, A., Kosman, J., Hazlett, T. L., & Jonas, A. (1999). Structural organization of the N-terminal domain of apolipoprotein A-I: Studies of tryptophan mutants. *Biochemistry*, *38*, 14387–14395.

Davidson, W. S., & Thompson, T. B. (2007). The structure of apolipoprotein A-I in high density lipoproteins. *The Journal of Biological Chemistry*, *282*, 22249–22253.

Dertinger, T., Pacheco, V., von der Hocht, I., Hartmann, R., Gregor, I., & Enderlein, J. (2007). Two-focus fluorescence correlation spectroscopy: A new tool for accurate and absolute diffusion measurements. *ChemPhysChem*, *8*, 433–443.

Dolinsky, T. J., Nielsen, J. E., McCammon, J. A., & Baker, N. A. (2004). Pdb2pqr: An automated pipeline for the setup of Poisson-Boltzmann electrostatics calculations. *Nucleic Acids Research*, *32*, W665–W667.

Edelstein, C., & Scanu, A. M. (1980). Effect of guanidine hydrochloride on the hydrodynamic and thermodynamic properties of human apolipoprotein A-I in solution. *The Journal of Biological Chemistry*, *255*, 5747–5754.

Frauenfelder, H., Nienhaus, G. U., & Johnson, J. B. (1991). Rate processes in proteins. *Berichte der Bunsengesellschaft für physikalische Chemie*, *95*, 272–278.

Garai, K., & Frieden, C. (2010). The association-dissociation behavior of the apoE proteins: Kinetic and equilibrium studies. *Biochemistry, 49*, 9533–9541.

Hatters, D. M., Budamagunta, M. S., Voss, J. C., & Weisgraber, K. H. (2005). Modulation of apolipoprotein E structure by domain interaction: Differences in lipid-bound and lipid-free forms. *The Journal of Biological Chemistry, 280*, 34288–34295.

Hatters, D. M., Peters-Libeu, C. A., & Weisgraber, K. H. (2006). Apolipoprotein E structure: Insights into function. *Trends in Biochemical Sciences, 31*, 445–454.

Haustein, E., & Schwille, P. (2007). Fluorescence correlation spectroscopy: Novel variations of an established technique. *Annual Review of Biophysics and Biomolecular Structure, 36*, 151–169.

He, X. M., & Carter, D. C. (1992). Atomic structure and chemistry of human serum albumin. *Nature, 358*, 209–215.

International Organization for Standardization, Geneva, Switzerland. *ISO ISO/TS 27687:2008 Nanotechnologies—Terminology and definitions for nano objects - nanoparticle, nanofibre and nanoplate.*

Jans, H., Liu, X., Austin, L., Maes, G., & Huo, Q. (2009). Dynamic light scattering as a powerful tool for gold nanoparticle bioconjugation and biomolecular binding studies. *Analytical Chemistry, 81*, 9425–9432.

Jansch, M., Stumpf, P., Graf, C., Rühl, E., & Müller, R. H. (2012). Adsorption kinetics of plasma proteins on ultrasmall superparamagnetic iron oxide (uspio) nanoparticles. *International Journal of Pharmaceutics, 428*, 125–133.

Jiang, X., Röcker, C., Hafner, M., Brandholt, S., Dörlich, R. M., & Nienhaus, G. U. (2010). Endo- and exocytosis of zwitterionic quantum dot nanoparticles by live Hela cells. *ACS Nano, 4*, 6787–6797.

Jiang, X., Weise, S., Hafner, M., Röcker, C., Zhang, F., Parak, W. J., et al. (2010). Quantitative analysis of the protein corona on FePt nanoparticles formed by transferrin binding. *Journal of the Royal Society Interface, 7*, S5–S13.

Kim, H., Andrieux, K., Delomenie, C., Chacun, H., Appel, M., Desmaële, D., et al. (2007). Analysis of plasma protein adsorption onto pegylated nanoparticles by complementary methods: 2-DE, CE and protein lab-on-chip® system. *Electrophoresis, 28*, 2252–2261.

Koyama, M., Tanaka, M., Dhanasekaran, P., Lund-Katz, S., Phillips, M. C., & Saito, H. (2009). Interaction between the N- and C-terminal domains modulates the stability and lipid binding of apolipoprotein A-I. *Biochemistry, 48*, 2529–2537.

Lemarchand, C., Gref, R., Passirani, C., Garcion, E., Petri, B., Müller, R. H., et al. (2006). Influence of polysaccharide coating on the interactions of nanoparticles with biological systems. *Biomaterials, 27*, 108–118.

Lunov, O., Syrovets, T., Beil, J., Delacher, M., Tron, K., Nienhaus, G. U., et al. (2011). Differential uptake of functionalized polystyrene nanoparticles by human macrophages and a monocytic cell line. *ACS Nano, 5*, 1657–1669.

Lunov, O., Syrovets, T., Buchele, B., Jiang, X., Röcker, C., Tron, K., et al. (2010). The effect of carboxydextran-coated superparamagnetic iron oxide nanoparticles on c-Jun N-terminal kinase-mediated apoptosis in human macrophages. *Biomaterials, 31*, 5063–5071.

Lunov, O., Syrovets, T., Röcker, C., Tron, K., Nienhaus, G. U., Rasche, V., et al. (2010). Lysosomal degradation of the carboxydextran shell of coated superparamagnetic iron oxide nanoparticles and the fate of professional phagocytes. *Biomaterials, 31*, 9015–9022.

Lunov, O., Zablotskii, V., Syrovets, T., Röcker, C., Tron, K., Nienhaus, G. U., et al. (2011). Modeling receptor-mediated endocytosis of polymer-functionalized iron oxide nanoparticles by human macrophages. *Biomaterials, 32*, 547–555.

Maffre, P., Nienhaus, K., Amin, F., Parak, W. J., & Nienhaus, G. U. (2011). Characterization of protein adsorption onto FePt nanoparticles using dual-focus fluorescence correlation spectroscopy. *Beilstein Journal of Nanotechnology, 2*, 374–383.

Monopoli, M. P., Walczyk, D., Campbell, A., Elia, G., Lynch, I., Bombelli, F. B., et al. (2011). Physical-chemical aspects of protein corona: Relevance to in vitro and in vivo biological impacts of nanoparticles. *Journal of the American Chemical Society*, *133*, 2525–2534.

Nel, A., Xia, T., Mädler, L., & Li, N. (2006). Toxic potential of materials at the nanolevel. *Science*, *311*, 622–627.

Nienhaus, G. U. (2006). Exploring protein structure and dynamics under denaturing conditions by single-molecule FRET analysis. *Macromolecular Bioscience*, *6*, 907–922.

Nienhaus, G. U., Müller, J. D., McMahon, B. H., & Frauenfelder, H. (1997). Exploring the conformational energy landscape of proteins. *Physica D*, *107*, 297–311.

Parak, F., Heidemeier, J., & Nienhaus, G. U. (1988). Protein structural dynamics as determined by mössbauer spectroscopy. *Hyperfine Interactions*, *40*, 147–158.

Pellegrino, T., Manna, L., Kudera, S., Liedl, T., Koktysh, D., Rogach, A. L., et al. (2004). Hydrophobic nanocrystals coated with an amphiphilic polymer shell: A general route to water soluble nanocrystals. *Nano Letters*, *4*, 703–707.

Röcker, C., Pötzl, M., Zhang, F., Parak, W. J., & Nienhaus, G. U. (2009). A quantitative fluorescence study of protein monolayer formation on colloidal nanoparticles. *Nature Nanotechnology*, *4*, 577–580.

Saito, H., Dhanasekaran, P., Nguyen, D., Holvoet, P., Lund-Katz, S., & Phillips, M. C. (2003). Domain structure and lipid interaction in human apolipoproteins A-I and E, a general model. *The Journal of Biological Chemistry*, *278*, 23227–23232.

Schenk, A., Ivanchenko, S., Röcker, C., Wiedenmann, J., & Nienhaus, G. U. (2004). Photodynamics of red fluorescent proteins studied by fluorescence correlation spectroscopy. *Biophysical Journal*, *86*, 384–394.

Schuler, B. (2007). Application of single molecule Förster resonance energy transfer to protein folding. *Methods in Molecular Biology*, *350*, 115–138.

Shang, L., Brandholt, S., Stockmar, F., Trouillet, V., Bruns, M., & Nienhaus, G. U. (2012). Effect of protein adsorption on the fluorescence of ultrasmall gold nanoclusters. *Small*, *8*, 661–665.

Shang, L., Wang, Y., Jiang, J., & Dong, S. (2007). pH-dependent protein conformational changes in albumin–gold nanoparticle bioconjugates: A spectroscopic study. *Langmuir*, *23*, 2714–2721.

Shao, M., Lu, L., Wang, H., Luo, S., & Duo Duo Ma, D. (2009). Microfabrication of a new sensor based on silver and silicon nanomaterials, and its application to the enrichment and detection of bovine serum albumin via surface-enhanced Raman scattering. *Microchimica Acta*, *164*, 157–160.

Silva, R. A., Hilliard, G. M., Fang, J., Macha, S., & Davidson, W. S. (2005). A three-dimensional molecular model of lipid-free apolipoprotein A-I determined by cross-linking/mass spectrometry and sequence threading. *Biochemistry*, *44*, 2759–2769.

Thomas, M. J., Bhat, S., & Sorci-Thomas, M. G. (2008). Three-dimensional models of HDL apo A-I: Implications for its assembly and function. *Journal of Lipid Research*, *49*, 1875–1883.

Treuel, L., Malissek, M., Gebauer, J. S., & Zellner, R. (2010). The influence of surface composition of nanoparticles on their interactions with serum albumin. *ChemPhysChem*, *11*, 3093–3099.

Treuel, L., & Nienhaus, G. U. (2012). Toward a molecular understanding of nanoparticle-protein interactions. *Biophysical Reviews*, *4*, 137–147. http://dx.doi.org/10.1007/s12551-12012-10072-12550.

Wally, J., Halbrooks, P. J., Vonrhein, C., Rould, M. A., Everse, S. J., Mason, A. B., et al. (2006). The crystal structure of iron-free human serum transferrin provides insight into inter-lobe communication and receptor binding. *The Journal of Biological Chemistry*, *281*, 24934–24944.

Wang, T., Bai, J., Jiang, X., & Nienhaus, G. U. (2012). Cellular uptake of nanoparticles by membrane penetration: A study combining confocal microscopy with FTIR spectroelectrochemistry. *ACS Nano*, *6*, 1251–1259.

Wilson, C., Wardell, M. R., Weisgraber, K. H., Mahley, R. W., & Agard, D. A. (1991). Three-dimensional structure of the LDL receptor-binding domain of human apolipoprotein E. *Science*, *252*, 1817–1822.

Zhang, J., & Yan, Y. B. (2005). Probing conformational changes of proteins by quantitative second-derivative infrared spectroscopy. *Analytical Biochemistry*, *340*, 89–98.

CHAPTER FIVE

# Studying Antibody–Antigen Interactions with Fluorescence Fluctuation Spectroscopy

**Sergey Y. Tetin[1], Qiaoqiao Ruan, Joseph P. Skinner**
Diagnostics Research, Abbott Diagnostics Division, Abbott Park, Illinois, USA
[1]Corresponding author: e-mail address: sergey.tetin@abbott.com

## Contents



## Abstract

Antibodies are excellent binding proteins that have found numerous applications in biological research, biotechnology, and medicine. Characterization of their ligand binding properties has long been, and continues to be, the focus of many researchers. Antibodies are also perfect test systems which can be used for the evaluation of newly introduced biophysical techniques. Working with many different antibodies, we continuously implement the growing arsenal of methods offered by fluorescence fluctuation spectroscopy (FFS) and apply them for antibody research. In this chapter, we will describe applications of FFS for antibody binding characterization and also provide examples how studying of antibodies helps to develop and enhance the tool set offered by FFS technology. In addition to traditional affinity evaluations, we will describe how resolving molecular populations enables determinations of the binding stoichiometry and provides further information about the system. Even though all our examples include antibodies, the same experimental procedures can also serve well for characterizing various proteins and other ligand binding systems.

*Methods in Enzymology*, Volume 519
ISSN 0076-6879
http://dx.doi.org/10.1016/B978-0-12-405539-1.00005-1

# 1. INTRODUCTION

Antibody-related topics have remained in the mainstream of biological research for the past six decades. Understanding the structure–function relationship in immunoglobulin molecules led to discoveries of new concepts in protein chemistry and molecular genetics (Edelman & Gally, 1968; Porter, 1967; Tonegawa, 1979). Antibodies are also known as irreplaceable tools in numerous research and diagnostic applications (Tetin & Stroupe, 2004). The progress in implementation of recombinant immunoglobulin libraries and antibody engineering technologies has resulted in rapid escalation of the development of antibodies and their derivatives for therapeutic uses (Gilliland, Luo, Vafa, & Almagro, 2012).

Immunoglobulins are stable and robust binding proteins with the widest range of binding affinities. A typical immunoglobulin G (IgG) antibody (MW ~ 150 kDa) is a symmetric molecule that contains two identical independent binding sites. Completely functional monovalent Fab fragments (MW ~ 50 kDa) are produced by proteolytic digestion of immunoglobulins (Karush & Hornick, 1973; Porter, 1967). Antibodies are the simplest systems with a higher order of stoichiometry which makes them very useful for testing techniques that have molecular resolution. At certain concentrations of the ligand, antibody solutions contain two different bound species as well as free ligand and/or free antibody. Substituting antibodies with respective Fab fragments eliminates the complexity and serves well as a proper first step in characterizing both the method and the system.[1]

The need to characterize various antibody–antigen systems triggers the development of new methods suitable for determination of the wide range of equilibrium and kinetic binding parameters. Reciprocally, antibodies serve as highly reliable test systems for characterization and validation of newly introduced biophysical tools. Using fluorescence quenching, anisotropy and lifetime measurements have been widely applied in antibody research and was previously reviewed (Tetin & Hazlett, 2000). Introduction of FFS techniques extends binding characterization to the molecular level and provides additional information on heterogeneity of antibody systems.

In the examples below, we share our experience in studying antibody systems and also discuss how antibody studies help to characterize currently

[1] In immunological literature, small antigenic molecules that trigger immune response only when tethered to a larger protein carrier are referred to as haptens. In this chapter, we are using the term ligand that could be applied for either antigen or hapten.

available and newly introduced FFS methods. The readers will find detailed discussions of FFS methods in other chapters of these two volumes written by inventors of the technology. In the sections below, we will present several practical examples of using FFS in binding studies, which will contain a brief description of the biochemical task together with related experiments. Each example will have specific equations used for analyzing the data.

The first method under discussion, fluorescence correlation spectroscopy (FCS), utilizes fluorescence intensity autocorrelation function analysis (Elson & Magde, 1974; Magde, Elson, & Webb, 1974). FCS can resolve two molecular species if their diffusion coefficients are about twofold different (Meseth, Wohland, Rigler, & Vogel, 1999). We will demonstrate applications of this technique for determination of the equilibrium dissociation constant, $K_D$, and also for antigenic epitope mapping performed in a competitive binding format.

Dual color fluorescence cross-correlation spectroscopy (DC-FCCS) was introduced to spectrally separate two species and characterize interactions by cross-correlating data measured in two separate channels (Schwille, Meyer-Almes, & Rigler, 1997). DC-FCCS has its advantages over FCS when studying molecules of similar size. In addition to $K_D$ determinations, we use this method for identifying trimolecular complexes, in which two different antibodies bind one antigen molecule, and also for measurements of the kinetic dissociation rate.

As an alternative to DC-FCCS, higher-order complexes can be studied using statistical information from the photon counts detected in a single channel (Palmer & Thompson, 1989; Qian & Elson, 1990a,b). These methods no longer rely on calculation of the temporal autocorrelation. Instead, the photon count distribution and its moments are used to analyze molecular heterogeneity (Chen, Müller, So, & Gratton, 1999; Kask, Palo, Ullmann, & Gall, 1999). Moment analysis measures the brightness and concentration of the fluorescent molecules in the sample and has been applied to determine antibody affinity (Chen, Müller, Tetin, Tyner, & Gratton, 2000). Closely related to moments are cumulants, which can also be used to characterize molecular heterogeneity (Müller, 2004). Time integrated fluorescence cumulant analysis (TIFCA) is the last technique under discussion. TIFCA resolves molecular populations using both the diffusion rate and brightness of each species. We applied moment analysis and TIFCA to resolve bound populations in a mixture of bivalent antibody and fluorescently labeled ligand at stoichiometric conditions.

## 2. BINDING MODEL AND EXPERIMENTAL CONSIDERATIONS

Solution phase measurements at equilibrium provide two important characteristics of the antibody system under study: the stoichiometry and the affinity. The stoichiometry defines the number of ligand molecules that bind to one antibody, and the affinity characterizes the strength of the binding interaction. These two basic properties are determined using the binding model presented in this section. Any applications of antibodies depend on the system's stoichiometry and equilibrium dissociation constant.

A typical IgG antibody contains two equivalent noninteracting binding sites (Edelman & Gally, 1968; Porter, 1967). Solution phase binding studies are usually performed by incubating various amounts of one of the reagents with the second reagent kept at constant concentration in the titration series. For instance, when using fluorescently labeled antigen, its concentration is usually kept constant, while antibody concentration varies. A dynamic equilibrium between the free antibody binding sites S and the free ligand L is described by the reaction

$$\mathrm{S} + \mathrm{L} \Leftrightarrow \mathrm{LS}, \qquad [5.1]$$

where the free species are related to the formed complex LS. The dissociation constant of the equilibrium

$$K_{\mathrm{D}} = \frac{\mathrm{L} \cdot \mathrm{S}}{\mathrm{LS}} \qquad [5.2]$$

is expressed in concentration units. The equilibrium association constant, often referred to as the binding affinity,

$$K_{\mathrm{A}} = \frac{\mathrm{LS}}{\mathrm{L} \cdot \mathrm{S}}, \qquad [5.3]$$

is expressed in reciprocal concentration units and directly relates to the free energy $\Delta G$, enthalpy $\Delta H$, and entropy $\Delta S$ of the binding interaction:

$$\Delta G = -RT \ln K_{\mathrm{A}} = \Delta H - T\Delta S. \qquad [5.4]$$

In Eq. (5.4), $R$ is the gas constant and $T$ is the absolute temperature. This expression (Eq. 5.2) is the simplest representation of the general ligand-binding concept formulated by Adair (1925). More detailed ligand-binding theory has been discussed in several classical reviews (Klotz, 1985; Weber, 1975, 1992).

According to the mass conservation law, the concentrations of the free and bound species are related to the total concentrations of the binding sites $S_t$ and the ligand $L_t$

$$\begin{aligned} S_t &= S + LS \\ L_t &= L + LS \end{aligned}. \quad [5.5]$$

Direct determination of the bound and free molecular species is often difficult, and researchers are limited to measuring an observable parameter that is proportional to the fraction of bound ligand or filled binding sites, $F_b$:

$$\begin{aligned} F_b^{\text{ligand}} &= \frac{LS}{L_t} = \frac{LS}{L + LS} \\ F_b^{\text{sites}} &= \frac{LS}{S_t} = \frac{LS}{S + LS} \end{aligned}. \quad [5.6]$$

The relationship between the fraction of bound complex with total concentrations of the mixed reagents and $K_D$ is given by the following relations:

$$\begin{aligned} F_b^{\text{ligand}} &= \frac{L_t + S_t + K_D - \sqrt{L_t^2 + S_t^2 + K_D^2 - 2S_tL_t + 2L_tK_D + 2S_tK_D}}{2L_t} \\ F_b^{\text{sites}} &= \frac{L_t + S_t + K_D - \sqrt{L_t^2 + S_t^2 + K_D^2 - 2S_tL_t + 2L_tK_D + 2S_tK_D}}{2S_t} \end{aligned}. \quad [5.7]$$

Usually antibodies bind one ligand molecule per binding site. Thus, concentrations of free ligand and free sites can be calculated from their total concentrations and fractions bound:

$$\begin{aligned} L &= L_t - S_t F_b^{\text{sites}} \\ S &= S_t - L_t F_b^{\text{ligand}} \end{aligned}. \quad [5.8]$$

Combining Eqs. (5.2) and (5.6) results in equations that can be used to fit data and determine the dissociation constant $K_D$:

$$\begin{aligned} F_b^{\text{ligand}} &= \frac{S}{K_D + S} \\ F_b^{\text{sites}} &= \frac{L}{K_D + L} \end{aligned}. \quad [5.9]$$

A typical binding curve shows the fraction of the bound reagent plotted as a function of the free concentration of the second reagent that was titrated into the mixture. Often, when performing a binding titration experiment, the labeled ligand is kept at a fixed concentration, while the concentration of the second reagent, an antibody or its fragment, varies covering several orders of concentration units. Binding curves are not symmetric and the apparent $K_D$ is biased to the higher values when random error in fraction bound increases. As a result, the random errors in fraction bound may lead to a systematic error in $K_D$ determinations. Therefore, experimental conditions and concentrations of the reagents should be thoroughly optimized in order to accurately measure the signal related to the bound species. It was shown that the precision in $K_D$ determination varies across the fraction bound and is highest when $F_b = 0.5$ (Weber, 1965). Hence, the concentration of the labeled ligand is expected to be kept close to or below the $K_D$ value. Fitting all titration data together minimizes effects caused by the random errors at each point in the titration series. Finding the right balance between the sensitivity of the detection method and optimal reagent concentration requires careful experimental design (Tetin & Hazlett, 2000).

On the contrary to equilibrium binding titrations used for affinity measurements, determination of the binding stoichiometry must be performed at the reagent concentration that is much higher than the equilibrium dissociation constant. In this case, the plot of fraction bound as a function of total sites (or ligand) depends on concentrations and stoichiometry rather than the equilibrium dissociation constant. When using traditional fluorescence techniques that do not have molecular resolution, we are limited to follow the signal that is proportional to fraction bound. FFS expands fluorescence techniques allowing direct measurement of the concentration of free and bound species. Hence, the binding stoichiometry and the affinity of the system can be calculated from the same set of experimental data.

# 3. STUDYING ANTIBODIES WITH FFS

## 3.1. Determination of equilibrium dissociation constants with FCS

In our experiments, fluorescent intensity $F(t)$ was used to calculate the autocorrelation of fluorescence fluctuations

$$G(\tau) = \frac{\langle \Delta F(t) \cdot \Delta F(t+\tau) \rangle}{\langle F \rangle^2}, \qquad [5.10]$$

where $\Delta F(t) = F(t) - \langle F \rangle$ is the fluorescence fluctuation at time $t$, $\langle F \rangle$ is the average fluorescence, and $\tau$ is the lag time. The autocorrelation function describes the average temporal decay of the fluctuations. For fluctuations caused by Brownian motion, decay of the autocorrelation function depends on the diffusion coefficient of the fluorescent species. The amplitude $G(0)$ is reciprocal to the average number of fluorescent molecules $\langle N \rangle$ in the observation volume:

$$G(0) = \frac{\langle \Delta F(t)^2 \rangle}{\langle F \rangle^2} = \frac{\gamma}{\langle N \rangle}. \qquad [5.11]$$

A geometric factor $\gamma$ is determined by the shape of the observation volume, referred to as the point spread function (PSF) (Thompson, 1991). For two-photon excitation experiments, the PSF can be approximated with a squared three-dimensional Gaussian (3DG) profile.

The functional form of the autocorrelation for each individual molecular species is

$$G(\tau) = G(0) \cdot g(\tau) = G(0) \cdot \left(1 + \frac{8D\tau}{w^2}\right)^{-1} \left(1 + \frac{8D\tau}{z^2}\right)^{-1/2}, \qquad [5.12]$$

where $w$ and $z$ are the beam waist and axial dimension of the PSF, respectively, and $D$ is the diffusion coefficient of the fluorescent species.

For multiple fluorescent species, the resulting autocorrelation function is the sum of the autocorrelation functions of each species multiplied by the corresponding fractional intensity squared. A binding system can be modeled using two species where the fluorescently labeled ligand is in either a free or a bound state, and the autocorrelation function is

$$G(\tau) = g'_{\mathrm{L}}(0) \cdot g_{\mathrm{L}}(\tau) + g'_{\mathrm{LS}}(0) \cdot g_{\mathrm{LS}}(\tau). \qquad [5.13]$$

In this equation, $g'_{\mathrm{L}}(0)$ is the fluctuation amplitude for the free ligand weighted by its brightness, and $g'_{\mathrm{LS}}(0)$ is the fluctuation amplitude for the bound ligand

$$g'_{\mathrm{L}}(0) = \left(\frac{\lambda_{\mathrm{L}} \langle N_{\mathrm{L}} \rangle}{\langle F \rangle}\right)^2 \frac{\gamma}{\langle N_{\mathrm{L}} \rangle} = \gamma \frac{\lambda_{\mathrm{L}}^2 \langle N_{\mathrm{L}} \rangle}{\langle F \rangle^2}. \qquad [5.14]$$

In Eq. (5.14), $\lambda_{\mathrm{L}}$ is the molecular brightness of free ligand expressed as the number of photon counts emitted by one molecule per second and $N_{\mathrm{L}}$ is the number of molecules of the free ligand. The ratio of the free and bound ligand is

$$\frac{N_{\mathrm{L}}}{N_{\mathrm{LS}}} = \left(\frac{\lambda_{\mathrm{LS}}}{\lambda_{\mathrm{L}}}\right)^2 \frac{g'_{\mathrm{L}}(0)}{g'_{\mathrm{LS}}(0)}. \quad [5.15]$$

If the brightness of both species is the same, the weighting is not necessary. Fitting the autocorrelation curves with the two-species model (Eqs. 5.12 and 5.13) permits calculation of the fraction ligand bound

$$F_{\mathrm{b}} = \frac{N_{\mathrm{LS}}}{N_{\mathrm{L}} + N_{\mathrm{LS}}}. \quad [5.16]$$

Plotting the fraction of ligand bound as a function of free binding sites leads to a binding plot and fitting with Eq. (5.9) provides $K_{\mathrm{D}}$.

### 3.1.1 Antibody binding to a small ligand

A mouse monoclonal antibody (mAb C11-10) was raised against the core protein of Hepatitis C virus (HCV) and used for diagnostic purpose. It binds a relatively short peptide QIV**GGVYL**LPRRGPRC in which the actual epitope sequence **GGVYL** is indicated in bold. The molecular weight of the antibody–peptide complex (152 kDa) is 63 times larger than that of the unbound peptide (2.4 kDa, including the fluorescent tag). Thus, the diffusion coefficient of the fluorescently labeled peptide should decrease approximately four times upon binding to the antibody ($D \propto 1/\sqrt[3]{\mathrm{MW}}$) which makes FCS application straightforward.

*Experiment*: Antibody solutions with concentrations ranging from 0.05 n$M$ to 5 μ$M$ were prepared in 16 serial dilution steps. The HCV peptide conjugated at the N-terminus with Alexa488 carboxylic acid, succinimidyl ester (Invitrogen, Carlsbad, CA), was added to each sample at a final concentration of 3 n$M$. The samples were kept in the dark at room temperature for 30 min before measurements. The calculated autocorrelation curves were fit to a one- or two-species model using Globals software (Laboratory of Fluorescence Dynamics, Irvine, CA).

Figure 5.1A shows the autocorrelation curves of the labeled peptide at various antibody concentrations. Shifting of the curves to the right indicates slower diffusion due to antibody binding. Initially, autocorrelation curves for the free and completely saturated peptide were independently fit to Eq. (5.12) to determine the diffusion coefficients of free and bound ligand. The fits yielded diffusion coefficients of 170 and 45 $\mu\mathrm{m}^2/\mathrm{s}$, respectively. Then, the curves from all titration data were fit to the two-component model (Eq. 5.13) to obtain $g'_{\mathrm{L}}(0)$ and $g'_{\mathrm{LS}}(0)$ for each concentration of

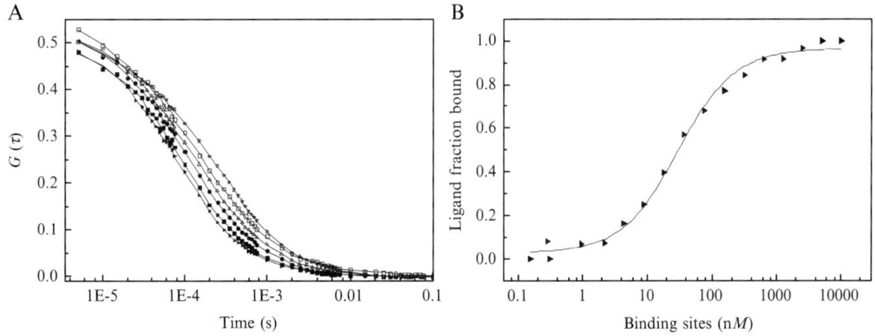


**Figure 5.1** (A) Examples of autocorrelation curves of 3 n*M* of Alexa488-HCV peptide in the presence of various antibody concentrations. (B) Equilibrium binding curve of anti-HCV peptide and mAb C11-10. The equilibrium dissociation constant is 27 ± 5 n*M*.

antibody by fixing the diffusion coefficients to the previously measured values. Fluorescence of the ligand decreases 25% upon binding, affecting the ratio of free and bound molecules such that $\lambda_{LS} = 0.75\lambda_L$ and

$$\frac{N_L}{N_{LS}} = \left(\frac{\lambda_{LS}}{\lambda_L}\right)^2 \frac{g'_L(0)}{g'_{LS}(0)} = 0.75^2 \frac{g'_L(0)}{g'_{LS}(0)}. \qquad [5.17]$$

The fraction of ligand bound at each antibody concentration can be calculated using Eq. (5.16). Calculations of the concentration of free antibody-binding sites at each titration point were performed with Eq. (5.8) and used in the equilibrium-binding plot shown in Fig. 5.1B. Fitting the binding data to the binding model in Eq. (5.9) resulted in $K_D = 27 \pm 5$ n*M*.

### *3.1.2 Binding of similar size proteins*

If protein complexes under study cannot be resolved by FCS due to similar size, the size of one of the reagents can be increased by premixing it with a specific antibody that does not interfere with the interaction of the primary binding partners. For instance, Alexa488-labeled human placenta growth factor (hPlGF) is a 29 kDa protein and the molecular weight of its receptor, human vascular endothelial growth factor receptor 1 (VEGFR1), is 45 kDa. The size of the PlGF–VEGR1 complex is not large enough for identification by its diffusion coefficient in the presence of unbound Alexa488-PlGF. Therefore, we formed an antibody-VEGFR1 complex, by addition of high affinity anti-VEGFR1 antibody, to increase the apparent molecular weight to 195 kDa. Thus, free and bound PlGF can be resolved by FCS.

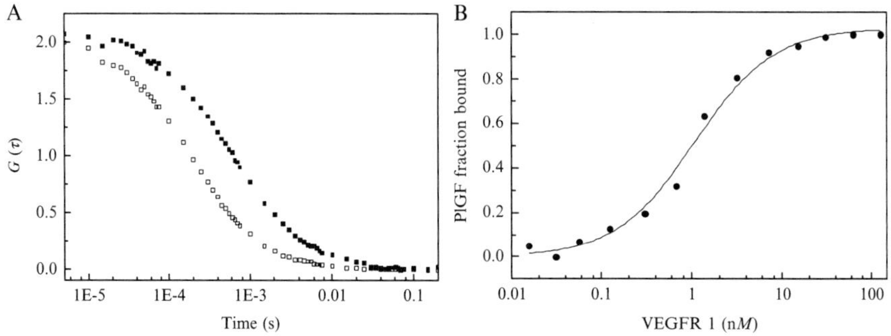


**Figure 5.2** (A) Autocorrelation curves of free Alexa488-PlGF (open square) and its complex (solid square) with VEGFR1 and anti-VEGFR1-antibody. (B) Equilibrium binding curve of PlGF and VEGFR1. The equilibrium dissociation constant is $1.2 \pm 0.2$ n*M*.

*Experiment*: A 100 n*M* solution of recombinant hVEGFR1 (R&D Systems, Minneapolis, MN, USA) in HEPES buffer was serially diluted in 16 samples to the final concentration 4 p*M*. An excess concentration of 100 n*M* anti-VEGFR1 mAb was added to each solution, completely saturating hVEGFR1 with antibody. Recombinant hPlGF (R&D Systems) was labeled at an 8:1 dye to protein ratio using Alexa488 yielding a labeling efficiency of 1.2 dye molecules per protein. Alexa488-PlGF was then added to each sample at a final concentration of 0.25 n*M* and incubated in the dark at room temperature for 30 min before measurement.

Figure 5.2A shows the calculated autocorrelation curves of free and bound Alexa488-PlGF. Formation of the antibody–VEGFR1 complex effectively increases the size of the receptor making binding of PlGF detectable, indicated by the shift of the autocorrelation curve (solid square). The data were processed by first determining diffusion coefficients of Alexa488-PlGF before and after saturation with antibody–VEGFR1 complex and then fit as described in the previous section. Figure 5.2B shows the fraction bound with the fit (solid line) to the first equality in Eq. (5.9) yielding $K_D = 1.2$ n*M*. In conclusion, it is important to indicate that approach should be applied cautiously, and other experiments are needed to confirm that adding the antibody has no effect on the primary binding interaction.

## 3.2. Antigenic epitope mapping

FCS-based binding studies can also be performed in competitive format and are especially useful when other solution phase methods cannot be applied because binding of the antibody does not affect fluorescence properties of the target antigen. In this section, we describe the use of FCS for determining antigenic epitopes in human brain natriuretic peptide (BNP) which is a valuable diagnostic cardiac marker (Sagnella, 1998). Human BNP is a 32-amino acid cyclic peptide with a ring structure confined between cysteines in the positions 10 and 26 (Nakao, Ogawa, Suga, & Imura, 1992). BNP is a relatively short and flexible molecule. Therefore, the residues that directly interact with antibodies (antigenic epitopes) can be identified by alanine scanning (Cunningham & Wells, 1989).

*Experiment*: We first prepared and purified N-terminus-labeled Alexa488-BNP. It was critical to label only the N-terminal amine because incorporating the probe in other positions may affect the binding epitope. Binding of N-terminus-labeled Alexa488-BNP to antibodies does not change its fluorescence intensity and results in very little change in polarization due to the flexibility of the molecule. Therefore, fluorescence polarization and quenching based assays cannot be used to measure binding parameters. However, FCS is an excellent tool for this experiment since the diffusion coefficients of the free and antibody-bound peptide are easily resolved.

Thirty mutated BNP peptides in which each amino acid residue was sequentially replaced with alanine, except residues Cys 10 and Cys 24, were prepared for competitive binding experiments. Alexa488-BNP tracer (4 n*M*) was mixed with mAb BC203 (30 n*M*), which allows at least 90% of the labeled peptide to be bound to the antibody. Serial titrations using BNP mutants and Alexa488-BNP–antibody complex kept at constant concentration were performed. A titration with native BNP was also done as a control. In this format, unlabeled peptides compete to displace Alexa488-BNP bound to the antibody.

FCS measurements were performed and data were processed to generate a binding plot for each mutant using the method described in Section 3.1.1. Figure 5.3A shows the competitive binding curves. The fractions of free Alexa488-BNP were plotted as a function of mutants concentration. The control curve using native BNP serves as a reference (solid line). Several mutants yielded binding curves similar to native BNP indicating that the alanine substitution did not affect the epitope. A number of BNP mutants cannot displace or only partially displace Alexa488-BNP from the antibody-binding site at high mutant concentration (1 μ*M*). Fig. 5.3B shows the fraction of Alexa488-BNP bound to the antibody in the presence of 400 n*M* BNP mutants. At 400 n*M*, the control solution (unlabeled native BNP) has less than 10% of bound Alexa488-BNP. On the other hand, at 400 n*M*, the H32A mutant could not displace any Alexa488-BNP from the antibody binding site. Figure 5.3B indicates that residue L29, R30, and H32 are the three most critical residues for BNP recognition by mAb BC203.

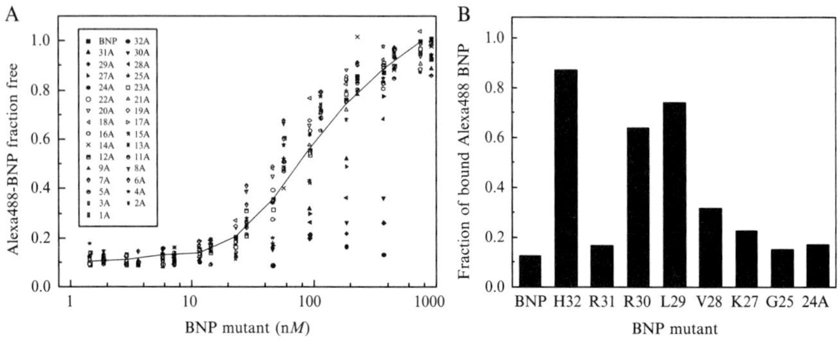


**Figure 5.3** Competitive-binding curves of alanine-substituted BNP mutants: (A) Fractions of free Alexa488-BNP were plotted as a function of BNP mutants concentrations calculated from autocorrelation curves. The positive control is shown as a line in the graph. (B) Fraction of bound Alexa488-BNP at 400 n*M* BNP mutants. Higher values indicate more involvement of the mutated residue in antibody recognition. Mutations outside of the epitope sequences did not affect the competition and therefore were excluded from the figures.

## 3.3. Cross-correlation analysis of antibody complexes

DC-FCCS is another FFS method specifically introduced to study complex formation (Kim, Heinze, Bacia, Waxham, & Schwille, 2005; Schwille et al., 1997; Weidemann, Wachsmuth, Tewes, Rippe, & Langowski, 2002). In DC-FCCS, two fluorescence signals, $F_G$ and $F_R$, are collected as a function of time by two separate detectors. Two independent autocorrelation curves are calculated following Eq. (5.10) for both channels. Additionally, the temporal cross-correlation function is calculated

$$G_x(\tau) = \frac{\langle \Delta F_G(t) \cdot \Delta F_R(t+\tau) \rangle}{\langle F_G \rangle \langle F_R \rangle}. \quad [5.18]$$

The auto- and cross-correlation curves can be fit with Eqs. (5.12) and (5.13).

At the decay time $\tau = 0$, the extrapolated value of $G(0)$ reflects the amplitude of the auto- or cross-correlation function and the number of molecules of each species:

$$\begin{aligned} G_G(0) &= \gamma \cdot \frac{1}{(\langle N_G \rangle + \langle N_{GR} \rangle)} \\ G_R(0) &= \gamma \cdot \frac{1}{(\langle N_R \rangle + \langle N_{GR} \rangle)} \\ G_x(0) &= \gamma \cdot \frac{\langle N_{GR} \rangle}{(\langle N_G \rangle + \langle N_{GR} \rangle)(\langle N_R \rangle + \langle N_{GR} \rangle)}, \end{aligned} \quad [5.19]$$

where $N_G$ represents molecules tagged with the green fluorophore, $N_R$ represents molecules tagged with the red fluorophore, $N_{GR}$ represents the complex. The fraction of bound species can then be determined by taking the ratios of the cross- and autocorrelation amplitudes:

$$\begin{aligned}\frac{G_x(0)}{G_R(0)} &= \frac{\langle N_{GR}\rangle}{\langle N_G\rangle + \langle N_{GR}\rangle} \\ \frac{G_x(0)}{G_G(0)} &= \frac{\langle N_{GR}\rangle}{\langle N_R\rangle + \langle N_{GR}\rangle}\end{aligned}. \qquad [5.20]$$

Calculation of the fraction bound using the above ratios has been used to determine the equilibrium dissociation constant of BNP antibody (Ruan & Tetin, 2008). Here, we focus on applying DC-FCCS for measuring higher-order complex formation and dissociation rate constants.

### *3.3.1 Antibody sandwich identification*

Antibody sandwiches are trimolecular complexes which form upon the binding of two different antibody molecules to a single antigen molecule. Sandwich formation is preferable for diagnostic immunoassays targeting low analyte concentrations (Tetin & Stroupe, 2004). Therefore, selection of antibody sandwich pairs is a critical task in immunoassay development. Human neutrophil gelatinase-associated lipocalin (NGAL), also known as Lipocalin-2 or LCN2, is used as an early protein marker for acute renal injury (Grenier et al., 2010). The molecular weight of NGAL is 25 kDa. In order to identify sandwich-forming pairs, we evaluated six mouse monoclonal antibodies which recognize different NGAL epitopes (Olejniczak et al., 2010).

*Experiment*: Three of six antibodies were labeled with Alexa488, while the other three were labeled with TexasRed. Initially, each Alexa488-labeled antibody was mixed with TexasRed-labeled antibodies at equal concentrations of 10 n*M*. This mixture was then measured using DC-FCCS before and after incubation with 15 n*M* NGAL.

Figure 5.4 shows examples of the cross-correlation curves of two antibody pairs before and after incubation with NGAL. In Fig. 5.4A, the result from a mixture of mAb ADD105 and mAb ADD104 is shown. The antibodies do not form a sandwich in the absence of NGAL (solid squares), and the $G_x(0)$ value only reflects spectral cross talk between the channels. The open squares show the cross-correlation curve after incubation with NGAL, which causes sandwich formation yielding an increased amplitude

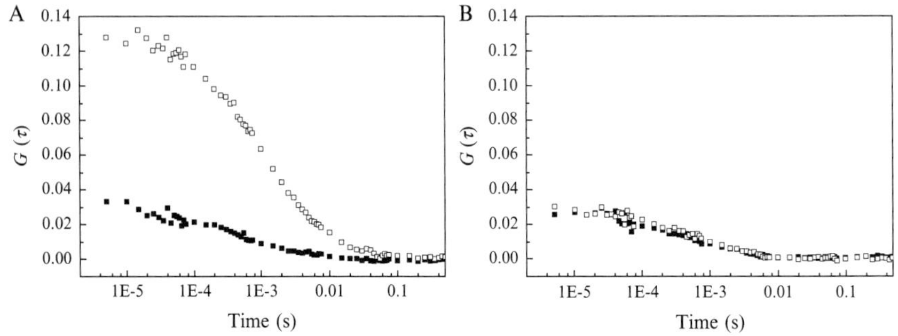


**Figure 5.4** Cross-correlation curves for antibody pairs before (solid square) and after (open square) NGAL addition: (A) mAb ADD105 and mAb ADD104; (B) mAb ADD103 and mAb ADD107.

$G_x(0)$ directly proportional to the concentration of the green–red complex (Eq. 5.19). Another result using mAb ADD103 and mAb ADD107 is shown in Fig. 5.4B. Addition of NGAL did not increase the cross-correlation amplitude indicating the sandwich did not form.

### 3.3.2 Dissociation kinetics

DC-FCCS can also be used for studying slow binding kinetics such as dissociation rates. The protein tissue inhibitor of metalloproteinases 1 (TIMP-1) is an inhibitor of matrix metalloproteinases (Visse & Nagase, 2003). Human TIMP-1 is a 30 kDa protein that has been identified as a promising biomarker for early detection of colorectal cancer and prediction of the postsurgical survival of treated patients (Holten-Andersen et al., 2002). In the following example, DC-FCCS is used to measure the dissociation rate $k_{\mathrm{off}}$ of one of the antibodies in a sandwich complex.

*Experiment*: Two anti-TIMP-1 monoclonal antibodies, mAb 63515 (R&D systems) and mAb VT-4 (Moller Sorensen et al., 2005), were labeled with Alexa488 and TexasRed, respectively. A total of 20 n*M* of labeled mAb 63515 and labeled mAb VT-4 were incubated with 30 n*M* TIMP-1 to form an antibody sandwich. Then, a 200-fold molar excess of the unlabeled mAb 63515 was added to compete with labeled antibody. Cross-correlation curves were collected every 15 s for the first 500 s, and then the collection interval was increased to 3 min.

Figure 5.5 shows the calculated cross-correlation amplitude as a function of time. As the unlabeled mAb 63515 replaces the Alexa488-labeled antibody in the sandwich complex, the $G_x(0)$ value gradually decreased. Fitting

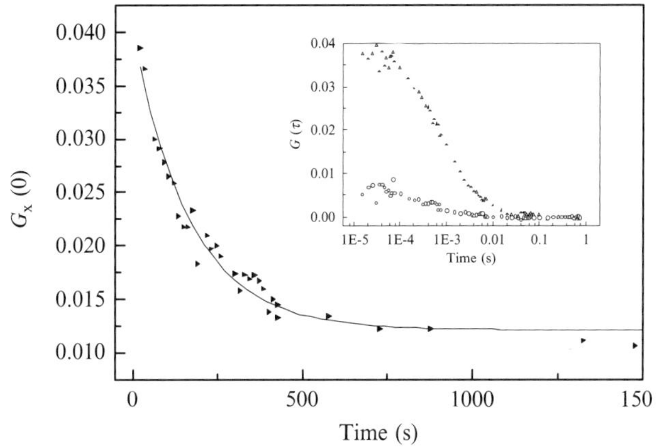


**Figure 5.5** Dissociation kinetics of TIMP-1 sandwiched with mAb 63515 and mAb VT-4. Labeled antibodies at an equal concentration (20 n*M*) were premixed with 30 n*M* TIMP-1 to achieve maximal sandwich formation. A 200-fold excess of the unlabeled mAb 63515 was used to displace the labeled mAb 63515; the off-rate is $6 \times 10^{-3}$ $s^{-1}$. The inset shows the cross-correlation in the presence (higher amplitude) and absence (low amplitude) of TIMP-1.

the dissociation curve to a single-exponential kinetics model resulted in $k_{off} = 6 \times 10^{-3}$ $s^{-1}$. For reference, the cross-correlation in the presence (higher amplitude) and absence (low amplitude) of TIMP-1 is shown in the inset to illustrate sandwich formation. Repeating this experiment with 300- and 400-fold molar excess of the unlabeled mAb 63515 confirmed the measured off-rate.

DC-FCCS is a powerful tool to study antibody–antigen binding interactions and antibody sandwich formations. However, it is difficult to achieve $G_x(0)/G(0) = 1$ using two interacting molecules mixed at equal concentration due to nonideal sample and instrument conditions. These effects have been the focus of a recent study (Foo, Naredi-Rainer, Lamb, Ahmed, & Wohland, 2012). Therefore, we use the relative changes of $G_x(0)$ or ratios of $G_x(0)/G(0)$ without extensive mathematical corrections for determination of the equilibrium-binding constants, dissociation rates, and optimization of antibody pairs that sandwich with the antigen.

## 3.4. Antibody stoichiometry

Bivalency is another important property of antibody molecules. In the experiments above, conditions were chosen to minimize the fraction of doubly bound antibody and analysis was performed with respect to the total number of sites present. Detailed discussions of the effect of bivalency on FCS data can be found elsewhere (Hazlett, Ruan, & Tetin, 2005). Here, we apply FFS tools which can resolve this bivalency and directly measure higher-order complexes.

We first expand the description of bound states to include three species by rewriting Eq. (5.5) from above in terms of total antibody concentration $A_t(S_t = 2\ A_t)$. Three observable states for the ligand are defined: the free ligand L and two bound forms (LA and L2A). The species LA specifies antibody with only a single binding site occupied, while L2A represents antibody with both sites occupied by ligand. The total concentrations of the ligand and antibody are

$$\begin{aligned} L_t &= L + LA + 2 \cdot L2A \\ A_t &= A + LA + L2A \end{aligned} \qquad [5.21]$$

The right-hand side of the first equality in Eq. (5.21) indicates the three ligand concentrations we wish to resolve experimentally. The probability $P_b = (LA + 2 \cdot L2A)/S_t$ that a given site is occupied is equal to the fraction of bound sites presented in Eq. (5.7), $P_b = F_b$. Thus, the concentration of free ligand L is calculated using Eq. (5.8). Because the sites are independent, we use equal probability for the ligand to bind to either site. The probabilities of only a single site occupied $P_{one}$ and both sites occupied $P_{two}$ are

$$\begin{aligned} P_{one} &= 2(1 - P_b)P_b \\ P_{two} &= P_b{}^2 \end{aligned} \qquad [5.22]$$

Thus, the concentrations of antibody complexes are

$$\begin{aligned} LA &= A_t \cdot P_{one} \\ L2A &= A_t \cdot P_{two} \end{aligned} \qquad [5.23]$$

Experimentally, FFS resolves the average number of molecules within the observation volume in each state. A conversion factor $\phi$ is required to express concentrations as the number of molecules in the observation volume yielding

$$N_{LA} = \phi S_t P_{one}, \qquad [5.24]$$

$$N_{L2A} = \phi A_t P_{two}, \qquad [5.25]$$

and

$$N_L = \phi(L_t - LS) + N_L'. \qquad [5.26]$$

In Eq. (5.26), a parameter $N'_L$ is added to account for any inactive ligand that may be present in the sample. To be consistent with its physical meaning, such an offset must not be included in the bound state described by Eqs. (5.24) and (5.25). These equations are applied below to generate models and fit experimental data.

### 3.4.1 Moment analysis

When a fluorophore is attached to a molecule of interest, its molecular brightness may be used to track complex formation (Chen & Müller, 2007). The molecular brightness is the number of photons emitted per molecule per unit time and proportional to the amplitude of the intensity fluctuations. Therefore, calculating the fluctuation amplitude $G(0)$ allows stoichiometry measurement by monitoring the correlation amplitude alone. We applied a very simple method of determining the amplitude of the autocorrelation function directly from the raw photon count data.

$G(0)$ is the ratio of the second and first moment of the photon count distribution. The moments of the discrete photon counts per sampling time $k$ can be calculated and used to determine the fluctuation amplitude

$$G(0) = \frac{\langle \Delta k^2 \rangle - \langle k \rangle}{\langle k \rangle^2}, \qquad [5.27]$$

which accounts for the discrete nature of photon counting by correcting for shot noise (Chen et al., 2000; Qian & Elson, 1990a,b).

Upon introduction of two or more species into the system, the value for $G(0)$ becomes the sum of the amplitudes which corresponds to each species weighted by the square of their fractional intensities. Thus, the fluctuation amplitude in terms of the molecular brightness of each species $\lambda_j$ is given by (Thompson, 1991)

$$G(0) = \frac{\gamma \sum_{j=1}^{J} \lambda_j^2 N_j}{\left( \sum_{j=1}^{J} \lambda_j N_j \right)^2}. \qquad [5.28]$$

For a mixture of ligand and antibody at stoichiometric conditions, the amplitude is determined by the concentrations of the three fluorescent species and their brightnesses:

$$G(0) = \frac{\gamma\left(\lambda_L^2 N_L + \lambda_{LA}^2 N_{LA} + \lambda_{L2A}^2 N_{L2A}\right)}{\left(\lambda_L N_L + \lambda_{LA} N_{LA} + \lambda_{L2A} N_{L2A}\right)^2}. \quad [5.29]$$

The values $N_L$, $N_{LA}$, and $N_{L2A}$ are directly proportional to the concentrations L, LA, and L2A.

Experimentally, the brightness of the fluorophore may change upon binding, which must be considered in the analysis. For antigen binding to antibody, we assume the following relations:

$$\begin{aligned} \lambda_{LA} &= q\lambda_L \\ \lambda_{L2A} &= 2q\lambda_L \end{aligned}. \quad [5.30]$$

The parameter $q$ is a quenching factor to account for a change in the brightness upon binding. Note that the relation also assumes that the doubly liganded antibody has two times the brightness of the singly liganded antibody. Using the above relations, Eq. (5.29) simplifies to a brightness-independent expression

$$G(0) = \frac{\gamma(N_L + q^2(N_{LA} + 4N_{L2A}))}{(N_L + q(N_{LA} + 2N_{L2A}))^2}. \quad [5.31]$$

A plot of Eq. (5.31) as a function of total antibody concentration is shown in Fig. 5.6 with a quenching factor of $q = 0.3$. Two conditions are shown in the

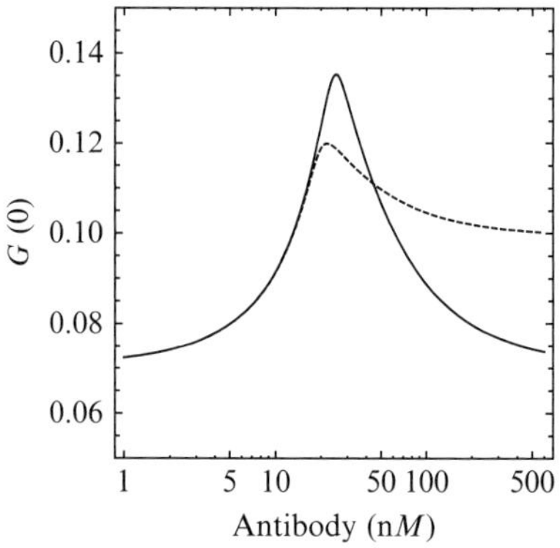


**Figure 5.6** Effect of inactive ligand in a titration. The figure shows the expected value of *G(0)* as a function of antibody concentration if the ligand is 100% active (solid line) and 80% active (dashed line). For this model, the dissociation constant is 0.1 n*M* and the ligand concentration is 25 n*M*. The antigen is quenched 70% upon binding in both cases.

plot. The first shows the expected results when all of the ligand in the titration is active, $N'_L = 0$ (solid line). The second curve indicates the effect of inactive ligand, $N'_L > 0$ (dashed line). This curve shows that $G(0)$ does not return to its initial value when a fraction of ligand is not active. This effect will be relevant in the experiments described below. To generate the plot, a dissociation constant of 0.1 n$M$ was used to calculate the number of molecules in each bound state.

*Experiment*: To perform a stoichiometric titration, anti-BNP mAb 106.3 was titrated with a constant concentration of 25 n$M$ BNP (5–13, C10A) labeled with Alexa488. The BNP ligand was described previously in Section 3.2. BNP (5–13, C10A) is the epitope region recognized by mAb 106.3. (Ruan and Tetin, 2008; Tetin et al. 2006; Longenecker et al., 2009).

Antibody was diluted serially in 12 samples starting with a concentration of 200 n$M$ in the first sample. After incubation for 20 min in the dark at room temperature, the samples were measured. The fluctuation amplitude $G(0)$ was calculated for each sample from the photon count data using Eq. (5.27). The results were fit using the model presented in Eq. (5.31). The uncertainty was determined by segmenting raw data into five sections and calculating $G(0)$ for each segment. The standard deviation of the five values was used as the uncertainty in fitting.

The calculated $G(0)$ from the titration data is presented in Fig. 5.7. As the concentration of antibody with two occupied sites increases, the total number of fluorescent entities decreases resulting in a higher fluctuation amplitude. Upon reaching a maximum, the amplitude decreases as singly liganded antibody dominates the fraction bound (solid circles). However, the amplitude does not return to its initial value as would be predicted when only singly liganded antibody is present. This observation can be attributed to inactive ligand. Upon inclusion of an inactive fraction, the data were fit to estimate the dissociation constant and the amount of inactive ligand present (solid line in the figure). The fit returns $K_D = 1.0 \pm 0.4$ n$M$ and an inactive fraction of $15 \pm 3\%$ (reduced $\chi^2 = 1.2$). This result is confirmed below by resolving all three species using another method of analysis.

### 3.4.2 Time Integrated Fluorescence Cumulant Analysis (TIFCA)

TIFCA is a method which utilizes both diffusion and brightness to resolve species. The theory behind TIFCA can be found elsewhere (Müller, 2004; Wu & Müller, 2005), including chapters in these volumes. In this

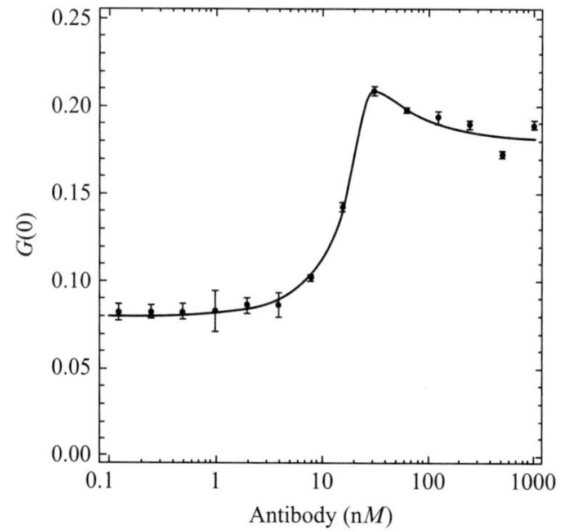


**Figure 5.7** Titration of BNP (5-13, C10A) with mAb 106.3. The fluctuation amplitude (solid circles) and the fit to the three-species model (solid line) are shown. The fit yielded an inactive fraction of 15 +/− 3% and a dissociation constant of 1.0 +/− 0.4 *nM*.

subsection, we describe application of TIFCA to resolve the mixture of three species present in a binding titration using a single color. We also show how TIFCA may be used to determine the proper binding model by comparing fits using single-site and two-site models.

TIFCA was applied to the same raw photon count data from the FFS experiments described in section 3.4.1. The factorial cumulants $\kappa_{[r]}$ of order $r$ are calculated at different sampling times $T$. These rebinned cumulants have been shown to obey the following relationship (Müller, 2004):

$$\kappa_{[r]} = \gamma_r \lambda^r \langle N \rangle B_r(T, \tau_D), \qquad [5.32]$$

where $\lambda$ is the molecular brightness, $\langle N \rangle$ is the average number of molecules and $\gamma_r$ is a PSF-dependent shape factor described previously (Müller, 2004; Thompson, 1991). Eq. (5.32) introduces the $r$-th binning function $B_r(T,\tau_D)$, which depends on the diffusion time $\tau_D$ and the sampling time $T$ (Wu & Müller, 2005). As the bin time increases, fluctuations are undersampled, which decreases the fluctuation amplitude. The binning function is derived from the higher-order autocorrelation function and includes the effect of undersampling. For a sample containing M species, the factorial cumulant of a mixture is given by the sum of the cumulants of each species:

$$\kappa_{[r]}^{M} = \sum_{j=1}^{M} \kappa_{[r]_j} = \gamma_r \sum_{j=1}^{M} \lambda_j^{r} \langle N_j \rangle B_r(T, \tau_{Dj}). \qquad [5.33]$$

This additive property makes cumulants very useful for resolving heterogeneity in the mixture.

*Experiment*: The same titration data presented in Section 3.4.1 were analyzed using TIFCA to determine the number of ligand molecules in each bound state within the mixture. The cumulants calculated for each sample in the titration experiment were globally fit using a three-species model representing free ligand, singly liganded antibody and doubly liganded antibody. The diffusion time and brightness of each species were linked. Only the number of molecules of each species was allowed to vary across the titration points. As an additional constraint, the brightness of the doubly liganded antibody was fixed to twice that of the singly liganded antibody. Furthermore, the diffusion times of these species (LA and L2A) were linked as they are predominantly determined by the size of antibody. Uncertainty was estimated by segmenting data as described in Section 3.4.1.

Data from this analysis are shown in Fig. 5.8 along with a global fit to the three expressions for the ligand populations given by Eqs. (5.24)–(5.26). By examining the amount of the free ligand as a function of antibody concentration (diamonds), the inactive species becomes evident. The

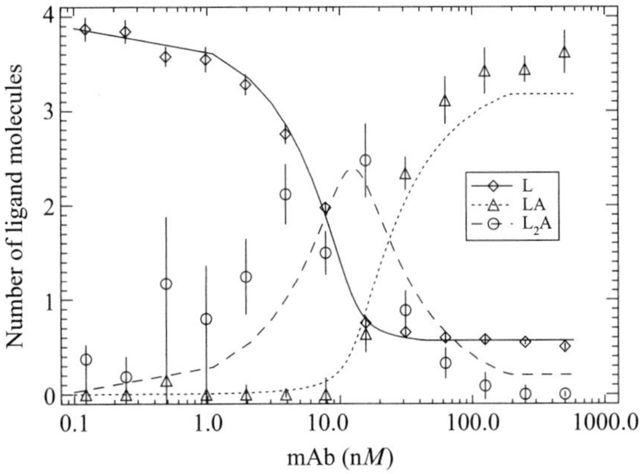


**Figure 5.8** Titration of 25 n*M* A-BNP (5–13) with mAb 106.3. The number of molecules of free ligand (diamond), ligand associated with single-bound antibody (triangle), and ligand associated with double-bound antibody (circles) versus antibody concentration is shown. The lines represent the result of a global fit (reduced $\chi^2 = 2.3$) of the titration, which recovered a dissociation constant of $0.7 \pm 0.2$ n*M*.

binding curve (solid line) of the free ligand has a nonzero offset ($N'_L > 0$) due to this inactive fraction. However, as expected, the increase and decrease of the doubly liganded antibody is seen (circles). The fit yielded a dissociation constant $K_D = 0.7 \pm 0.2$ n*M* and a fraction of free ligand of ~14% (reduced $\chi^2 = 2.3$). These values agree well with the results from moment analysis presented above. However, the uncertainty in the dissociation constant is reduced by twofold, perhaps due to globally fitting all three binding curves. TIFCA was also applied to a loading study in which the antibody concentration was fixed and labeled antigen was added to the solution (Skinner, Wu, Mueller, & Tetin, 2011). The result yielded a similar Kd indicating flexibility of the method.

### *3.4.3 Model selection using TIFCA*

We have found that TIFCA enables selection of the correct binding model. To establish the quality of fit to a particular model, we examined both a global reduced $\chi^2$ and a local reduced $\chi^2$ for each cumulant in the titration. It became evident that the second cumulant is particularly sensitive to the proper binding model.

As an example, we compared analysis of the data presented in Fig. 5.8 using both two-species and three-species models. Globally fitting TIFCA data to the two-species model (only one binding site) yielded a global reduced $\chi^2 = 4.0$, while the three-species model returned a reduced $\chi^2 = 1.9$. Comparing these values is not sufficient to rule out the two-species model since the difference in global reduced $\chi^2$ is only twofold. Upon inspection of the individual fits to each cumulant, however, we found that the second cumulant became quite sensitive to the proper model. Therefore, we defined a local reduced $\chi^2$ to yield quantitative verification of the goodness of fit to the second cumulant through all titration points. A detailed definition of local reduced $\chi^2$ has been presented elsewhere (Skinner et al., 2011).

Model selection based on the second cumulant can be observed graphically by examining the fit of the second cumulant function at a given titration point using the globally determined parameters. For example, the fit to the second cumulant for the antibody titration experiment (Fig. 5.8) at 16 n*M* antibody is shown in Fig. 5.9 as a solid line. The normalized residuals are shown in the bottom panel and aid in judging the goodness of fit. Figure 5.9A shows the fit to the second cumulant using a two-species model. In order to include all the normalized residuals,

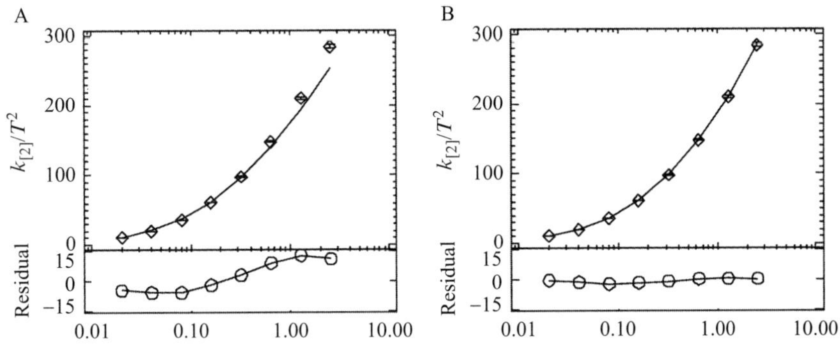


**Figure 5.9** Fits to the second cumulant for the antibody titration with 16 n*M* antibody. (A) The fit to the two-species model leads to large normalized residuals. (B) The fit to the three-species model, which includes two binding sites, results in a large reduction of the residuals.

the *y*-axis has been scaled to ±15. The local reduced $\chi^2 = 78$. Figure 5.9B displays the fit to the second cumulant using a three-species model, which includes two antibody-binding sites. The *y*-axis of the residuals has the same scale as in Fig. 5.9A to show the difference. The local reduced $\chi^2 = 2.4$, and the magnitude of the residuals has decreased significantly. By taking an average of all the local reduced $\chi^2$ values for the second cumulant, a definite model can be obtained. For the two-species model, the average local reduced $\chi^2 = 9.1$. For a three-species model, this value becomes 1.5. Thus, the local reduced $\chi^2$ of the second cumulant has ruled out the two-species, single-site model.

## 4. INSTRUMENTATION

All experiments described in this chapter were performed using an Alba dual-channel fluorescence correlation spectrometer (ISS, Champaign, IL) connected to an inverted Nikon Eclipse TE300 microscope (Nikon InsTech Co., Ltd., Kanagawa, Japan) equipped with a Nikon Plan Apo 60X/1.2 NA water immersion objective.

A mode-locked Tsunami Titanium-Sapphire laser (Spectra-Physics, Mountain View, CA) was used as a two-photon excitation light source. The Alba spectrometer contains two SPCM-AQR-15-Si avalanche photo diodes (Perkin Elmer, Inc. Fremont, CA) aligned perpendicularly. When performing DC-FCCS measurements, a dichroic mirror Q565lp and two band-pass filters HQ535/50 and HQ645/75 (all from Chroma Technology

Corp., Rockingham, VT) were placed in the light path before the detectors. These filters and dichroic mirror were moved out of the light path for single channel measurements. We used 384 microwell optical bottom plates in all measurements (Nalge Nunc International, Rochester, NY).

## 4.1. Instrument calibration for FCS measurements

Before any experiments can be performed, the instrument must be calibrated to allow quantification of the data. The calibration parameters sought depend on the particular FFS technique being performed.

For single color FCS, calibration of the instrument was performed using a 20 n*M* solution of Alexa488 (Life Technologies, Grand Island, NY) in 10 m*M* HEPES buffer and 0.15 *M* NaCl, pH 7.4, containing 3 m*M* EDTA and 0.005% surfactant P20. We have found that this buffer works well in preventing sample loss. It is commercially available from GE Healthcare under the name HBS-EP.

The laser was set at 780 nm, and 10 million data points were collected at 200 kHz frequency. The raw photon count data were processed with Vista (version 3.36) software (ISS) to calculate and fit the autocorrelation function to a squared 3DG PSF. The volume parameters $w$ and $z$ were calculated from the fit to Eq. (5.16) using a fixed diffusion coefficient $D = 340\ \mu\text{m}^2/\text{s}$. Typical volume parameters for $w$ and $z$ were 0.3 and 1.2 μm, respectively.

Due to uneven flatness of the glass bottom plate, the axial focal position of the objective relative to the inner surface of glass may shift as the stage translates from well to well. Therefore, it is important to adjust the $z$ position from one well to the next. We have found it is best to lower the objective down from the sample until very few photon counts are observed, indicating focus on the glass. Then, the objective is translated up ~100 μm into the sample using the markings on the microscope focus dial as an approximate position. This method yields very stable intensity values when moving across wells on the sample plate.

## 4.2. Instrument calibration for DC-FCCS measurements

Two-photon DC-FCCS requires careful selection of the excitation wavelength and dichroic filter sets to permit similar brightness for the chosen fluorophores in each channel with minimum spectral cross talk. A fluorescent dye which emits in both channels can be used for alignment. The auto- and cross-correlation curves should completely overlap when proper alignment is achieved. We use Alexa488 and TexasRed for labeling antibodies and an excitation wavelength of 810 nm to simultaneously excite both fluorophores. Using these dyes, spectral cross-talk was less than 1% and 5% in the green and red channels, respectively.

## 4.3. Instrument calibration for TIFCA measurements

Moment analysis and TIFCA are related techniques, and therefore, calibration is also similar. It involves determining shape factors and relating molar concentration to the number of molecules. For TIFCA, higher-order shape factors dependent on the PSF must be determined. It is known that the theoretical PSF used for modeling FFS data only approximately describes the experimental PSF (Hess & Webb, 2002; Huang, Perroud, & Zare, 2004; Wu & Müller, 2005). Deviations become more pronounced as higher-order moments are calculated. To calibrate, a solution of Alexa488 at a concentration of 16 n$M$ was measured using a laser power of ~3.5 mW. Raw photon count data were acquired at 50 kHz for 1 min and exported for analysis using programs written in IDL version 6.4 (ITT Corp., Boulder CO). The $G(0)$ value was determined by calculation of moments. A factor of $\gamma=0.3535$ for a 3DG is used to determine the number of molecules $N=2.8$. The calibration parameter $\phi$ which relates concentration and number of molecules in the observation volume is given by $\phi=2.8/16$ n$M$=0.18 molecules/n$M$.

To determine the higher-order shape factors necessary for TIFCA, the data were binned serially by a factor of two corresponding to sampling times ranging from 20 μs to 2.56 ms. The first four factorial cumulants along with their variances were calculated for the eight bin times $T$. The four cumulants at each sampling time were fit to determine the brightness, diffusion time, and the number of molecules (Wu & Müller, 2005). This calibration step provides experimentally determined values of the PSF-dependent shape factors $\gamma_3$ and $\gamma_4$. To do so, $\gamma_1=1$ and $\gamma_2=0.3535$ are fixed based on a 3DG PSF, while $\gamma_3$ and $\gamma_4$ vary.

The first four cumulants calculated for each sampling time are shown in the four panels in Fig. 5.10 along with a global fit for all sampling times to Eq. (5.32). Each panel displays a single normalized cumulant $\kappa_r/T^r$ plotted as a function of sampling time $T$ on a log scale. The fitted brightness of Alexa488 is 44 kHz, the diffusion time is 0.047 ms, and $N=2.8$ molecules. The PSF calibration parameters returned by the fit are $\gamma_3=0.25$ and $\gamma_4=0.22$.

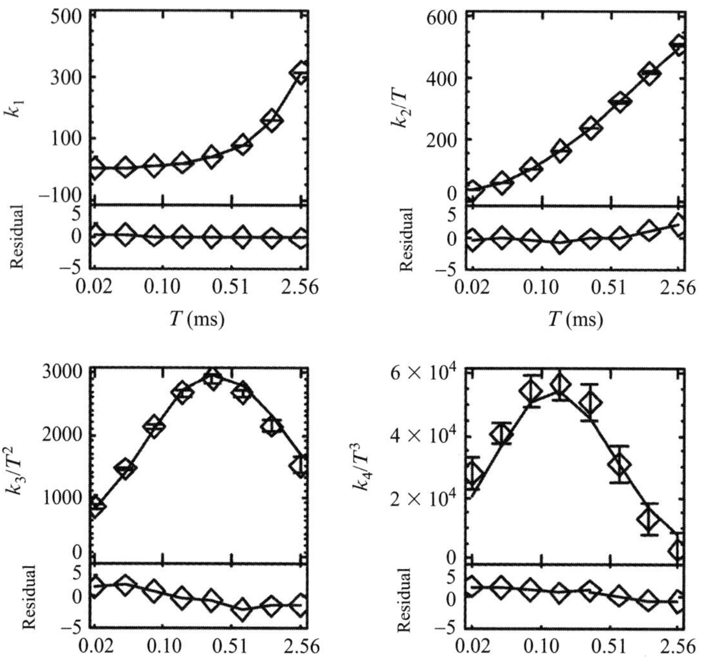


**Figure 5.10** Instrument calibration. The first four cumulants and the resulting fit from a measured solution of 16 n*M* Alexa488. The brightness determined by the fit is 44 kHz, the number of molecules is 2.8, and the diffusion time is 0.047 ms (reduced $\chi^2 = 1.0$). The calibrated beam shape parameters are $\gamma_3 = 0.25$ and $\gamma_4 = 0.22$ with $\gamma_2 = 0.3535$ fixed during the fit.

## ACKNOWLEDGMENTS

We are thankful to many of our Abbott colleagues for providing well characterized reagents and valuable discussions. We are especially thankful to Sylvia Saldana for her support in the laboratory.

## REFERENCES

Adair, G. S. (1925). The hemoglobin system. VI. The oxygen dissociation curve of hemoglobin. *The Journal of Biological Chemistry*, *63*, 529–545.

Chen, Y., & Müller, J. D. (2007). Determining the stoichiometry of protein heterocomplexes in living cells with fluorescence fluctuation spectroscopy. *Proceedings of the National Academy of Sciences of the United States of America*, *104*, 3147–3152.

Chen, Y., Müller, J. D., So, P. T., & Gratton, E. (1999). The photon counting histogram in fluorescence fluctuation spectroscopy. *Biophysical Journal*, *77*, 553–567.

Chen, Y., Müller, J. D., Tetin, S. Y., Tyner, J. D., & Gratton, E. (2000). Probing ligand protein binding equilibria with fluorescence fluctuation spectroscopy. *Biophysical Journal*, *79*, 1074–1084.

Cunningham, B. C., & Wells, J. A. (1989). High-resolution epitope mapping of hGH-receptor interactions by alanine-scanning mutagenesis. *Science*, *244*, 1081–1085.

Edelman, G. M., & Gally, J. A. (1968). Antibody structure, diversity, and specificity. *Brookhaven Symposia in Biology*, *21*, 328–344.

Elson, E. L., & Magde, D. (1974). Fluorescence correlation spectroscopy: I. Conceptual basis and theory. *Biopolymers*, *13*, 29–61.

Foo, Y. H., Naredi-Rainer, N., Lamb, D. C., Ahmed, S., & Wohland, T. (2012). Factors affecting the quantification of biomolecular interactions by fluorescence cross-correlation spectroscopy. *Biophysical Journal*, *102*, 1174–1183.

Gilliland, G. L., Luo, J., Vafa, O., & Almagro, J. C. (2012). Leveraging SBDD in protein therapeutic development: Antibody engineering. *Methods in Molecular Biology*, *841*, 321–349.

Grenier, F. C., Ali, S., Syed, H., Workman, R., Martens, F., Liao, M., et al. (2010). Evaluation of the ARCHITECT urine NGAL assay: Assay performance, specimen handling requirements and biological variability. *Clinical Biochemistry*, *43*, 615–620.

Hazlett, T. L., Ruan, Q., & Tetin, S. Y. (2005). Application of fluorescence correlation spectroscopy to hapten-antibody binding. *Methods in Molecular Biology*, *305*, 415–438.

Hess, S. T., & Webb, W. W. (2002). Focal volume optics and experimental artifacts in confocal fluorescence correlation spectroscopy. *Biophysical Journal*, *83*, 2300–2317.

Holten-Andersen, M. N., Christensen, I. J., Nielsen, H. J., Stephens, R. W., Jensen, V., Nielsen, O. H., et al. (2002). Total levels of tissue inhibitor of metalloproteinases 1 in plasma yield high diagnostic sensitivity and specificity in patients with colon cancer. *Clinical Cancer Research*, *8*, 156–164.

Huang, B., Perroud, T. D., & Zare, R. N. (2004). Photon counting histogram: One-photon excitation. *ChemPhysChem*, *5*, 1523–1531.

Karush, F., & Hornick, C. L. (1973). Multivalence and affinity of antibody. *International Archives of Allergy and Applied Immunology*, *45*, 130–132.

Kask, P., Palo, K., Ullmann, D., & Gall, K. (1999). Fluorescence-intensity distribution analysis and its application in biomolecular detection technology. *Proceedings of the National Academy of Sciences of the United States of America*, *96*, 13756–13761.

Kim, S. A., Heinze, K. G., Bacia, K., Waxham, M. N., & Schwille, P. (2005). Two-photon cross-correlation analysis of intracellular reactions with variable stoichiometry. *Biophysical Journal*, *88*, 4319–4336.

Klotz, I. M. (1985). Ligand–receptor interactions: Facts and fantasies. *Quarterly Reviews of Biophysics*, *18*, 227–259.

Longenecker, K. L., Ruan, Q., Fry, E. H., Saldana, S. C., Brophy, S. E., Richardson, P. L., et al. (2009). Crystal structure and thermodynamic analysis of diagnostic mAb 106.3 complexed with BNP 5-13 (C10A). *Proteins*, *76*, 536–547.

Magde, D., Elson, E. L., & Webb, W. W. (1974). Fluorescence correlation spectroscopy. II. An experimental realization. *Biopolymers*, *13*, 29–61.

Meseth, U., Wohland, T., Rigler, R., & Vogel, H. (1999). Resolution of fluorescence correlation measurements. *Biophysical Journal*, *76*, 1619–1631.

Moller Sorensen, N., Dowell, B. L., Stewart, K. D., Jensen, V., Larsen, L., Lademann, U., et al. (2005). Establishment and characterization of 7 new monoclonal antibodies to tissue inhibitor of metalloproteinases-1. *Tumour Biology*, *26*, 71–80.

Müller, J. D. (2004). Cumulant analysis in fluorescence fluctuation spectroscopy. *Biophysical Journal*, *86*, 3981–3992.

Nakao, K., Ogawa, Y., Suga, S., & Imura, H. (1992). Molecular biology and biochemistry of the natriuretic peptide system. I: Natriuretic peptides. *Journal of Hypertension*, *10*, 907–912.

Olejniczak, E. T., Ruan, Q., Ziemann, R. N., Birkenmeyer, L. G., Saldana, S. C., & Tetin, S. Y. (2010). Rapid determination of antigenic epitopes in human NGAL using NMR. *Biopolymers*, *93*, 657–667.

Palmer, A. G., 3rd, & Thompson, N. L. (1989). High-order fluorescence fluctuation analysis of model protein clusters. *Proceedings of the National Academy of Sciences of the United States of America*, *86*, 6148–6152.

Porter, R. R. (1967). The structure of immunoglobulins. *Essays in Biochemistry, 3,* 1–24.

Qian, H., & Elson, E. L. (1990a). Distribution of molecular aggregation by analysis of fluctuation moments. *Proceedings of the National Academy of Sciences of the United States of America, 87,* 5479–5483.

Qian, H., & Elson, E. L. (1990b). On the analysis of high order moments of fluorescence fluctuations. *Biophysical Journal, 57,* 375–380.

Ruan, Q., & Tetin, S. Y. (2008). Applications of dual-color fluorescence cross-correlation spectroscopy in antibody binding studies. *Analytical Biochemistry, 374,* 182–195.

Sagnella, G. A. (1998). Measurement and significance of circulating natriuretic peptides in cardiovascular disease. *Clinical Science (London, England), 95,* 519–529.

Schwille, P., Meyer-Almes, F. J., & Rigler, R. (1997). Dual-color fluorescence cross-correlation spectroscopy for multicomponent diffusional analysis in solution. *Biophysical Journal, 72,* 1878–1886.

Skinner, J. P., Wu, B., Mueller, J. D., & Tetin, S. Y. (2011). Determining antibody stoichiometry using time-integrated fluorescence cumulant analysis. *The Journal of Physical Chemistry. B, 115,* 1131–1138.

Tetin, S. Y., & Hazlett, T. L. (2000). Optical spectroscopy in studies of antibody-hapten interactions. *Methods, 20,* 341–361.

Tetin, S. Y., Ruan, Q., Saldana, S. C., Pope, M. R., Chen, Y., Wu, H., et al. (2006). Interactions of two monoclonal antibodies with BNP: High resolution epitope mapping using fluorescence correlation spectroscopy. *Biochemistry, 45,* 14155–14165.

Tetin, S. Y., & Stroupe, S. D. (2004). Antibodies in diagnostic applications. *Current Pharmaceutical Biotechnology, 5,* 9–16.

Thompson, N. L. (1991). *Fluorescence correlation spectroscopy.* New York: Plenum.

Tonegawa, S. (1979). Somatic recombination and mosaic structure of immunoglobulin genes. *Harvey Lectures, 75,* 61–83.

Visse, R., & Nagase, H. (2003). Matrix metalloproteinases and tissue inhibitors of metalloproteinases: Structure, function, and biochemistry. *Circulation Research, 92,* 827–839.

Weber, G. (1965). The binding of small molecules to proteins. In B. Pullman (Ed.), *Molecular Biophysics: Proceedings of an international summer school held in Squaw Valley, California, August 17–28, 1964* (pp. 369–396). New York, Squaw Valley, California: Academic Press.

Weber, G. (1975). Energetics of ligand binding to proteins. *Advances in Protein Chemistry, 29,* 1–83.

Weber, G. (1992). *Protein interactions.* New York, London: Chapman and Hall.

Weidemann, T., Wachsmuth, M., Tewes, M., Rippe, K., & Langowski, J. (2002). Analysis of ligand binding by two-color fluorescence cross-correlation spectroscopy. *Single Molecules, 3,* 49–61.

Wu, B., & Müller, J. D. (2005). Time-integrated fluorescence cumulant analysis in fluorescence fluctuation spectroscopy. *Biophysical Journal, 89,* 2721–2735.

CHAPTER SIX

# Fluorescence Fluctuation Approaches to the Study of Adhesion and Signaling

**Alexia I. Bachir[1], Kristopher E. Kubow, Alan R. Horwitz**
Department of Cell Biology, University of Virginia, Charlottesville, Virginia, USA
[1]Corresponding author: e-mail address: ab8su@virginia.edu

## Contents



## Abstract

Cell–matrix adhesions are large, multimolecular complexes through which cells sense and respond to their environment. They also mediate migration by serving as traction points and signaling centers and allow the cell to modify the surroucnding tissue. Due to their fundamental role in cell behavior, adhesions are germane to nearly all major human health pathologies. However, adhesions are extremely complex and dynamic structures that include over 100 known interacting proteins and operate over multiple space (nm–μm) and time (ms–min) regimes. Fluorescence fluctuation techniques are well suited for studying adhesions. These methods are sensitive over a large

*Methods in Enzymology*, Volume 519
ISSN 0076-6879
http://dx.doi.org/10.1016/B978-0-12-405539-1.00006-3

spatiotemporal range and provide a wealth of information including molecular transport dynamics, interactions, and stoichiometry from a single time series. Earlier chapters in this volume have provided the theoretical background, instrumentation, and analysis algorithms for these techniques. In this chapter, we discuss their implementation in living cells to study adhesions in migrating cells. Although each technique and application has its own unique instrumentation and analysis requirements, we provide general guidelines for sample preparation, selection of imaging instrumentation, and optimization of data acquisition and analysis parameters. Finally, we review several recent studies that implement these techniques in the study of adhesions.

# 1. INTRODUCTION

Cells sense and interact with other cells and the extracellular matrix (ECM) through multimolecular assemblies called adhesions. The bidirectional transduction of stimuli through adhesions allows cells, for example, to sense the stiffness of their microenvironment, migrate, and remodel the ECM by proteolysis, contraction, and fibrillogenesis (Geiger, Spatz, & Bershadsky, 2009). Due to their pivotal role in migration and other cell behaviors, adhesions are a focal point in the study of such diverse topics as inflammation, wound healing, tumor progression, embryonic morphogenesis, and tissue engineering and regeneration (Ridley et al., 2003).

Adhesions serve in two general capacities: as a continuous physical connection between the substratum and the actin cytoskeleton, and as signaling centers that initiate and coordinate a complex network of interrelated signal transduction pathways. ECM components bind to integrins (cell adhesion receptors), which span the cell membrane and connect to actin through a system of linked proteins (Vicente-Manzanares, Choi, & Horwitz, 2009). Through this connection, forces from actomyosin contraction and membrane resistance to actin polymerization are transferred to the substratum (Brown et al., 2006). Concurrently, integrin clustering, ligand binding, and the forces exerted on adhesions produce biochemical signals that, among other things, feedback to alter actin polymerization, myosin contractility, and adhesion (Geiger et al., 2009). Through this loop, physical and chemical stimuli transduced through adhesions modify the cytoskeleton and force development, which in turn feedback to affect adhesion formation and the composition, organization, and function of the cell and its extracellular microenvironment.

While this conceptual framework is well established, the mechanisms through which the physical and biochemical signals are bidirectionally transduced are not well understood. This is due in part to the molecular complexity of adhesions, which can contain over 160 different molecules that participate in over 700 putative interactions, most of which have been primarily identified and characterized by *in vitro* biochemical assays (Zaidel-Bar, Itzkovitz, Ma'ayan, Iyengar, & Geiger, 2007). The components that comprise adhesions serve several general functions; for example, as scaffolds that organize signaling complexes, force-bearing connections, enzymatic de-/activators (e.g., kinases, phosphatases, and proteases), and combinations of these functions. Although recent studies have begun to provide static snapshots of adhesion structure *in situ* (Kanchanawong et al., 2010), we have little knowledge of and few tools to investigate how the functions of adhesion components are organized and function in space and time in living, dynamic cells.

A major challenge to studying adhesions is the large spatiotemporal range that must be spanned. In a "typical" migrating cell, small, diffraction-limited "nascent adhesions" assemble and disassemble within a thin (~2–3 μm) band at the leading edge at rates as fast as tens of seconds (Choi et al., 2008). Under the influence of nonmuscle myosin II, nascent adhesions grow and elongate as they mature initially into focal complexes (~1 μm diameter) and subsequently into focal adhesions and fibrillar adhesions, which can span many micrometers and remain stable for minutes to hours (Geiger et al., 2009). While some adhesions can be static over long time periods, individual adhesion components can exchange on and off the adhesion over a broad timescale: from subsecond to minutes, depending on the adhesion (Goetz, 2009; Lele et al., 2006; Wolfenson, Bershadsky, Henis, & Geiger, 2011). Presumably, these dynamics are intrinsic to the organization and function of the adhesion. Therefore, it is not enough to simply study a single adhesion in a single place and time, since they are in a continuous process of structural and functional change and thus heterogeneous throughout the cell. Understanding the organization and functions of these different adhesions, including their formation and interconversion, requires monitoring potentially short-lived molecular associations in adhesions that are themselves moving and changing size at slower rates and over a larger area. Due to these challenges, we still lack such basic information about adhesions as, for example, the numbers and stoichiometry of components in different adhesions, how adhesions assemble (e.g., whether components exchange as complexes or individual molecules), and when and where signaling complexes form and become active. Ideally, one would like to generate association maps that relate molecular events such as molecular number and clustering, the formation and properties of the integrin–actin linkage, and the specific signals generated by adhesions within the context of large-scale cell processes and ultimately cell movement, itself. However, current mainstream techniques are not well suited for such a multi-scale challenge.

Traditional techniques used to study the interactions and dynamics of adhesion components suffer from limited spatial and/or temporal resolution. The interactions among adhesion components have been characterized primarily by co-immunoprecipitation and fluorescence colocalization. The former provides high temporal but poor spatial resolution and requires an *in vitro* setting; the latter provides *in situ* localization but low spatiotemporal resolution. Fluorescence recovery after photobleaching (FRAP) provides *in situ* information about the exchange rates (apparent affinity) of adhesion components (Wehrle-Haller, 2007; Wolfenson et al., 2011) but is limited to relatively large areas and long timescales (slow-moving objects and slow exchange rates). Förster resonance energy transfer (FRET) is used to study molecular associations in adhesions (e.g., Ballestrem et al., 2006); but it suffers from a low dynamic range: FRET is sensitive to changes in distance between 0.5 and 2 times the Förster radius (i.e., between $\sim$2 and 10 nm) and the magnitude of the differences are small. In adhesions, molecules can be associated indirectly in large complexes (distances $>$10 nm), and therefore may be missed by FRET. Conversely, molecules that are not in common complexes can nevertheless be in juxtaposition and show FRET. Moreover, FRET experiments suffer from artifacts (e.g., multiple donor and acceptor interactions) and thus require extensive controls for every FRET pair studied. Finally, none of these methods can determine the number of molecules of a certain component and stoichiometries within an adhesion.

The need for techniques that can provide data on molecular dynamics and associations in adhesions of living cells has driven the development of a toolbox of fluorescence fluctuation techniques. These methods can derive multiple pieces of information from a single data set and at high spatiotemporal resolution. From a single time series, for example, fluctuation analyses can provide molecular numbers, aggregation states, exchange rates, diffusion and flow rates, and interactions with their stochiometries. These methods operate over multiple spatiotemporal regimes with high resolution. In this chapter, we discuss how fluorescence fluctuation techniques described elsewhere in this volume are being applied to study adhesions in living cells. Although we limit our discussion to cell–matrix adhesions, the techniques, methods, and considerations outlined in this chapter are generally applicable to other kinds of adhesions as well as the study of other biological processes.

## 2. A FLUORESCENCE FLUCTUATION TOOLBOX

The machines (e.g., cell protrusion, adhesion, and retraction) and regulatory systems that drive cell migration are dynamic, spatially restricted, and integrated in the migrating cell. They appear to operate using a plethora of transient, localized molecular interactions (Parsons, Horwitz, & Schwartz, 2010; Vicente-Manzanares & Horwitz, 2011). Understanding the molecular machines and interactions underlying cell migration would benefit greatly from high-resolution mapping of the dynamics and associations of adhesion molecules in space and time. In response to this need, a "toolbox" of complementary fluorescence fluctuation techniques has been developed. This toolbox promises to enable the in-depth study of the associations and dynamics that contribute to nearly every aspect of cell migration.

In this section, we introduce the underlying principles of fluorescence fluctuation microscopy beginning with its original development as a single-point temporal correlation approach for solution measurements (fluorescence correlation spectroscopy, FCS). We then describe the recent advances that have extended FCS to migrating cells (image correlation spectroscopy, ICS). The ICS approaches use spatiotemporal correlations from an image time series instead of a single point, thereby extending the scope of the original FCS from a single point to high-resolution cellular maps. Finally, we introduce a variant of intensity variance analysis (number and brightness, N&B) that provides information on molecular aggregation and interactions at pixel resolution.

### 2.1. Fluorescence correlation spectroscopy

FCS was originally developed to measure diffusion coefficients and chemical rate constants of biomolecules in solution (Magde, Elson, & Webb, 1974). With the development of genetically encoded fluorescence labeling (i.e., fluorescent proteins, FPs) (Miyawaki, 2011; Shaner, Steinbach, & Tsien, 2005), better fluorescence microscopes (Axelrod, 2008), and sensitive fluorescence detectors, this technique and its variants have become established biophysical approaches for studying biomolecular aggregation, dynamics, and interactions with a broad range of applications both *in vitro* and in cells (Haustein & Schwille, 2007).

FCS is based on measuring fluorescence intensity fluctuations that arise from the transport or movement of fluorophores in and out of an optically defined focal volume ($\sim$1 fL), that is, the point spread function (PSF) which characterizes the focused laser beam. These fluctuations arise from a number of processes including diffusion, directed movement (flow), and binding and unbinding (exchange) phenomena (Fig. 6.1). In addition to these processes, fluorophore blinking and bleaching also produce intensity fluctuations. Intensity fluctuations are usually analyzed using the autocorrelation function (ACF), the decay rate of which reflects the rates of these fluctuations. Assuming a quasi-equilibrium, the ACF can be modeled and fit to estimate diffusion coefficients, flow speeds, and rate constants (Elson & Magde, 1974; Magde et al., 1974). Fluorescence cross-correlation analysis is also used to quantify molecular interactions through the presence of correlated fluctuations in the intensity of two molecules labeled with different color fluorescent probes (Bacia, Kim, & Schwille, 2006; Berland, 2004).

The amplitude of the ACF can be used to determine the average number of fluctuating units (i.e., individual molecules or complexes) within the beam

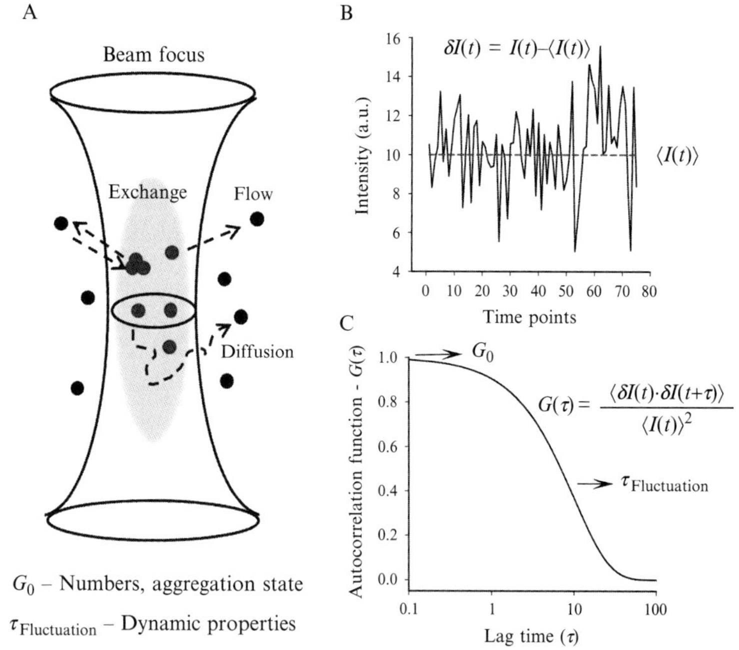


**Figure 6.1** Principles underlying fluorescence correlation spectroscopy (FCS). (A) Examples of dynamic processes (diffusion, flow, and exchange) that give rise to fluorescence intensity fluctuations in a focal volume. (B) Fluorescence intensity fluctuations over time. (C) Autocorrelation analysis of the intensity fluctuations quantifies the transport processes underlying the fluorescence fluctuations and the particle number densities and aggregation states. (See Color Insert.)

volume. In most cases, the number of particles follows a Poisson distribution, where the root mean square fluctuation (amplitude of ACF) is inversely proportional to the number of fluctuating units (Fig. 6.1). When the fluctuating units are homogeneous (e.g., all monomers or all dimers), the number of fluctuating units within the observation volume can be accurately determined from the ACF amplitude. When the units are heterogeneous, the ACF gives the average aggregate size. In this case, other fluctuation analysis methods, such as fluorescence intensity distribution analysis (FIDA) or the photon-counting histogram (PCH), can be used to determine the number of molecules in the different size complexes (monomers, dimers, etc.) (Chen, Müller, So, & Gratton, 1999; Chen, Wei, & Müller, 2003; Kask, Palo, Ullmann, & Gall, 1999).

The ability to record fluorescence counts from a single, defined focal volume for a long period of time (1–10 s) is an advantage of FCS. This enables the study of a broad range of processes, from fast photophysical dynamics such as fluorophore blinking (1–10 μs) to the slower diffusion of molecules in solution or the cytoplasm. The even-slower transport properties of cell-membrane-associated molecules such as lipids and integrins can also be monitored by FCS; however, due to membrane crowding effects and clustering at cellular structures such as cell adhesion sites, reduced characteristic fluctuation times and even immobile populations are often reported. Therefore, the dynamics of adhesions are at or beyond the sensitivity FCS (see Table 6.1). In principle, one can image for a longer period of time to capture these slower

**Table 6.1** Characteristic timescales for biological processes and the fluorescence fluctuation techniques used to study them

| Dynamic process | Characteristic timescale | Fluctuation correlation technique |
|---|---|---|
| EGFP fluorescence blinking | 0.1 μs–0.1 ms | FCS |
| EGFP diffusion in solution | ~0.05–0.1 ms | FCS-sFCS-RICS |
| EGFP diffusion in cytoplasm | 0.5–1 ms | FCS-RICS-sFCS |
| Lipid diffusion in cell membrane | 0.01–1 s | FCS-sFCS-TICS-RICS |
| Protein receptor diffusion in cell membrane | 1–10 s | sFCS-TICS |
| Retrograde flow of adhesion and cytoskeletal molecules | ~1–100 s | STICS |
| Adhesion molecules exchange rates | 500 ms to ~min | TICS-RICS |

dynamics; however, photobleaching, cell toxicity due to continuous laser illumination, and macroscopic membrane fluctuations over time limit its use.

Scanning-FCS-a 2D extension of classical FCS-overcomes some of these limitations (Ruan, Cheng, Levi, Gratton, & Mantulin, 2004). In this approach, multiple points are monitored by moving the observation volume to different points (e.g., a circle) within the sample. Scanning-FCS has been used to study slower membrane protein dynamics (Petrasek et al., 2008; Ries, Yu, Burkhardt, Brand, & Schwille, 2009); however, its applicability to study cell-migration-related processes is limited due to the restricted spatial observation area.

## 2.2. Image correlation spectroscopy techniques

ICS is an extension of FCS; it was originally developed to measure molecular number densities and aggregation states from individual images (Petersen, Hoddelius, Wiseman, Seger, & Magnusson, 1993). It now consists of a family of analysis methods that combines spatial and temporal correlations of intensity fluctuations in an image time series and, like FCS, estimates molecular numbers, aggregation states, and dynamics. However, it is capable of capturing a wide range of dynamics that includes the slow transport properties of membrane receptors and adhesion components. ICS can also reveal protein interactions by measuring cross-correlations between intensity fluctuations from pairs of molecules labeled with different color fluorescent markers (Fessart et al., 2007; Wiseman & Petersen, 1999). A major advantage of ICS is that multiple analysis methods can be applied to the same image time series to extract complementary pieces of information (Fig. 6.2). Here, we briefly introduce ICS methods and outline the information each provides. For a more detailed overview of the background theory and implementation of the ICS techniques, refer to other chapters in this volume.

### *2.2.1 Image correlation spectroscopy*

ICS measures spatial correlations within a single confocal image to determine *number densities* and *molecular aggregation states*. The spatial ACF (s-ACF) is fit to a 2D Gaussian function, and the recovered fit amplitude is used to estimate the number density and aggregation state (Petersen et al., 1993). Changes in these parameters can be followed over time by applying the analysis to individual images in an image time series (Wiseman & Petersen, 1999). This approach is best suited for slow moving or stationary proteins since they do not move over the image scan time ($\sim$seconds).

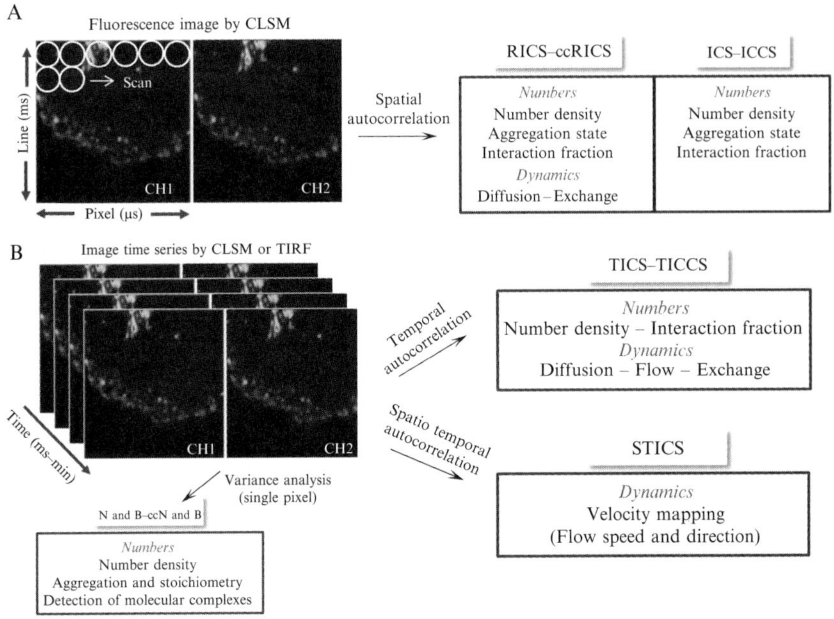


**Figure 6.2** Schematic illustration of the fluorescence fluctuation toolbox. (A) CLSM dual-color fluorescence images of a cell expressing two proteins, each tagged with a different color fluorescent probe visualized in channels 1 and 2 (CH1 and CH2). In a confocal setup, fluorescence excitation is achieved as a laser beam is scanned across a sample with microsecond pixel dwell time and millisecond line-scan time. Image correlation spectroscopy (ICS) and raster-image correlation spectroscopy (RICS) are implemented on CLSM images (or selected regions in an image) to measure molecular numbers and fast dynamics using spatial autocorrelation analysis. (B) Image time series acquired using either CLSM or TIRFM are used for temporal image correlation spectroscopy (TICS) and spatiotemporal image correlation analysis (STICS) analysis to study transport dynamics (diffusion, flow) and exchange kinetics. Number density information can also be recovered from TICS analysis. Single pixel variance analysis of fluorescence fluctuation in the time series provides high-resolution mapping of number densities and stoichiometry of molecular aggregates (N&B). All of these techniques can be applied to two fluorescence detection channels to detect and quantify molecular interactions in complexes. (See Color Insert.)

### *2.2.2 Temporal-image correlation spectroscopy*

Temporal-image correlation spectroscopy (TICS) is used to measure *transport coefficients* and *exchange rates* of biomolecules from image time sequences acquired using confocal laser scanning microscopy (CLSM) or camera-based microscopes (Digman, Brown, Horwitz, Mantulin, & Gratton, 2008; Wiseman, Squier, Ellisman, & Wilson, 2000). A temporal ACF (t-ACF) is calculated from region-to-region temporal intensity fluctuations and then fit to appropriate models for diffusion, directed motion, or exchange kinetics. For slow dynamics, for example, membrane receptors, TICS generally requires spatial averaging of temporal intensity fluctuations to improve sensitivity. Similar to ICS, the amplitude of the temporal ACF can also be used to estimate number densities.

### 2.2.3 Raster-image correlation spectroscopy

Raster-image correlation spectroscopy (RICS) is a complement to TICS and is designed to measure the *rapid diffusion* and *exchange* characteristic of molecules in the cytoplasm (Digman et al., 2005; Digman, Sengupta, et al., 2005). It is based on the temporal information intrinsic to an image acquired using CLSM. The laser scans across in a single line (in *x*-direction) with an ~2–20 μs pixel-to-pixel scan speed (dwell time) and then moves down (*y*-direction) to scan the next line with ~millisecond line-to-line scan speed. Fast diffusion and exchange occur on these timescales and will manifest in the spatial autocorrelation function (s-ACF), from which diffusion coefficients and exchange rates are computed. Number densities can also be recovered from the amplitude of the s-ACF.

### 2.2.4 Spatiotemporal image correlation spectroscopy

Spatiotemporal image correlation spectroscopy (STICS) measures the *velocity of directed movement* from image time series acquired using CLSM or TIRF. Unlike TICS, which averages single-point temporal autocorrelations of intensity fluctuations over a region in a time series, STICS uses combined spatiotemporal autocorrelation to detect directed motion over time (Hebert, Costantino, & Wiseman, 2005). The directed motion manifests as a translocation in the peak of the sp-ACF over time. Temporal tracking of the peak will provide an estimate of the magnitude and direction of flow.

Spatial and temporal mapping of adhesion protein dynamics and interactions can be achieved by applying the image correlation toolbox to different temporal segments of an image time series as well as subregions at different locations within the cell. Small defined regions (e.g., 10 × 10 pixels) and time segments (~10–20 frames) provide a more detailed picture of the spatiotemporal coordination of cell migration processes. The size of the spatiotemporal "window" is, however, limited by the need to sample a sufficient number of fluctuations to yield adequate statistics. Computer simulations have tested the limits of these techniques under various experimental acquisition settings (Brown et al., 2008; Costantino, Comeau, Kolin, & Wiseman, 2005;

Hebert et al., 2005; Kolin, Costantino, & Wiseman, 2006). In this context, the ICS techniques do not have sufficient spatiotemporal resolution to study molecular interactions and dynamics in diffraction-limited nascent adhesions. The need for greater spatiotemporal resolution drove the development of the N&B variance analysis.

## 2.3. Number and brightness variance analysis

The N&B analysis is a fluctuation variance method that provides pixel-resolution mapping of number densities and aggregation states of adhesion proteins without sacrificing temporal resolution (i.e., long imaging times) (Digman, Dalal, Horwitz, & Gratton, 2008). N&B analyzes the variance in the intensity fluctuations. In contrast, the correlation approach requires many images to capture the entire spectrum of kinetic information and thereby fit the ACF and is therefore not practical for the rapidly turning over adhesions in migrating cells. In the N&B approach, two parameters are computed independently: *N*, which is the average number of fluctuating units (monomers or aggregates) and *B*, the brightness, which is an estimate of the aggregate size. For example, if a molecular complex consists of trimers, the measured *B* parameter will be three times the value measured for a monomer. Therefore, using appropriate monomer calibration, one can estimate the aggregation size (see Section 3). Furthermore, associations of different adhesion molecules within complexes can be detected by measuring a cross-brightness parameter, Bcc (Choi, Zareno, Digman, Gratton, & Horwitz, 2011; Digman, Wiseman, Choi, Horwitz, & Gratton, 2009). Dual-color calibration to monomer species tagged with two different fluorescent probes provides the aggregation state of the individual proteins within the complex (Choi et al., 2011).

# 3. EXPERIMENTAL IMPLEMENTATION

## 3.1. Fluorescent labeling

Fluorescence fluctuation techniques in living cells typically rely on ectopically expressed fluorescent proteins (FP) linked to the protein of interest (e.g., mGFP–paxillin). There are numerous resources and guides for selecting FPs for specific applications (Chudakov, Matz, Lukyanov, & Lukyanov, 2010; Shaner et al., 2005). Tables of relevant properties for most common FPs can be found online at the Molecular Probes (Invitrogen, http://www.invitrogen.com), Nikon microscopy (http://www.microscopyu.com/), Zeiss microscopy (http://zeiss-campus.magnet.fsu.edu/), and

Olympus microscopy (http://www.olympusmicro.com/) Web sites. Here, we present some considerations for choosing a particular FP (or combination of FPs) for fluorescence fluctuation analyses of adhesions.

### 3.1.1 Spectral properties

Select FPs that are optimally excited by the laser lines available in your system (e.g., mGFP with the 488 nm line) to maximize the signal-to-noise ratio. For cross-correlation analysis in which two different protein types are labeled with different color FPs ("dual-color" experiments), choose FPs that minimize spectral bleedthrough and the potential for FRET. For example, GFP and mKusabira-Orange have large areas of spectral overlap that may contribute to FRET and/or bleedthrough (Fig. 6.3A). Alternatively, we often use mGFP and mCherry for dual-color experiments because they are spectrally well separated; however, this combination requires a 568-nm laser as mCherry is only weakly excited by the 543-nm HeNe laser usually included on commercial microscope setups (Fig. 6.3B).

### 3.1.2 Brightness

Fluorescence brightness depends on two intrinsic fluorophore parameters: the extinction coefficient (efficiency of light absorption) and quantum yield (efficiency of light emission). For optimal signal to noise, use FPs with high brightness values.

### 3.1.3 Photostability

Repeated excitation of FPs can generate photo-oxidation products (free radicals) that can injure living cells (phototoxicity) and ultimately result in permanent loss of fluorescence (photobleaching). Consequently, photostable FPs should be selected. For example, when selecting a red FP, use mCherry instead of DsRed, as it is brighter and more photostable.

### 3.1.4 Oligomerization

While many FPs (e.g., DsRed) readily oligomerize, most FPs are available in monomeric form or can be made so by mutagenesis (Shaner et al., 2004). Monomeric FPs are required, for example, when quantifying the number density and aggregation state of molecular complexes using fluorescence fluctuation techniques. Transport studies are less sensitive to this, provided the oligomerization of the FP does not impair the function of the protein and thereby bias the recovered data.

In general, most fluorescently tagged proteins are fully functional; however, their biological activity needs to be tested in the context of the process

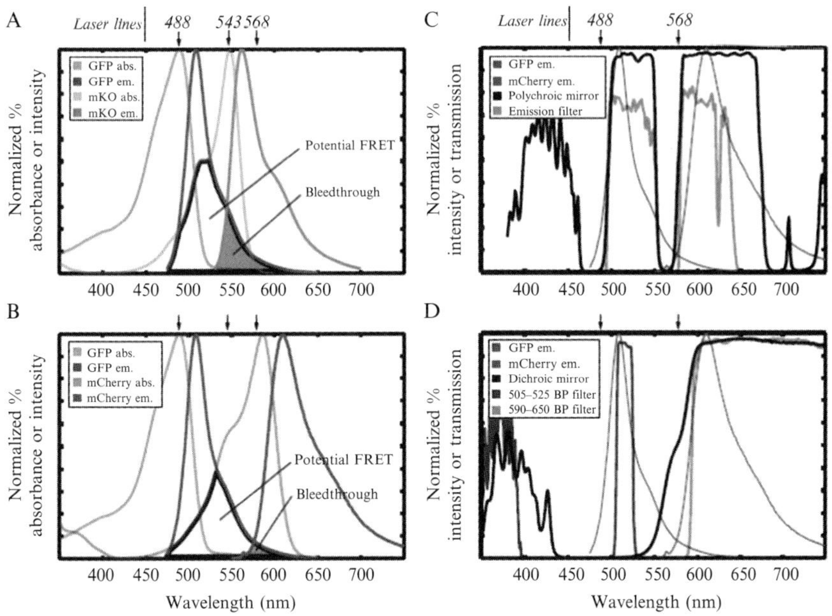


**Figure 6.3** (A) Absorbance and emission spectra of EGFP (green) and mKusabira-Orange (mKO; orange). Spectral overlaps potentially leading to FRET (black outlined area) or signal bleedthrough (gray area) are indicated. (B) Absorbance and emission spectra of EGFP (green) and mCherry (red) with spectral overlaps indicated as in (A). Note that the red-shifted mCherry spectra have less overlap with EGFP but also require a long-wavelength laser (568 nm) for optimal excitation. (C) Emission spectra of EGFP and mCherry, overlayed with the transmission spectra of the polychroic mirror (black) and emission filter (gray) used for TIRF microscopy. The two channels are further separated and projected onto different areas of a CCD chip using a Dual-View adapter (spectra of these mirrors not shown). (D) Emission spectra of EGFP and mCherry, overlaid with the transmission spectra of the dichroic mirror (black) and band-pass filters for the EGFP (dark gray) and mCherry (light gray) channels used in confocal microscopy. Notch filters at specific wavelengths remove reflected light from the lasers (spectra not shown). (See Color Insert.)

under study. For adhesion molecules, correct localization to adhesions can be verified by immunostaining. Other approaches involve testing whether the genetically tagged construct can properly rescue a knockdown of the endogenous protein and showing that the construct responds to known perturbations, mutations, and/or inhibitors.

## 3.2. Transfection optimization

The DNA plasmid encoding the FP construct is introduced to cells via transfection. There are three primary transfection methods: cationic lipids, electroporation, and viruses. The last two methods are often used for difficult-to-transfect cell types and are more difficult to implement: electroporation requires additional instrumentation (e.g., Nucleofector from Lonza) and viral transfection requires specialized techniques and can entail working with retroviruses. Lipid-based approaches (Felgner et al., 1987) require no additional equipment and have been successfully used in many cell lines used for migration and adhesion studies.

### 3.2.1 Lipid-based transfection protocol

The following is a sample procedure for the lipid-based transient transfection of mGFP–paxillin and FAK–mCherry in CHO.K1 cells using Lipofectamine (Invitrogen) (Choi et al., 2011). Detailed protocols and additional reagents can be found on the manufacturer's Web site (www.invitrogen.com).

Perform all steps under sterile conditions.

- **a.** Grow cells to 50–70% confluency in a six-well tissue culture dish (usually seeded the day before transfection).
- **b.** For each well, fill two sterile 1.5-ml microcentrifuge tubes with 100 μl of Opti-MEM (Invitrogen, 11058) each.
- **c.** In the first tube (for each well), add 5 μl of Lipofectamine reagent (Invitrogen, 18324) for every μg of DNA used (typically 1 μg).
- **d.** In the second tube, add 0.1 μg mGFP–paxillin, 0.1 μg FAK–mCherry, and 0.8 μg pBluescript ("blank" nonexpressing plasmid) to yield 1 μg total DNA.
- **e.** Combine the solution from the Lipofectamine tube with the DNA solution and incubate for 20 min at room temperature.
- **f.** Meanwhile, rinse cells twice with Opti-MEM, then add 1 ml of Opti-MEM to each dish.
- **g.** After the incubation period, add 800 μl Opti-MEM to each Lipofectamine-DNA solution (total is now 1 ml), aspirate Opti-MEM from the wells, then add the entire solution dropwise to the wells.
- **h.** Incubate cells for 3 h in the tissue culture incubator (37 °C, 8.5% $CO_2$).
- **i.** Rinse cells twice with PBS, then add normal culture medium.
- **j.** Cells can be imaged on the following 1–2 days.

### 3.2.2 Expression level optimization

After transfection, cells typically express multiple copies of the plasmid for a few days and then begin to lose the ones that are not incorporated into chromosomal structures. Cells can be made to stably express constructs by using antibiotic resistance markers on the plasmids or virus-based transfection methods.

Expressing transgenic FP constructs can result in a large, nonphysiological excess of the protein. This can perturb the biology (e.g., mislocalization, alter binding kinetics, and perturb normal structure), mask fluctuations, and increase the fluorescence background as excess protein accumulates in the cytoplasm. Expression levels can be controlled by varying the amount of plasmid DNA and using carrier DNA to ensure high overall efficiency (Webb et al., 2005). There are at least three additional means for controlling expression:

**a.** Limit transcription of your construct by using a weak promoter.
**b.** Deplete the cells' endogenous supply of the protein of interest by knocking down or knocking out the corresponding gene.
**c.** Generate stably transfected cell lines and/or select for low expressors.

The first method involves simply exchanging the strong promoter used in most expression vectors for a weak one (Watanabe & Mitchison, 2002)—a process that, for most plasmids, can be completed by a straightforward restriction digest and ligation. We have used this modification with great success in our laboratory (Kubow & Horwitz, 2011; Vicente-Manzanares, Newell-Litwa, Bachir, Whitmore, & Horwitz, 2011). When using reduced-expression constructs, larger DNA amounts (0.2–0.4 μg) may be necessary. The second and third approaches are recommended for fluctuation measurements of number densities and aggregation of adhesion proteins. For example, stable cell lines of CHO.B2 expressing α5integrin-mGFP have been used to characterize the aggregation state of integrins in cell adhesions (Wiseman et al., 2004). As CHO.B2 cells lack endogenous α5 integrin expression, the measured numbers from the transfected cells accurately reflect α5 aggregation in adhesions.

## 3.3. Sample preparation for imaging adhesions in migrating cells

The following is a list of major considerations for preparing samples for imaging adhesions in migrating cells. It is followed by a typical procedure used in our lab for CHO.K1 cells.

### *3.3.1 Migration promoting conditions*

The procedure mentioned below has been optimized to promote migration in CHO.K1 cells. The same basic procedure can be used for other cell types but may need to be reoptimized to promote migration. Cell-migration speed varies biphasically as a function of adhesion strength, which is itself a function of ligand density, cell adhesiveness (integrin density and avidity), and substrate stiffness (Engler et al., 2004; Lo, Wang, Dembo, & Wang, 2000; Palecek, Loftus, Ginsberg, Lauffenburger, & Horwitz, 1997). Titrate ligand density to obtain optimal migration (Palecek et al., 1997). Moreover, many cell types (e.g., CHO, fibroblasts) will, over time, secrete and organize matrix and/or become highly contractile and stop migrating (~2 h); obviously, imaging must be completed before this time. The culture medium will also influence cell migration (see Section 3.3.2).

### 3.3.2 Cell culture medium for imaging

Most normal cell culture media contain serum, which has a number of factors that affect migration, that is, fibronectin and chemokinetic molecules such as growth factors. To have a standardized, migration-promoting environment, serum-free medium containing promigratory growth factors (e.g., EGF, PDGF, or IGF) should be used. In addition, select media without phenol red (included as a pH indicator) to decrease background fluorescence. Finally, if the imaging system is not equipped with $CO_2$ control, the media will need to be formulated to have the correct pH at atmospheric $CO_2$ levels. The pH of normal media can be stabilized without supplemental $CO_2$ by adding 10–25 m*M* HEPES. Our laboratory routinely uses CCM1 (HyClone).

### 3.3.3 Sample heater

The sample must be maintained at 37 °C during imaging. There are basically two ways to accomplish this: encase the entire microscope body in an environmental control chamber, or use a sample holder with a heating element (e.g., DH-35 culture dish heater, Warner Instruments). The first option is expensive, but allows for $CO_2$ control and is more resistant to temperature fluctuations that result in focus drift. The second option is easier and cheaper to implement but requires a $CO_2$-independent medium and is more susceptible to temperature fluctuations between the sample and the instrument. An objective heater (e.g., from Bioptechs) can help stabilize the focus and local temperature.

### 3.3.4 Dishes for imaging

We use dishes with a coverglass bottom of the thickness specified by the objective (typically No. 1 or 1.5). Suitable dishes are commercially available (e.g., MatTek dishes; Lab-Tek chambers) or can be made in-house provided they are clean and sterile.

The following is a basic procedure to prepare CHO.K1 cells for imaging adhesions during migration (e.g., Choi et al., 2011).

Day before experiment
1. Transfect CHO.K1 cells with the desired constructs.
2. Incubate glass-bottomed dishes with 2 μg/ml fibronectin in PBS overnight at 5 °C.

Day of the experiment
1. Rinse the fibronectin-coated dishes with PBS.
2. Trypsinize cells, resuspend in CCM1 medium (HyClone), and seed into a rinsed dish.
3. Place the dish in the incubator and allow the cells to attach and spread (~20–30 min).
4. Transfer the dish to the microscope and begin imaging. Cells will migrate up to ~2 h after plating at which point they slow and become primarily contractile.

## 3.4. Instrumentation for image-based correlation measurements

The most common commercial instruments used for correlation studies with migrating cells are:
1. TIRF microscope equipped with a CCD camera (or EMCCD for fast dynamic studies and single molecule sensitivity). This setup is used for studying processes at or near adhesions, visualizing diffraction-limited nascent adhesions, and studying fast dynamic process (~100–500 ms). It can also be used for dual-color experiments using appropriate filter cubes and laser lines. We use a Dual View adaptor (Photometrics) for simultaneous acquisition of two channels.
2. CLSM (e.g., Olympus FV1000) equipped with multiple laser lines and PMT detectors that can operate in analog or pseudophoton-counting mode. Photon-counting PMTs allow for simplified analysis (see Section 3.4.5) and should be used, if possible. CLSM is usually implemented for easily visualized adhesions under conditions that do not require fast time acquisition. It is also used for protein dynamics and interaction studies in the cytoplasm (e.g., signaling complexes).

The following are some general considerations applicable to either optical configuration.

### *3.4.1 Laser power*

The selection of laser lines depends on the fluorescent probes used. When possible, select lines closest to the absorption maximum of the fluorescent probe. The laser power used for fluorescence excitation should be minimized to avoid photobleaching and phototoxicity while still providing an adequate signal-to-noise ratio. Optimization will also depend on the exposure settings as well as the expression levels of the fluorescently tagged proteins of interest. The extent of photobleaching can be estimated by an average intensity plot over time and should be less than 10%. Adverse cell response to intense light often manifests as blebbing and global cell retraction. Begin with the lowest laser power and exposure settings and increase gradually until a balance is achieved between the aforementioned parameters.

### 3.4.2 Laser alignment

Since different laser lines have characteristic critical TIRF angles, alignment of the optical fiber should be performed to obtain the optimal TIRF depth of field for a given wavelength. When a combiner is used for multiple laser lines, this process is usually automated and can be controlled via software. Otherwise, for dual-color experiments, an average critical angle for both laser lines is selected through manual alignment of the optical fiber. For CLSMs, laser alignment primarily affects the excitation intensity; but unlike TIRF, misalignment will not result in illumination inhomogeneities—only decreased intensity—and so is less critical.

### 3.4.3 Objectives

Select high magnification (60×, 100 ×), high NA objectives (> 1.2) to maximize resolution, optimize sample illumination intensity and sensitivity, and in the case of systems employing CCDs, enable high spatial sampling frequencies (pixel sizes < 0.1 μm). Ensure that the objective is corrected for chromatic aberrations for the wavelengths used. Live-cell imaging requires that the sample be held at 37 °C. The optimal immersion oil for the objective at 37 °C will likely be different than the optimal oil for room temperature. For example, our PlanApo 60 × 1.4 NA objective (Olympus) has an optimal PSF at room temperature when an oil with $n=1.514$ is used; but 37 °C, requires an oil with $n=1.522$. Test your oil by verifying that the PSF is symmetrical under typical experimental conditions (37 °C). The PSF can be viewed by taking high-resolution images of fluorescent beads with a diameter less than the diffraction limit (e.g., 0.1 μm Tetraspeck beads, Invitrogen) (see Cole, Jinadasa, & Brown, 2011 for detailed procedure).

### 3.4.4 Filter sets

This is primarily a concern for dual-color imaging. Select emission filters to minimize spectral bleedthrough while still ensuring sufficient sensitivity. For example, we use the following emission filter combination for mGFP and mCherry on the Olympus FV1000: SDM560 dichroic mirror; DM505-525; and DM590-650 band pass filters (Fig. 6.3D). In contrast to a CLSM, the TIRF configuration uses only one detector, usually a CCD camera (Cascade 512B or QuanntEM 512SC, Photometrics). Therefore, for simultaneous dual-color imaging on a TIRF microscope, we use a polychroic mirror (Z488/568rpc), a dual-pass emission filter (Z488/568m) (Chroma Technology; Fig. 6.3C), and a Dual-View simultaneous imaging system (Photometrics) to project the two channels onto separate areas of the chip. Alignment of the Dual View is critical to ensure accurate pixel overlap for both channels and should be checked over the course of the experiment as temperature or mechanical fluctuations can result in misalignment of the system. A protocol for the Dual-View alignment is found online (http://www.photometrics.com/support/pdfs/manuals/dv2_alignment.pdf).

### 3.4.5 Photon detectors

If possible, photon-counting detectors should be used for correlation measurements in lieu of analog detectors such as PMTs and CCDs. The shot noise of photon-counting detectors has a larger dynamic range and follows a Poisson distribution, which simplifies the fluctuation analysis. The pseudophoton counting mode of the Olympus FV1000 has been used successfully in lieu of true photon-counting PMTs (Choi et al., 2011; Digman & Gratton, 2009). However, not all commercial microscopes are equipped with photon counting detectors; therefore, additional theory has been developed for the N&B technique to account for the dark current of analog PMTs and CCDs (Dalal, Digman, Horwitz, Vetri, & Gratton, 2008; Unruh & Gratton, 2008). A crucial consideration for fluctuation techniques is the stability and sensitivity of the detectors over time. This can be tested by collecting a blank image time series (no light) and plotting a histogram of the digital counts for individual images within the series. The same time series can be used to determine the detector parameters used in the N&B analysis (see below). Correlation techniques like ICS, STICS, and TICS are less sensitive to these variations over time and do not require any detector parameter corrections in the analysis. To avoid camera gain saturation over time, use moderate gain settings. For QEM:512SC, we use 3 × multiplication gain and 500–700 EM gain settings. Exact settings will depend on the camera. PMT detectors should also be operated at moderate settings: 600–700 V with no multiplication gain and no offset. The latter two parameters are usually manipulated to improve image quality and should be avoided as they affect fluctuation analyses.

## 3.5. General considerations for image series acquisition

All the fluorescence fluctuation analyses mentioned in this chapter operate on individual images or image time series. The exact acquisition protocol, correlation analysis, and data fitting routines will vary depending on the technique (e.g., see Chapter 10 and others in this volume). In this section, we outline general considerations for selecting acquisition settings.

### *3.5.1 Sampling rate*

The sampling rate is critical for all fluorescence fluctuation techniques. It depends on the image exposure (TIRF-CCD) or scan (line scan and rescan time; CLSM) time and the user-defined delay between frames. The delay time between frames is limited by the camera readout time (~milliseconds) and the confocal scan time (~seconds), both of which scale with image size (number of pixels). The choice of these parameters will vary depending on the characteristic timescale of the dynamic process under study (e.g., molecular exchange, diffusion, and flow), the type of adhesion, and the particular adhesion molecule. For molecular aggregation and interaction studies of cell adhesion proteins like paxillin, a 100–200 ms camera exposure time with no delay time between frames (stream acquisition) has been used for N&B analysis (Choi et al., 2011). For retrograde flow studies of paxillin, images are acquired every few seconds (Brown et al., 2006). As the dynamics of adhesion proteins depend on the location (cytoplasm vs. adhesion) and type of adhesion (focal vs. nascent), the sampling time should be tailored accordingly. FRAP measurements provide an initial estimate of this parameter and can be used as a guideline. FRAP also identifies immobile populations which are commonly observed for adhesion molecules. Even though immobile populations do not contribute intensity fluctuations to the autocorrelation analysis, they do contribute to the total intensity and thereby can bias the correlation analysis. Immobile population removal methods were developed to account for these effects (see below). An alternative is to collect image time series over a range of timescales that span an order of magnitude (if possible) to determine the correct ranges. Sampling too slow or too fast relative to the characteristic timescale of the fluctuation will produce no temporal or spatial correlation function (for TICS, STICS, and RICS) and indicates that the selected sampling time is inappropriate. For variance analysis, sampling at the wrong timescale will simply bias the measured values.

### 3.5.2 Number of sampled fluctuations

The number of sampled fluctuations is also critical for fluctuation analyses. For N&B, this will depend on the total number of frames in a time series. For example, for analyses of paxillin dynamics and interactions in nascent adhesions, a typical time series will consist of 2000–4000 frames acquired with a 100 ms exposure time using streaming acquisition (i.e., no delay between frames) (Choi et al., 2011). We then divide the series into 200–500 frame segments for analysis. For spatiotemporal correlation approaches like TICS and STICS, the required number of frames will also depend on the size of the analyzed area, as increases in spatial average produce better statistics. Velocity mapping retrograde flow of adhesion proteins in adhesions has been reported from areas as small as $16^2$ pixels in time series consisting of a few hundred images, acquired every 2–5 s (Hebert et al., 2005). In the case of image correlation analysis (ICS, RICS), the size of the image and the pixel resolution are important parameters for determining the number of sampled fluctuations. Typically, $32^2$–$64^2$ pixel regions are used with a pixel size of $\sim$1/3 the beam radius size. Small, defined regions ($<10 \times 10$ pixels) and time segments ($\sim$10–20 frames) provide higher spatial resolution. However, one should be cautious when selecting small ranges as the accuracy of the parameters recovered by fluctuation correlation techniques depends on the number of sampled fluctuations. For example, smaller areas used for enhanced spatial resolution will require longer sampling times to achieve similar sampling statistics. Computer simulations have tested the limits of these techniques under various experimental acquisition settings and guide implementing the analyses in this toolbox (Brown et al., 2008; Costantino et al., 2005; Hebert et al., 2005; Kolin et al., 2006).

## 3.6. Post-acquisition image processing

Standard microscope imaging software (e.g., Metamorph, Olympus FluoView) typically saves images as 12- or 16-bit tiff files. An ImageJ plugin for ICS and ICCS is available through the Stower's Institute Web site (http://research.stowers.org/imagejplugins/). For complete fluctuation correlation toolbox analysis, the Globals Software for Images (SimFCS) is also available through the Laboratory for Fluorescence Dynamics free of charge for a trial period. Custom codes for selected correlation techniques have also been written in programming platforms such as Matlab and are available through the Cell Migration Consortium Web site (http://www.cellmigration.org/). For all correlation techniques, image time series should be processed to account for the following issues.

### 3.6.1 Background intensity

Background noise affects the accuracy of number density, interaction fraction, and aggregation estimates. This can be addressed by subtracting the average background intensity in each image. The average background intensity can be estimated from a selected adjacent background region in each image in the time series to account for any variations over time.

### 3.6.2 Analog detector

To account for analog detector (PMT, CCD) parameters like dark current and digital conversion on variance analysis approaches such as N&B, blank (no light) image time series are acquired under the same detector settings used for imaging the protein(s) of interest. A histogram count of the digital counts is generated from the image time series and used to estimate: (1) the Offset parameter (dark current in absence of photons) and (2) the *S*-factor (photon–analog current conversion factor). The Offset factor is estimated from a Gaussian fit of the digital counts distribution. This parameter is then subtracted from the image time series. The apparent brightness parameter *B* of a background region is then used to estimate the *S*-factor. *S*-values that yield a *B* distribution centered around "one" are used for the analysis. Refer to Dalal et al. (2008) and Unruh and Gratton (2008) for the theoretical framework and a detailed discussion of analog detector effects on N&B analysis.

### 3.6.3 Photobleaching and macroscopic fluctuations

Photobleaching and macroscopic fluctuations due to cell edge protrusion or slow movement of adhesion structures bias correlation analyses by introducing artifactual fluctuations. Theoretical models have been developed to account for photobleaching in temporal autocorrelation analyses (Kolin et al., 2006). A generalized approach applicable to most fluctuation techniques is to use a moving average filter on an image time series (Digman, Brown, et al., 2005). This eliminates low frequency fluctuations while maintaining the high-frequency molecular fluctuations. The cutoff frequency will depend on the length of the moving average window, which will vary for different data sets depending on the kinetics of the process under study. Typically, we have used a 20–40 s filter for cell migration related studies.

### 3.6.4 Immobile populations

Immobile molecular complexes are commonly observed in adhesions. These structures mask intensity fluctuations and thereby affect correlation analysis. Hebert et al., 2005 implemented an immobile filtering algorithm to eliminate intensity contributions in an image time series from static structures. In this algorithm, the image series is averaged over time to yield an "average image." As immobile structures will contribute constant fluorescence intensity at a given location over time, subtraction of the "average image" from each image in the series will eliminate the immobile population contributions (Hebert et al., 2005). Moving average filters (see Section 3.6.3) can also eliminate contributions from immobile populations.

## 3.7. Control samples

In addition to labeling your protein of interest(s), control constructs appropriate for the fluorescence fluctuation technique should be produced.

### 3.7.1 Bleedthrough control

For cross-correlation measurements, cells expressing only one fluorescent construct (e.g., either mGFP–paxillin or mCherry–FAK) are used to measure the extent of spectral bleedthrough or indirect excitation under the same experimental conditions used for the actual experiment (e.g., laser intensity, camera exposure time or confocal scan speed, detector gain, and filter combinations). For example, to measure the bleedthrough of mGFP into the mCherry detector channel (Fig. 6.3B), cells expressing mGFP–paxillin are excited using the 488 nm Ar ion laser line, and the fluorescence signal is collected in both the mGFP and mCherry detection channels. The ratio of the average background-corrected signal in the mCherry channel to the average background-corrected signal in the mGFP channel is a measure of the bleedthrough and should usually be $\lesssim 10\%$.

### 3.7.2 Positive and negative controls for cross-correlation

For cross-correlation studies, it is useful to have positive and negative controls. For example, in our cross-correlation studies of adhesions using mGFP–paxillin and mCherry–FAK (Choi et al., 2011), we created a dual-color paxillin (mGFP–paxillin–mCherry) and membrane targeted versions of mGFP and mCherry (GAP–mGFP, GAP–mCherry). The dual-color paxillin, which did not show detectable FRET, served as a positive control for maximal cross-correlation. The membrane-localized

GAP–mGFP and GAP–mCherry (GAP-43; growth-associated protein; Moriyoshi, Richards, Akazawa, O'Leary, & Nakanishi, 1996) were coexpressed as negative controls, as they do not interact and diffuse freely in the membrane.

### 3.7.3 Fluorescence intensity calibration controls

To determine the aggregation state (stoichiometry) of individual adhesion proteins in a complex using N&B analysis or ICS, controls of fluorescently tagged monomer species are needed (e.g., the membrane targeted versions mentioned above). Control samples are imaged under the same experimental conditions as the protein(s) of interest (e.g., laser intensity, camera exposure time or confocal scan speed, detector gain, and filter combinations). The following is a basic protocol for measuring aggregation states from:

**a.** TIRF-based N&B measurements.

- Collect ~1024–2048 images of GAP using stream acquisition settings and 50–100 ms camera exposure time (Cascade 512B or QuantEM:512SC, Photometrics).
- Calculate the average apparent brightness parameter $B_{\mathrm{GAP}}$ from a selected region of a cell expressing GAP–mGFP or/and GAP–mCherry (Choi et al., 2011).
- Calculate the apparent brightness parameter $B_{\mathrm{protein}}$ for the protein(s) of interest at selected adhesion locations and over the desired time segments (~20–40 s).
- For single-channel experiments, the aggregation state of a protein at a given pixel location is the ratio of $(B_{\mathrm{protein}}-1)/(B_{\mathrm{GAP}}-1)$.
- For dual-channel experiments, select the pixel locations that indicate the presence of complexes (positive Bcc value) to estimate the aggregation state of each protein in the complex. For example, use GAP–mGFP and GAP–mCherry for mGFP–paxillin and FAK–mCherry calibration, respectively. For more analysis details, refer to Choi et al. (2011) and Digman, Wiseman, Choi, Horwitz, and Gratton (2009).

**b.** ICS measurements.

- Collect confocal images ($512^2$ or $1024^2$ pixels) of GAP using 8 μs pixel dwell time and an optical zoom (~4–6×) that yields ~0.1 μm pixel resolution.
- Calculate the spatial ACF of a cell region expressing GAP–mGFP and fit the *x*-axis component of the spatial ACF to a 1D Gaussian function. From the fit amplitude ($G_0$) and the beam area ($\pi\omega^2$; $\omega$,

beam radius), calculate the number of particles (NP) per micron ($1/(G_0 \cdot \pi\omega^2)$).

- Calculate the cluster density (CD) from the fluorescence average intensity ($\langle I \rangle$) and the NP per micron ($CD = \langle I \rangle / NP$). For GAP-mGFP, CD should be proportional to the fluorophore parameters under the given imaging parameters.
- Calculate CD for the protein of interest in a selected cell region. Use $CD_{GAP}$ as a normalization factor to estimate the protein degree of aggregation. For more analysis details, refer to Wiseman et al. (2004) and references therein.

# 4. APPLICATIONS

Fluorescence fluctuation techniques have only recently been applied to adhesion and migration. So far the applications have focused on the dynamics and interactions of adhesion proteins and have provided critical information on the mechanisms of adhesion formation, signaling, and turnover. Here, we outline some of these applications:

1. Mapping integrin dynamics and aggregation in migrating cells (Wiseman et al., 2004).
2. Differential retrograde flow dynamics of adhesion components in the integrin–actin linkage (the molecular clutch model) (Brown et al., 2006).
3. Association of adhesion components in complexes occurs at adhesion sites (Paxillin–FAK (Choi et al., 2011) and Paxillin–Vinculin (Digman, Wiseman, Choi, Horwitz, & Gratton, 2009)).

## 4.1. Mapping of integrin aggregation and dynamics in migrating cells

Integrin receptors are critical components of adhesions, as they provide the mechanical link between the substratum and the actin cytoskeleton. Integrin clustering at adhesion sites reinforces this link and provides a hub for recruitment of numerous adaptor and signaling molecules (Vicente-Manzanares et al., 2009). Wiseman et al. (2004) used ICS to map the number density and aggregation state (number of molecules per aggregate) of integrins. In this study, image time series of CHO.B2 cells stably expressing the mGFP-α5 integrin subunit and plated on fibronectin-coated coverslips were acquired on a CLSM. Different areas (or adhesions) within the cells were selected from individual images in the time series for spatial autocorrelation analysis. The average number density and aggregation states were estimated from the amplitude of the spatial autocorrelation function. CHO.K1 cells expressing GAP-GFP (GFP with a membrane-targeting sequence) were used as a monomeric control. The analyses revealed that integrins are clustered throughout the cells—even at locations away from visible adhesions (with 3–4 molecules per cluster) where no visible structural organization was discerned. At the sites of large adhesions, the integrin number and aggregation increased by approximately four- to five-fold. These parameters changed as adhesions disassembled. The authors also used TICS and cross-TICS on selected regions of cells to characterize the dynamics of $\alpha5\beta1$ integrin and $\alpha$-actinin in and away from adhesion sites. Temporal image correlation functions were fit to 2D diffusion, 2D flow, or combined 2D diffusion and flow (for single or two populations) equations to estimate relevant transport parameters and their respective populations. Immobile fractions were also estimated from the fits (Fig. 6.4A). The cross-TICS functions revealed associations between integrin and $\alpha$-actinin in molecular complexes at adhesion sites and throughout the cell; interactions were not observed for integrin and paxillin (Fig. 6.4B). Similar maps of the dynamics and percent populations of the interacting integrin-$\alpha$-actinin fractions were also presented.

## 4.2. Retrograde flow of adhesion components in the integrin–actin linkage

Adhesions are sites of traction where the forces arising from membrane resistance to actin polymerization and myosin-mediated contraction are transmitted to the substratum. In the absence of a firm linkage between actin and the substratum, these forces lead to retrograde flow of actin (opposite the direction of polymerization) as new actin monomers are added or the filaments are contracted and moved toward the cell center. The presence of actin retrograde flow suggests that this linkage might not always be efficient and points to the present of a molecular "clutch" with varying efficiency (Mitchison & Kirschner, 1988; Ponti, Machacek, Gupton, Waterman-Storer, & Danuser, 2004). Brown et al. (2006) used spatiotemporal image correlation spectroscopy to address this issue. Single-channel spatiotemporal ACFs were calculated from image time series of cells coexpressing RFP-actin and GFP-tagged variants of paxillin, $\alpha5$-integrin, talin, FAK, vinculin, and $\alpha$-actinin. The magnitude and direction of flow of the individual molecules were estimated from the temporal displacement of the peak of the spatial autocorrelation.

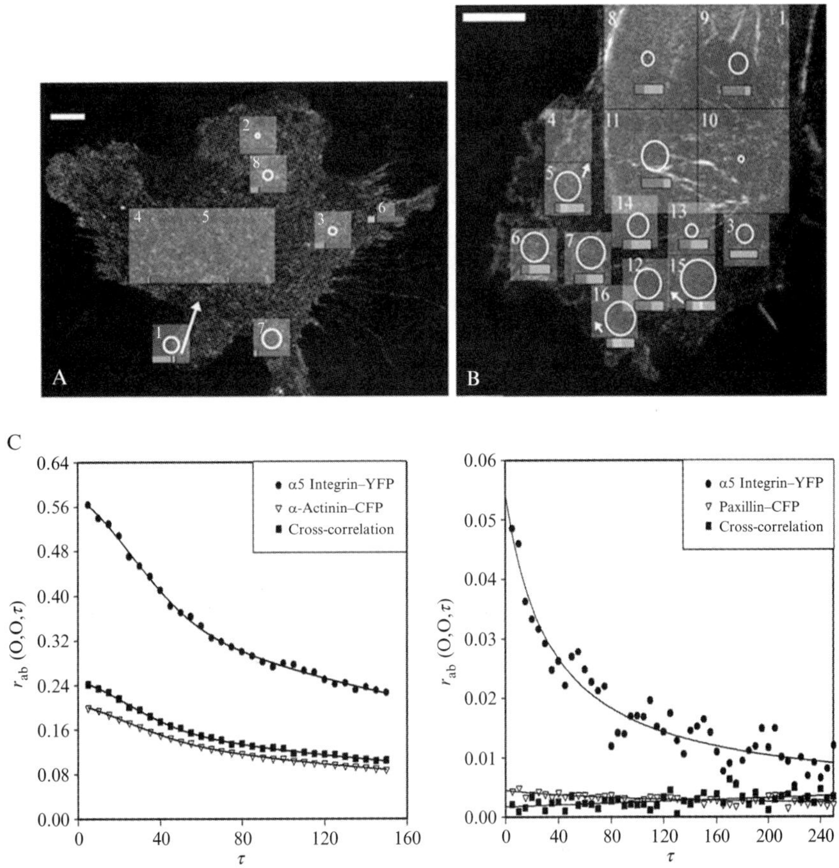


**Figure 6.4** Spatial map of the dynamics of (A) a5 integrin-GFP and (B) α-actinin GFP in CHO.B2 (A) and CHO.K1 (B) cells and plated on fibronectin. Image time series consisting of 100 images were captured every 5 s using (A) CLSM and (B) two-photon microscopy. Scale bars are 5 μm in (A) and 10 μm in (B). Highlighted regions ($32^2$–$128^2$ pixels) were selected for TICS analysis. The circles represent the root mean square average diffusion distance from the center of the circle for a 10-min period; it is derived from the average diffusion coefficient for each region. The vectors represent the mean translation distance and direction over a 10-min period based on the measured velocities for regions exhibiting flowing integrin populations. The colored bars depict the proportion of immobile (green), flowing (yellow), and diffusing (cyan) integrin or α-actinin within each region. (C) Cross-TICS reveals associations between $\alpha_5$ integrin–CFP and α-actinin–YFP in molecular complexes. Similar associations were absent in the case of $\alpha_5$ integrin–CFP and paxillin–YFP. The presented single-channel autocorrelation and cross-correlation functions are calculated from CHO.K1 image time series expressing the fluorescently labeled constructs and captured using two-photon microscopy. *Figure and caption reproduced from Wiseman et al. (2004).* (For interpretation of the references to color in this figure legend, the reader is referred to the online version of this chapter.)

An example of a spatial map of the flow of paxillin–GFP and RFP–actin is presented in Fig. 6.5A. The relative magnitude and direction of the velocity vectors for the various adhesion components relative to actin revealed a differential correlation between certain adhesion and actin retrograde velocities (Fig. 6.5B). The nature of the correlation was cell-type dependent: highly contractile cells, like MEF or 3T3 cells, showed a stronger coupling between adhesion molecules and actin, whereas CHO cells, which are less contractile, showed reduced coupling (Fig. 6.5B). Thus, it appears that the forces on adhesions are modulated by the

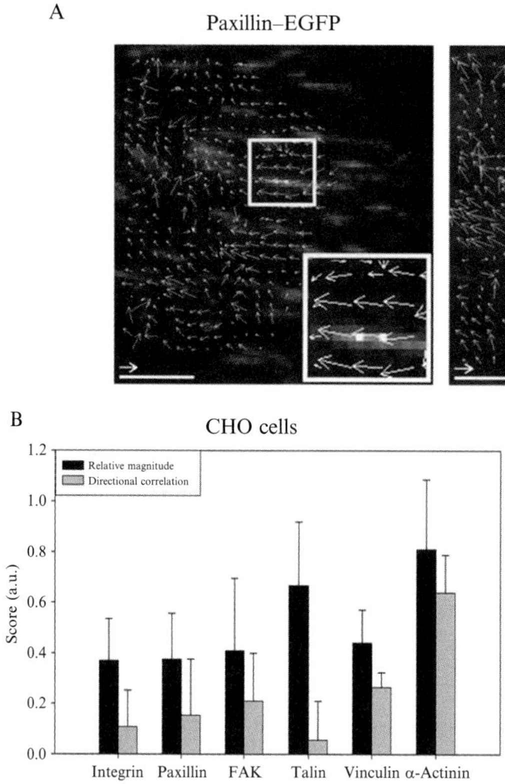

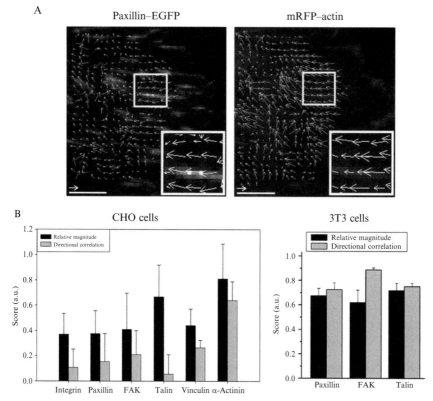


**Figure 6.5** Retrograde flow of adhesion proteins. (A) Velocity maps for paxilin–EGFP and mRFP–Actin coexpressed in NIH3T3 fibroblasts plated on fibronectin. The image time series consisted of 121 imaged captured every 5 s using TIRF. Scale bar is 5 μm. Velocity scale arrow is 1 μm/min$^{-1}$. (B) Comparison of the plots of the median relative magnitude and directional correlation coefficients of various adhesion proteins relative to actin in CHO cells and 3T3 cells. *Figure and caption reproduced from Brown et al. (2006).* (For interpretation of the references to color in this figure legend, the reader is referred to the online version of this chapter.)

efficiency of the integrin–actin linkage to retrograde forces. As the force on adhesions feeds back to dictate adhesion size, organization, and signaling, this clutch has major ramifications for the cell.

## 4.3. Association of adhesion components in complexes occurs at adhesion sites

The fluorescence fluctuation toolbox has been used to characterize the interactions between focal adhesion kinase (FAK) and paxillin, which constitute a key signaling complex that regulates cell migration (Choi et al., 2011). Paxillin and FAK organize and control signals that regulate protrusion and adhesion formation/disassembly through phosphorylation (FAK) and scaffolding functions (FAK and paxillin). As scaffolds, they organize, regulate, and generate signals that impinge on Rho GTPases, which in turn control actin polymerization and contraction (Deakin & Turner, 2008; Schaller, 2010). Paxillin and FAK are regulated by several kinases, including Src, a tyrosine kinase (Schaller & Parsons, 1995; Zaidel-Bar, Milo, Kam, & Geiger, 2007). Choi et al. (2011) investigated the role of tyrosine phosphorylation on two sites in paxillin on paxillin–FAK interactions in adhesions. Using ccN&B analysis of TIRF image time series of CHO cells expressing mGFP–paxillin and FAK–mCherry, they showed that wild-type paxillin and FAK exchange as molecular complexes in newly forming (nascent) adhesions. They also quantified the size and relative stoichiometry of the complex and showed that phosphomimetic mutants of paxillin (Y31E–Y118E) increased its size. In contrast, the non-phosphorylatable mutant of paxillin (Y31F–Y118F) resulted in few detectable paxillin–FAK complexes; those that existed had a reduced complex size (Fig. 6.6). In addition, using ccRICS analysis on CLSM image time series, they showed that paxillin–FAK interactions were not detected in the cytoplasm outside the immediate vicinity of adhesions, indicating that their association occurs once they are in adhesion sites. This confirms and extends earlier correlation studies by Digman, Wiseman, Choi, Horwitz, and Gratton (2009) that showed little, if any, vinculin–paxillin or paxillin–FAK association in the cytoplasm.

Digman et al. also investigated whether molecular complexes mediate the formation or disassembly of adhesions. Using temporal correlation analysis (FCS, scanning FCS, TIRF–TICS) and fluorescence variance analysis (PCH, ccN&B), they studied paxillin dynamics and interactions with FAK and vinculin. They observed heterogeneity in the dynamics and aggregation state of paxillin in different adhesions and in regions across the cell.

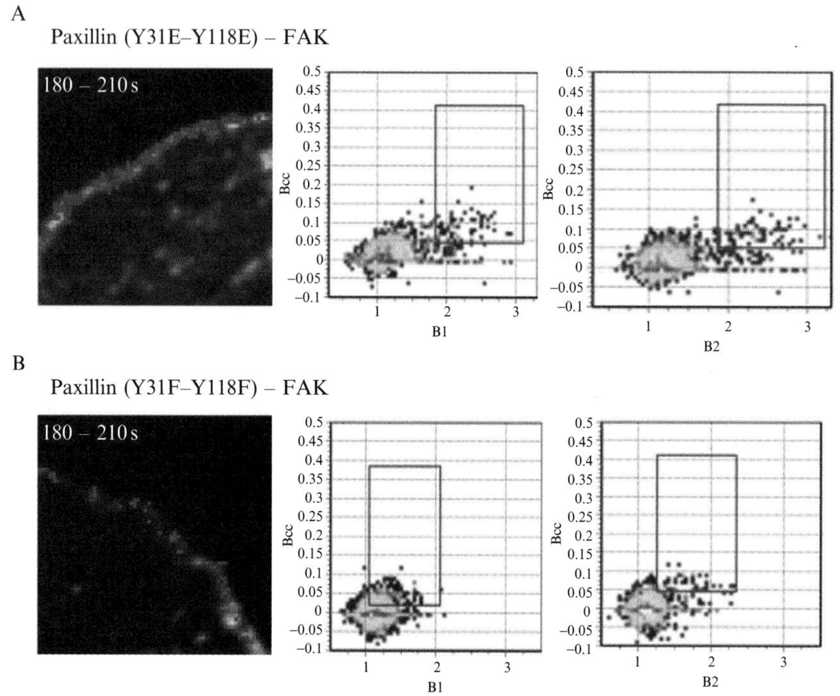


**Figure 6.6** Cross-number and -brightness (ccN&B) analysis of Paxillin–FAK interactions in nascent adhesions (Choi et al., 2011). Highlighted pixels (red) indicate the presence of complexes and are selected for positive cross-brightness (Bcc) parameter values as indicated in the highlighted box in the histogram plots of pixels with a given Bcc value and corresponding B1 or B2 for each channel, respectively. For pixels that show complexes, the B1 and B2 values will reflect the size of stoichiometry of the individual species in the complex. More molecular complexes are detected for the phosphomimetic (Y31E, Y118E) versus the nonphosphorylatable (Y31F, Y118F) mutants of paxillin. TIRF image time series of CHO.K1 cells expressing paxillin–GFP and mCherry–FAK were acquired with 100 ms exposure time using stream acquisition for ~3.5 min. *Figure and caption reproduced from Choi et al. (2011).* (See Color Insert.)

Whereas diffusion is the main mode of transport for adhesion molecules in the cytoplasm, exchange (binding–unbinding) kinetics with a broad range of rates (0.1–10 $s^{-1}$) dominate in the vicinity of adhesions (Fig. 6.7) (Digman, Wiseman, Choi, Horwitz, & Gratton, 2009). They observed large clusters and complexes exchanging relatively slowly in the vicinity of the disassembling regions of adhesions, whereas small aggregates (largely monomers) were observed exchanging rapidly in assembling adhesions (Fig. 6.8) (Digman, Wiseman, Horwitz, & Gratton, 2009). This suggests that

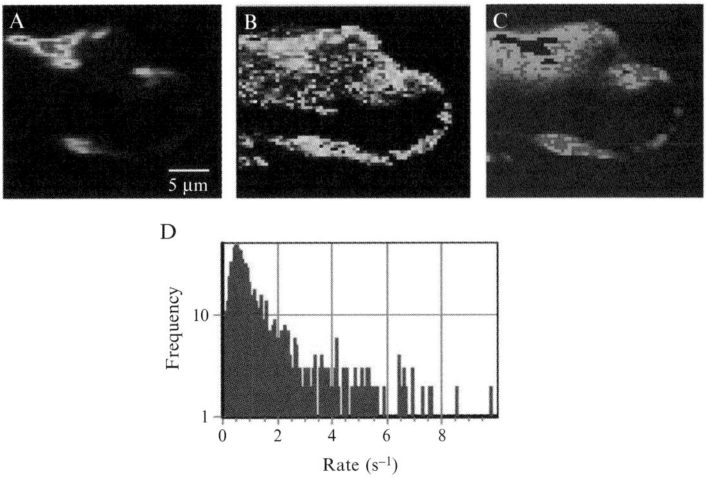


**Figure 6.7** Temporal pixel-autocorrelation analysis of CHO cells expression paxillin–EGFP. (A) Average fluorescence intensity of the image time series. Spatial mapping of the (B) amplitude ($G_0$) of the temporal autocorrelation function and (C) exchange rate constants (s$^{-1}$) obtained from the fits of the single pixel autocorrelation functions to an exponential decay function. Red and green pixels correspond to rates $>1$ and $<1$ s$^{-1}$, respectively. (D) Histogram plot of the rate constants obtained from the single pixel temporal autocorrelation analysis. *Figure and caption reproduced from Digman, Brown, Horwitz, Mantulin, and Gratton (2008).* (See Color Insert.)

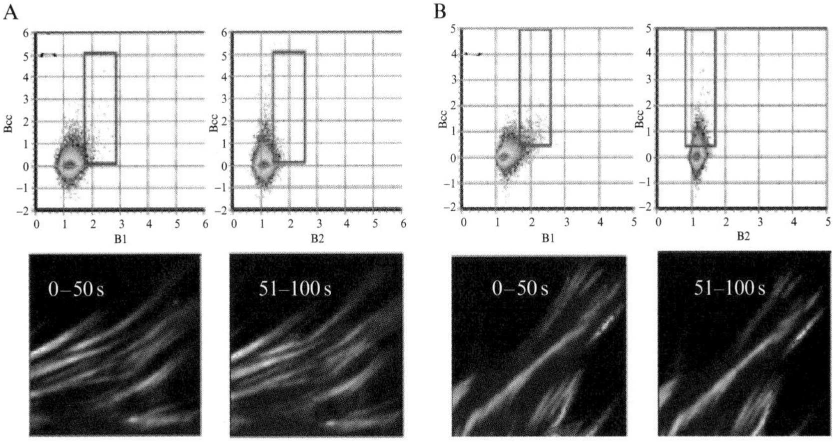


**Figure 6.8** Cross-number and brightness (ccN&B) analysis of paxillin–vinculin reveals the presence of (A) larger molecular aggregates between adhesions and smaller molecular aggregates in adhesions (B) as indicated by the B1 and B2 values observed for pixels that indicate the presence of molecular complexes. CLSM image time series of MEF cells expressing vinculin–GFP and paxillin–mCherry were acquired every 1.15 s for ~4 min. *Figure and caption reproduced from Digman, Wiseman, Choi, Horwitz, and Gratton (2009).* (For color version of this figure, the reader is referred to the online version of this chapter.)

adhesion assembly involves the addition of monomer units or small aggregates, whereas adhesion disassembly results from the dissociation of large clusters or complexes.

## 5. CONCLUSION

Until recently, methodological limitations have hindered the study of the molecular-scale interactions and dynamics of the molecules that comprise adhesions in living cells. Fluorescence fluctuation techniques have recently been developed that potentiate these studies in a way that senses fluctuations down to the level of single molecules, provides spatial maps across the entire cell, and enables analysis of multiple pieces of information with a single data set. Two major themes are critical in the implementation of these methods. First, when working with biological specimens, there are a number of potential artifacts that arise as one attempts to increase the signal-to-noise ratio: increasing excitation power increases photobleaching/phototoxicity; increasing FP construct expression results in higher background fluorescence and may also alter the biology. Second, the timescale of the process under study is critical: optimizing the sampling rate imaging times will have a profound effect on the detection of fluctuations. While the experimental implementation of most fluorescence fluctuation techniques is relatively straightforward and uses commercially available instrumentation, the subsequent analysis requires care.

## REFERENCES

Axelrod, D. (2008). Chapter 7: Total internal reflection fluorescence microscopy. *Methods in Cell Biology*, *89*, 169–221.

Bacia, K., Kim, S. A., & Schwille, P. (2006). Fluorescence cross-correlation spectroscopy in living cells. *Nature Methods*, *3*, 83–89.

Ballestrem, C., Erez, N., Kirchner, J., Kam, Z., Bershadsky, A., & Geiger, B. (2006). Molecular mapping of tyrosine-phosphorylated proteins in focal adhesions using fluorescence resonance energy transfer. *Journal of Cell Science*, *119*, 866–875.

Berland, K. M. (2004). Fluorescence correlation spectroscopy: A new tool for quantification of molecular interactions. *Methods in Molecular Biology*, *261*, 383–398.

Brown, C. M., Dalal, R. B., Hebert, B., Digman, M. A., Horwitz, A. R., & Gratton, E. (2008). Raster image correlation spectroscopy (RICS) for measuring fast protein dynamics and concentrations with a commercial laser scanning confocal microscope. *Journal of Microscopy*, *229*, 78–91.

Brown, C. M., Hebert, B., Kolin, D. L., Zareno, J., Whitmore, L., Horwitz, A. R., et al. (2006). Probing the integrin-actin linkage using high-resolution protein velocity mapping. *Journal of Cell Science*, *119*, 5204–5214.

Chen, Y., Müller, J. D., So, P. T., & Gratton, E. (1999). The photon counting histogram in fluorescence fluctuation spectroscopy. *Biophysical Journal*, *77*, 553–567.

Chen, Y., Wei, L. N., & Müller, J. D. (2003). Probing protein oligomerization in living cells with fluorescence fluctuation spectroscopy. *Proceedings of the National Academy of Sciences of the United States of America, 100*, 15492–15497.

Choi, C. K., Vicente-Manzanares, M., Zareno, J., Whitmore, L. A., Mogilner, A., & Horwitz, A. R. (2008). Actin and alpha-actinin orchestrate the assembly and maturation of nascent adhesions in a myosin II motor-independent manner. *Nature Cell Biology, 10*, 1039–1050.

Choi, C. K., Zareno, J., Digman, M. A., Gratton, E., & Horwitz, A. R. (2011). Cross-correlated fluctuation analysis reveals phosphorylation-regulated paxillin-FAK complexes in nascent adhesions. *Biophysical Journal, 100*, 583–592.

Chudakov, D. M., Matz, M. V., Lukyanov, S., & Lukyanov, K. A. (2010). Fluorescent proteins and their applications in imaging living cells and tissues. *Physiological Reviews, 90*, 1103–1163.

Cole, R. W., Jinadasa, T., & Brown, C. M. (2011). Measuring and interpreting point spread functions to determine confocal microscope resolution and ensure quality control. *Nature Protocols, 6*, 1929–1941.

Costantino, S., Comeau, J. W., Kolin, D. L., & Wiseman, P. W. (2005). Accuracy and dynamic range of spatial image correlation and cross-correlation spectroscopy. *Biophysical Journal, 89*, 1251–1260.

Dalal, R. B., Digman, M. A., Horwitz, A. F., Vetri, V., & Gratton, E. (2008). Determination of particle number and brightness using a laser scanning confocal microscope operating in the analog mode. *Microscopy Research and Technique, 71*, 69–81.

Deakin, N. O., & Turner, C. E. (2008). Paxillin comes of age. *Journal of Cell Science, 121*, 2435–2444.

Digman, M. A., Brown, C. M., Horwitz, A. R., Mantulin, W. W., & Gratton, E. (2008). Paxillin dynamics measured during adhesion assembly and disassembly by correlation spectroscopy. *Biophysical Journal, 94*, 2819–2831.

Digman, M. A., Brown, C. M., Sengupta, P., Wiseman, P. W., Horwitz, A. R., & Gratton, E. (2005). Measuring fast dynamics in solutions and cells with a laser scanning microscope. *Biophysical Journal, 89*, 1317–1327.

Digman, M. A., Dalal, R., Horwitz, A. F., & Gratton, E. (2008). Mapping the number of molecules and brightness in the laser scanning microscope. *Biophysical Journal, 94*, 2320–2332.

Digman, M. A., & Gratton, E. (2009). Analysis of diffusion and binding in cells using the RICS approach. *Microscopy Research and Technique, 72*, 323–332.

Digman, M. A., Sengupta, P., Wiseman, P. W., Brown, C. M., Horwitz, A. R., & Gratton, E. (2005). Fluctuation correlation spectroscopy with a laser-scanning microscope: Exploiting the hidden time structure. *Biophysical Journal, 88*, L33–L36.

Digman, M. A., Wiseman, P. W., Choi, C., Horwitz, A. R., & Gratton, E. (2009). Stoichiometry of molecular complexes at adhesions in living cells. *Proceedings of the National Academy of Sciences of the United States of America, 106*, 2170–2175.

Digman, M. A., Wiseman, P. W., Horwitz, A. R., & Gratton, E. (2009). Detecting protein complexes in living cells from laser scanning confocal image sequences by the cross correlation raster image spectroscopy method. *Biophysical Journal, 96*, 707–716.

Elson, E. L., & Magde, D. (1974). Fluorescence correlation spectroscopy. I. Conceptual basis and theory. *Biopolymers, 13*, 1–27.

Engler, A., Bacakova, L., Newman, C., Hategan, A., Griffin, M., & Discher, D. (2004). Substrate compliance versus ligand density in cell on gel responses. *Biophysical Journal, 86*, 617–628.

Felgner, P. L., Gadek, T. R., Holm, M., Roman, R., Chan, H. W., Wenz, M., et al. (1987). Lipofection: A highly efficient, lipid-mediated DNA-transfection procedure. *Proceedings of the National Academy of Sciences of the United States of America, 84*, 7413–7417.

Fessart, D., Simaan, M., Zimmerman, B., Comeau, J., Hamdan, F. F., Wiseman, P. W., et al. (2007). Src-dependent phosphorylation of beta2-adaptin dissociates the beta-arrestin-AP-2 complex. *Journal of Cell Science, 120*, 1723–1732.

Geiger, B., Spatz, J. P., & Bershadsky, A. D. (2009). Environmental sensing through focal adhesions. *Nature Reviews. Molecular Cell Biology, 10*, 21–33.

Goetz, J. G. (2009). Bidirectional control of the inner dynamics of focal adhesions promotes cell migration. *Cell Adhesion & Migration, 3*, 185–190.

Haustein, E., & Schwille, P. (2007). Fluorescence correlation spectroscopy: Novel variations of an established technique. *Annual Review of Biophysics and Biomolecular Structure, 36*, 151–169.

Hebert, B., Costantino, S., & Wiseman, P. W. (2005). Spatiotemporal image correlation spectroscopy (STICS) theory, verification, and application to protein velocity mapping in living CHO cells. *Biophysical Journal, 88*, 3601–3614.

Kanchanawong, P., Shtengel, G., Pasapera, A. M., Ramko, E. B., Davidson, M. W., Hess, H. F., et al. (2010). Nanoscale architecture of integrin-based cell adhesions. *Nature, 468*, 580–584.

Kask, P., Palo, K., Ullmann, D., & Gall, K. (1999). Fluorescence-intensity distribution analysis and its application in biomolecular detection technology. *Proceedings of the National Academy of Sciences of the United States of America, 96*, 13756–13761.

Kolin, D. L., Costantino, S., & Wiseman, P. W. (2006). Sampling effects, noise, and photobleaching in temporal image correlation spectroscopy. *Biophysical Journal, 90*, 628–639.

Kubow, K. E., & Horwitz, A. R. (2011). Reducing background fluorescence reveals adhesions in 3D matrices. *Nature Cell Biology, 13*, 3–5 author reply 5–7.

Lele, T. P., Pendse, J., Kumar, S., Salanga, M., Karavitis, J., & Ingber, D. E. (2006). Mechanical forces alter zyxin unbinding kinetics within focal adhesions of living cells. *Journal of Cellular Physiology, 207*, 187–194.

Lo, C. M., Wang, H. B., Dembo, M., & Wang, Y. L. (2000). Cell movement is guided by the rigidity of the substrate. *Biophysical Journal, 79*, 144–152.

Magde, D., Elson, E. L., & Webb, W. W. (1974). Fluorescence correlation spectroscopy. II. An experimental realization. *Biopolymers, 13*, 29–61.

Mitchison, T., & Kirschner, M. (1988). Cytoskeletal dynamics and nerve growth. *Neuron, 1*, 761–772.

Miyawaki, A. (2011). Proteins on the move: Insights gained from fluorescent protein technologies. *Nature Reviews. Molecular Cell Biology, 12*, 656–668.

Moriyoshi, K., Richards, L. J., Akazawa, C., O'Leary, D. D., & Nakanishi, S. (1996). Labeling neural cells using adenoviral gene transfer of membrane-targeted GFP. *Neuron, 16*, 255–260.

Palecek, S. P., Loftus, J. C., Ginsberg, M. H., Lauffenburger, D. A., & Horwitz, A. F. (1997). Integrin-ligand binding properties govern cell migration speed through cell-substratum adhesiveness. *Nature, 385*, 537–540.

Parsons, J. T., Horwitz, A. R., & Schwartz, M. A. (2010). Cell adhesion: Integrating cytoskeletal dynamics and cellular tension. *Nature Reviews. Molecular Cell Biology, 11*, 633–643.

Petersen, N. O., Hoddelius, P. L., Wiseman, P. W., Seger, O., & Magnusson, K. E. (1993). Quantitation of membrane receptor distributions by image correlation spectroscopy: Concept and application. *Biophysical Journal, 65*, 1135–1146.

Petrasek, Z., Hoege, C., Mashaghi, A., Ohrt, T., Hyman, A. A., & Schwille, P. (2008). Characterization of protein dynamics in asymmetric cell division by scanning fluorescence correlation spectroscopy. *Biophysical Journal, 95*, 5476–5486.

Ponti, A., Machacek, M., Gupton, S. L., Waterman-Storer, C. M., & Danuser, G. (2004). Two distinct actin networks drive the protrusion of migrating cells. *Science, 305*, 1782–1786.

Ridley, A. J., Schwartz, M. A., Burridge, K., Firtel, R. A., Ginsberg, M. H., Borisy, G., et al. (2003). Cell migration: Integrating signals from front to back. *Science*, *302*, 1704–1709.

Ries, J., Yu, S. R., Burkhardt, M., Brand, M., & Schwille, P. (2009). Modular scanning FCS quantifies receptor-ligand interactions in living multicellular organisms. *Nature Methods*, *6*, 643–645.

Ruan, Q., Cheng, M. A., Levi, M., Gratton, E., & Mantulin, W. W. (2004). Spatial-temporal studies of membrane dynamics: Scanning fluorescence correlation spectroscopy (SFCS). *Biophysical Journal*, *87*, 1260–1267.

Schaller, M. D. (2010). Cellular functions of FAK kinases: Insight into molecular mechanisms and novel functions. *Journal of Cell Science*, *123*, 1007–1013.

Schaller, M. D., & Parsons, J. T. (1995). pp125FAK-dependent tyrosine phosphorylation of paxillin creates a high-affinity binding site for Crk. *Molecular and Cellular Biology*, *15*, 2635–2645.

Shaner, N. C., Campbell, R. E., Steinbach, P. A., Giepmans, B. N., Palmer, A. E., & Tsien, R. Y. (2004). Improved monomeric red, orange and yellow fluorescent proteins derived from Discosoma sp. red fluorescent protein. *Nature Biotechnology*, *22*, 1567–1572.

Shaner, N. C., Steinbach, P. A., & Tsien, R. Y. (2005). A guide to choosing fluorescent proteins. *Nature Methods*, *2*, 905–909.

Unruh, J. R., & Gratton, E. (2008). Analysis of molecular concentration and brightness from fluorescence fluctuation data with an electron multiplied CCD camera. *Biophysical Journal*, *95*, 5385–5398.

Vicente-Manzanares, M., Choi, C. K., & Horwitz, A. R. (2009). Integrins in cell migration—The actin connection. *Journal of Cell Science*, *122*, 199–206.

Vicente-Manzanares, M., & Horwitz, A. R. (2011). Cell migration: An overview. *Methods in Molecular Biology*, *769*, 1–24.

Vicente-Manzanares, M., Newell-Litwa, K., Bachir, A. I., Whitmore, L., & Horwitz, A. R. (2011). Myosin IIA/IIB restrict adhesive and protrusive signaling and generate front-back polarity in migrating cells. *The Journal of Cell Biology*, *193*, 381–396.

Watanabe, N., & Mitchison, T. J. (2002). Single-molecule speckle analysis of actin filament turnover in lamellipodia. *Science*, *295*, 1083–1086.

Webb, D. J., Schroeder, M. J., Brame, C. J., Whitmore, L., Shabanowitz, J., Hunt, D. F., et al. (2005). Paxillin phosphorylation sites mapped by mass spectrometry. *Journal of Cell Science*, *118*, 4925–4929.

Wehrle-Haller, B. (2007). Analysis of integrin dynamics by fluorescence recovery after photobleaching. *Methods in Molecular Biology*, *370*, 173–202.

Wiseman, P. W., Brown, C. M., Webb, D. J., Hebert, B., Johnson, N. L., Squier, J. A., et al. (2004). Spatial mapping of integrin interactions and dynamics during cell migration by image correlation microscopy. *Journal of Cell Science*, *117*, 5521–5534.

Wiseman, P. W., & Petersen, N. O. (1999). Image correlation spectroscopy. II. Optimization for ultrasensitive detection of preexisting platelet-derived growth factor-beta receptor oligomers on intact cells. *Biophysical Journal*, *76*, 963–977.

Wiseman, P. W., Squier, J. A., Ellisman, M. H., & Wilson, K. R. (2000). Two-photon image correlation spectroscopy and image cross-correlation spectroscopy. *Journal of Microscopy*, *200*, 14–25.

Wolfenson, H., Bershadsky, A., Henis, Y. I., & Geiger, B. (2011). Actomyosin-generated tension controls the molecular kinetics of focal adhesions. *Journal of Cell Science*, *124*, 1425–1432.

Zaidel-Bar, R., Itzkovitz, S., Ma'ayan, A., Iyengar, R., & Geiger, B. (2007). Functional atlas of the integrin adhesome. *Nature Cell Biology*, *9*, 858–867.

Zaidel-Bar, R., Milo, R., Kam, Z., & Geiger, B. (2007). A paxillin tyrosine phosphorylation switch regulates the assembly and form of cell-matrix adhesions. *Journal of Cell Science*, *120*, 137–148.

CHAPTER SEVEN

# Interactions in Gene Expression Networks Studied by Two-Photon Fluorescence Fluctuation Spectroscopy

**Nathalie Declerck**[*,†], **Catherine A. Royer**[*,1]
[*]Centre de Biochimie Structurale, INSERM U1054, CNRS UMR5048, Université Montpellier 1 and 2, Montpellier, France
[†]Département de Microbiologie, Institut National pour la Recherche Agronomique, Paris, France
[1]Corresponding author: e-mail address: catherine.royer@cbs.cnrs.fr

## Contents



## Abstract

Fluorescence fluctuation techniques have proven to be extremely useful in the characterization of biomolecular interactions. Here an overview of recent applications of two-photon fluorescence fluctuation to the study of protein–nucleic acid complexes implicated in translational and transcriptional regulation is presented. In particular, the issue of the stoichiometry of the complexes is addressed using fluorescence (cross) correlation spectroscopy (F(C)CS) for the *in vitro* studies of one RNA-binding protein (the bacterial ribosomal protein L20) and two transcriptional repressors (CggR and CcpN, implicated in the control of the central carbon metabolism in *Bacillus subtilis*). Then, the application of two-photon scanning number and brightness (2psN&B) analysis of fluorescence microscopy measurements in single cells is presented. Multiple technical aspects related to the adaptation of this method to live bacteria are discussed.

*Methods in Enzymology*, Volume 519
ISSN 0076-6879
http://dx.doi.org/10.1016/B978-0-12-405539-1.00007-5

This approach was used to count the number of fluorescent protein molecules produced from different inducible promoters in *B. subtilis* reporter strains, in hundreds of individual cells under both permissive and repressive conditions. We present a case study in which the stochastic activity of glycolytic and gluconeogenic gene promoters could be quantified *in vivo* by 2psN&B and be related to the repression mechanisms proposed from *in vitro* studies.

# 1. INTRODUCTION

The power of fluorescence fluctuation techniques to measure biomolecular interactions and dynamics has long been recognized in biophysics. Indeed, even before the advent of stable, coherent, and tunable light sources and ultrasensitive, low-noise detectors, the community was very excited by the perspectives offered when fluorescence correlation spectroscopy (FCS) was first introduced nearly 40 years ago (Magde, Elson, & Webb, 1974). Since that time, an enormous number of studies and reviews (see these recent reviews: Chiantia, Ries, & Schwille, 2009; Digman & Gratton, 2009; Sahoo & Schwille, 2011; Weiss, 2008 and references therein) have appeared in the literature. Numerous useful innovations and improvements to the original fluorescence correlation approach have been presented as well, many of the latest of which are discussed in this volume. Therefore, the purpose of this review is not to provide an overview of the field, or even its applications to the study of biomolecular interactions implicated in the regulation of gene expression. Rather, it will present a summary of recent work from our group and collaborators, using two-photon fluorescence fluctuation spectroscopy (FFS) or microscopy (FFM) approaches to address issues of affinity, stoichiometry, and activity of regulatory proteins interacting with specific DNA or RNA targets both *in vitro* and *in vivo*.

# 2. *IN VITRO* INTERACTIONS BETWEEN PROTEINS AND NUCLEIC ACIDS USING FLUCTUATION APPROACHES

## 2.1. L20: Protein–RNA interactions implicated in translational control

L20 is a protein of the large ribosomal subunit in bacteria which not only recognizes a specific site on the 23S ribosomal RNA but is also responsible for translational repression of L35, and of its own expression through translational coupling (Lesage et al., 1992). L20 from *Escherichia coli* recognizes two sites in a rather long operator RNA region (660 nucleotides) of the L35 leader mRNA. The first site is formed by a pseudo-knot joining an upstream sequence with a short sequence just prior to the L35 cistron Shine-Dalgarno sequence. The second site is located in between the sequences involved in the long range interaction (Guillier et al., 2005, 2002). All three specific recognition sites (23S rRNA, S1, and S2) share a common motif containing a bulged pair of adenosines, which when mutated to uracil decrease the affinity for L20 to nonspecific binding levels. L20 is a small protein exhibiting an N-terminal domain forming a long positively charged helix in the context of the ribosome, but which is unstructured in solution, and a 6-kD C-terminal domain (L20-Cter) adopting a globular structure containing three α-helices that form the specific RNA-binding surface (Ban, Nissen, Hansen, Moore, & Steitz, 2000). The group of Mathias Springer has shown that mutations in either one of the two S1 or S2 sites abolish translational control *in vivo* (Guillier et al., 2002). Using fluorescence anisotropy assays titrating an N-terminal Alexa488-labeled L20-Cter (Molecular Probes, Eugene, OR) with full-length operator RNA, we showed that mutating either S1 or S2 had a little or no effect on the affinity of the protein for the operator RNA (Allemand, Haentjens, Chiaruttini, Royer, & Springer, 2007), whereas mutating both sites decreased the affinity to nonspecific levels.

The contradiction between the *in vivo* and *in vitro* results prompted us to investigate the stoichiometry of L20-operator interactions. We wondered in particular whether two molecules of L20 could simultaneously interact with the operator RNA (presumably at both S1 and S2). To test this, we used florescence cross-correlation spectroscopy (FCCS), which provides a very powerful means for detecting specific interactions between differently labeled fluorescent partners. For this purpose, we labeled L20-Cter with either a green dye (Alexa-488 as above for the anisotropy measurements) or a red dye (Atto-647N, ref supplier, Attotech, Seigen Germany). Protein samples (at a concentration between 25 and 100 μ*M* in a total volume of 500 μL) were incubated with a 10-fold molar excess of NHS ester derivatives of the fluorescent dye at pH 8 in phosphate buffer for 20 min at room temperature. Under these conditions, well below the p$K_a$ of lysine and arginine residues, the protein is expected to be primarily labeled on its N-terminus, although this was not verified. Protein was separated from free unreacted dye using a PD-10 desalting column (GE Healthcare Life Sciences), equilibrated in 50 m*M* Tris buffer, by depositing 500 μL of the labeling reaction mixture on the column and eluting with $3 \times 750$ of

Tris-containing buffer. Labeling ratios obtained under these conditions were near 1 dye molecule per L20 molecule as assessed by Bradford assay.

We carried out first two-photon FCCS measurements (Fig. 7.1) on equimolar concentrations of the two L20-Cter preparations (150 n*M* nominal concentrations) to determine whether the protein (at 300 n*M*) remained monomeric or formed dimers or higher order oligomers. In two-photon cross-correlation measurements, the larger two-photon cross sections allow for simultaneous excitation of both fluorophores with a single laser line, in this case 850 nm (typically $<60$ mW at the microscope entrance). Differential detection of the red- and green-emitting dyes is provided by a set of appropriate filters (a 565-nm dichroic and 675 (red channel, CH1), or 525 (green channel, CH2) $\pm$50-nm bandpass filters). The actual concentration of the freely diffusing labeled proteins (extrapolated from the time-zero amplitudes of the autocorrelation functions ($G(0)$) measured for each channel) was lower for Atto-647N-labeled L20-Cter (CH1) than that of the Alexa-488-labeled L20-Cter (CH2) due to a greater loss of protein on the slide surface (despite pegylation). In any case, the cross-correlation signal detected simultaneously in both channels is within shot noise levels,

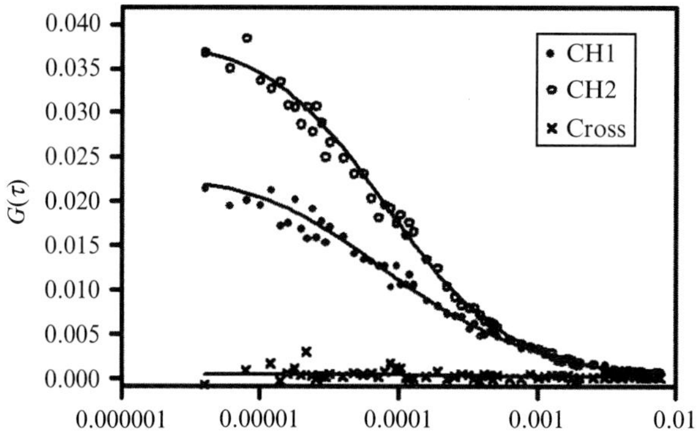


**Figure 7.1** Oligomerization state of L20: FCS measurements. Open circles correspond to the autocorrelation profiles obtained from the data collected in channel 2 (corresponding to L20-Cter-Alexa-488 emission) and dots correspond to the autocorrelation profiles calculated from the data collected in channel 1 (corresponding to L20-Cter-Atto-647N emission). The crosses correspond to the cross-correlation profile between the fluctuations in channel 1 and channel 2. L20-Cter-Alexa-488 and L20-Cter-Atto-647N were at a nominal concentration of 150 n*M*. *Figure taken from Allemand et al. (2007).*

demonstrating no interaction between L20-Cter monomers under these conditions. It is of note that our setup allows a complete separation of the green and red signal, evidencing the absence of cross talk between the detection channels. This is very important for FCCS because cross talk seriously contaminates the cross-correlation signal given that it is 100% correlated. We show further on, using similar dyes (fluorescein and Atto-647N) with identical two-photon excitation properties, that significant cross-correlation is observed for positive controls using doubly labeled oligonucleotides under identical excitation and detection conditions (Figure 7.5B).

We next sought to determine whether dimers or higher order L20-Cter species are formed upon association with the operator RNA. First, we carried out FCS measurements on Alexa-488-labeled L20-Cter in absence and presence of operator RNA containing both S1 and S2 recognition sites (Fig. 7.2A). Protein and RNA molecules were mixed at equimolar 150 n$M$ concentration, about 10-fold above the $K_D$ as determined from the anisotropy measurements. The autocorrelation curve (Fig. 7.2A) shifted to longer timescale in presence of the RNA, indicating the formation of a complex with lower diffusion rate compared to that of the free protein. Additional RNA produced no further shift, indicating saturation of the L20 protein under these conditions. We performed the same experiment using Atto-647N-labeled L20-Cter and obtained similar results (Fig. 7.2B). Then for both situations we added 300 n$M$ (saturating) L20-Cter with the complementary probe, and still detected in the same channel. The autocorrelation curve for the Alexa-488 L20-Cter (detected in channel 2—green) bound by the operator shifted back to shorter times when Atto-647N-labeled L20-Cter (red) was added, and vice versa. These observations indicate that the added protein competes with the protein present in the complex for binding, and hence that a 2:1 L20:operator RNA complex does not form. FCCS experiments (Fig. 7.2C and D) confirmed this conclusion. Absolutely no cross-correlation was observed for solutions containing equimolar concentration of each component (150 n$M$ operator RNA, Alexa-488 L20-Cter and Atto-647N L20-Cter). Thus we come to the surprising conclusion that despite the existence of two specific S1 and S2 sites on the operator RNA (which can each bind the L20-Cter protein when the other is mutated), only one monomer of L20-Cter interacts with the operator mRNA. Such a situation can hold only if there is strong negative cooperativity between the two sites, as suggested by chemical probing of the secondary structure (Allemand et al., 2007).

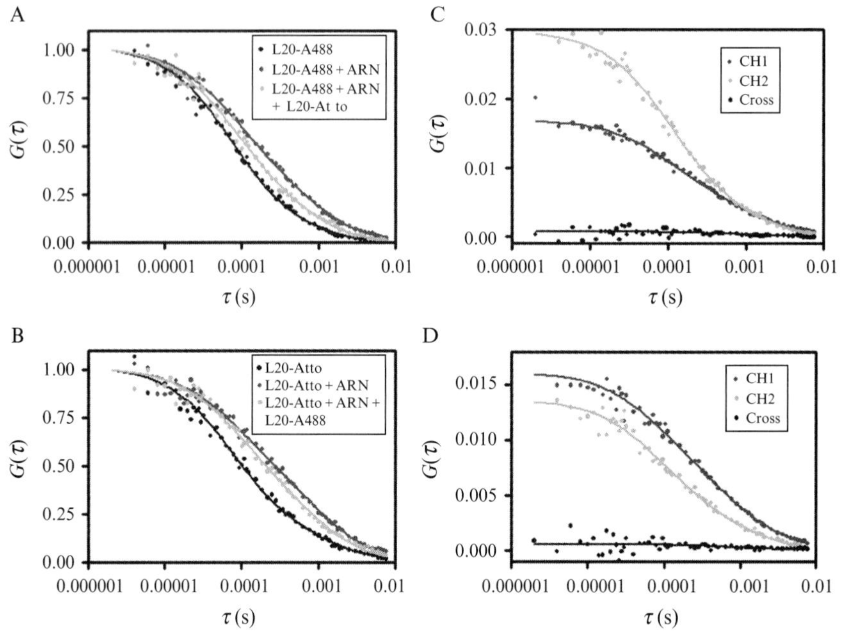


**Figure 7.2** FCS measurements on the L20-operator RNA complex. Combinations of L20-Cter-Alexa488, L20-Cter-Atto-647, and operator RNA at a nominal concentration of 150 n*M* were used. (A, B) Normalized autocorrelation profiles for channel 2 (L20-Cter-Alexa488) and channel 1 (L20-Cter-Atto-647). Dots correspond to the protein alone in (A) L20-Cter Alexa488 and (B) L20-Cter-Atto-647. Triangles correspond to data obtained after the addition of the RNA. Crosses correspond to the autocorrelation profile obtained after addition of 150 n*M* of (A) L20-Cter-Atto-647 to the L20-Cter-Alexa488/RNA complex and (B) L20-Cter-Alexa488 to the L20-Cter-Atto-647/RNA complex. (C, D) Auto correlation profiles from channel 1 (dots), channel 2 (open circles), and cross-correlation profiles (crosses) obtained for the solutions in which (C) 150 n*M* of L20C-ter-Atto-647 was added to the L20-Cter-Alexa488/RNA complex at a concentration of 150 n*M* and (D) 300 n*M* of L20-Cter-Atto-647 was added to the L20-Cter-Alexa488/RNA complex at a concentration of 300 n*M*. *Figure taken from Allemand et al. (2007).* (For color version of this figure, the reader is referred to the online version of this chapter.)

## 2.2. Transcriptional regulators of the central carbon metabolism in *Bacillus subtilis*

Our longstanding collaboration with the group of Stephane Aymerich has led to the application of fluorescence fluctuation methods to unravel the molecular mechanisms underlying the transcriptional regulation of the central carbon metabolism (CCM) in *Bacillus subtilis*. In this model organism of Gram-positive bacteria, the control of the switch between glycolysis and gluconeogenesis upon a change in carbon source is under the transcriptional control of two repressors, CggR and CcpN (Doan & Aymerich, 2003;

Fillinger et al., 2000; Servant, Le Coq, & Aymerich, 2005; Tännler et al., 2008). CggR controls the expression of the *gapA* operon encoding central glycolytic enzymes whose synthesis is stimulated in the presence of glucose through the binding of fructose 1,6 bis-phosphate (FBP) to CggR (Doan & Aymerich, 2003). Inversely, CcpN strongly represses the *gapB* and *pckA* genes encoding gluconeogenic enzymes that are produced only when cells are grown on noncarbohydrate carbon sources such as malate (Fig. 7.3). Both repressors recognize operator DNA of about 45–50 bp and made of two directly repeated half-sites. In the case of CggR, the operator sequence lies between the promoter and the ribosome binding site (Doan & Aymerich, 2003), whereas in the case of CcpN, at least one of the half-sites overlaps with the *gapB* or the *pckA* promoter (Licht & Brantl, 2006; Servant et al., 2005), suggesting different repression mechanism.

### 2.2.1 FCS measurements of CggR repressor oligomerization

Insights into the stoichiometry and thermodynamics of the interactions involving the CggR repressor were first provided by fluorescence anisotropy, combined with analytical ultracentrifugation. Using fluorescein-labeled oligonucleotides comprising the full-length *gapA* operator or only one of the half-sites, we showed that CggR interacts as a tetramer with its full-length DNA target, with each direct repeat constituting the binding site for one CggR dimer. The interaction of tetrameric CggR with DNA is very tight ($K_D < 0.5$ n$M$) and highly cooperative in the absence of the inducer FBP, whereas the cooperativity, but not the affinity of the CggR-operator interaction, is abolished upon inducer binding (Zorrilla, Chaix, et al., 2007). FCS experiments were first carried out on the CggR protein alone labeled with tetramethylrhodamine (TMR) (according to the same protocol as above using TMR succinimidyl ester). Comparison of the autocorrelation curves obtained as a function of increasing FBP and decreasing CggR concentration (Fig. 7.4) revealed an increase in the translational diffusion coefficient, consistent with the existence of higher order oligomers of CggR, confirmed by analytical ultracentrifugation, that were destabilized by the binding of FBP (Zorrilla, Doan, et al., 2007). These protein aggregates completely disappeared upon interaction of CggR with DNA, as observed by size-exclusion chromatography as well as FCS. Indeed, the autocorrelation curves recovered for a 45-bp fluorescein-labeled oligonucleotide corresponding to the *gapA* full-length operator were only marginally shifted in the presence of a 1000-fold excess of protein, the size change of the DNA probe upon CggR binding being insufficient to be monitored by this method (Zorrilla, Lillo, et al., 2008).

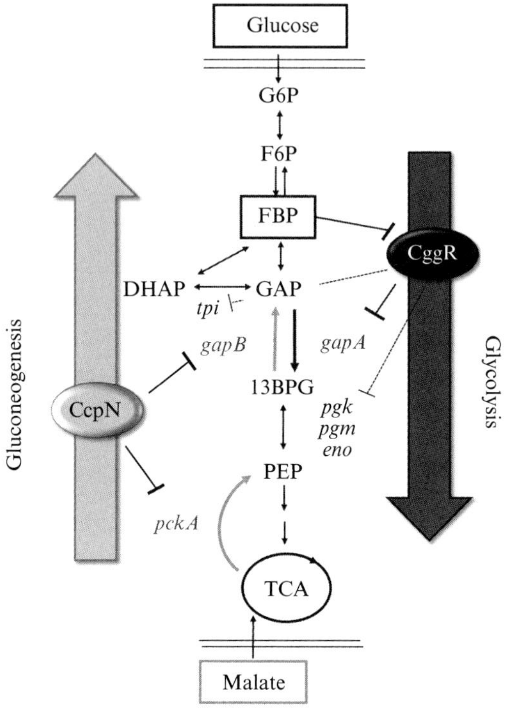


**Figure 7.3** Schematic of the central carbon metabolism of *B. subtilis*. Highlighted is the switch between glycolysis and gluconeogenesis controlled by the repressors CggR and CcpN. Regulatory proteins are in ellipses, and the genes coding for the enzymes are in small italic letters. When glucose is available for cell growth, fructose-1,6-biphosphate (FBP) accumulates and blocks the repressive action that CggR exerts on the transcription of *gapA* and four other central glycolytic genes (*pgk*, *pgm*, *eno*, and *tpi*). As CggR is transcribed from the same *gapA* operon that it represses, it is also an autorepressor. Inversely, when cells are grown on malate or other nonglycolytic carbon sources, the CcpN repressor is inhibited by an unknown mechanism involving YqfL, allowing expression of the essential gluconeogenic genes *gapB* and *pckA*. *Figure adapted from Ferguson et al. (2012)*. G6P: glucose-6-phosphate, F6P: fructose-6-phosphate, FBP: fructose-1,6-bis-phosphate, DHAP: di-hydroxy-acetone-phosphate, GAP: D-glyceraldehyde-3-phosphate, 13BPG: 1,3-di-phosphoglycerate, PEP: phosphoenol pyruvate; TCA: tri-carboxylic acid cycle. (For color version of this figure, the reader is referred to the online version of this chapter.)

### *2.2.2 FCCS measurements of CggR and CcpN repressor-operator interactions*

We thus turned to FCCS, coupled with other biophysical methods including SAXS and noncovalent mass spectrometry, to further characterize CggR–DNA complexes and investigate on the origin of the FBP-dependent

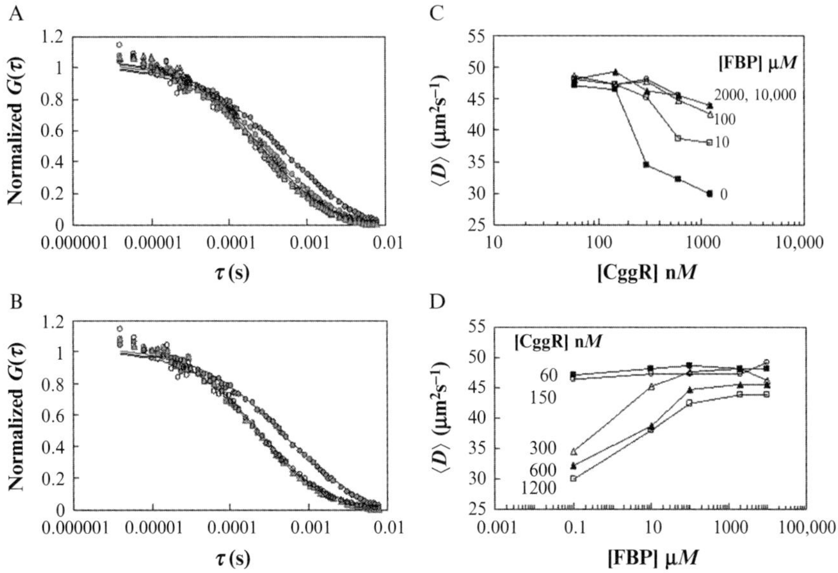


**Figure 7.4** Fluorescence correlation profiles of CggR as a function of FBP concentration. Experiments were carried out using TMR-labeled CggR. (A) Normalized autocorrelation curves recorded at 1.2 μ*M* CggR in absence (closed blue circles) and in presence of 10 μ*M* (red circles), 100 μ*M* (blue triangles), 2 m*M* (red triangles), and 10 m*M* (open blue circles) FBP. (B) Normalized autocorrelation curves registered at 1.2 μ*M* CggR in absence (closed blue circles) and in presence (open red circles) of 10 m*M* FBP and at 60 n*M* CggR in absence (open blue circles) and in presence (full red triangles) of 10 m*M* FBP. The fitting functions shown in the graphs correspond to best fitting of the experimental data to two independent diffusing species models. (C) The diffusion coefficient ⟨D⟩ as a function of repressor concentration in absence (closed circle) and in presence of 10 μ*M* (open circles), 100 μ*M* (closed triangles), 2 m*M* (open triangles), and 10 m*M* (closed squares) FBP. (D) ⟨D⟩ as a function of FBP concentration at 1.2 μ*M* (open squares), 600 n*M* (closed triangles), 300 n*M* (open triangles), 150 n*M* (open circles), and 60 n*M* (closed squares) CggR. *Figure taken from Zorrilla, Chaix, et al. (2007).* (For interpretation of the references to color in this figure legend, the reader is referred to the online version of this chapter.)

cooperativity (Chaix et al., 2010). Oligonucleotides of 23 bp corresponding to the half-site operator and labeled with either Atto-647N (red) or fluorescein (green) were mixed at equimolar concentration (60 n*M*) in absence and presence of CggR and FBP (Fig. 7.5). In absence of CggR, no cross-correlation between the oligonucleotides was observed (Fig. 7.5A), while FCCS on a doubly labeled hybrid DNA (Fig. 7.5B) led to the maximal level of cross-correlation expected given the labeling ratios of the two oligonucleotides (about 80% for both DNA probes). It is important to try to maximize

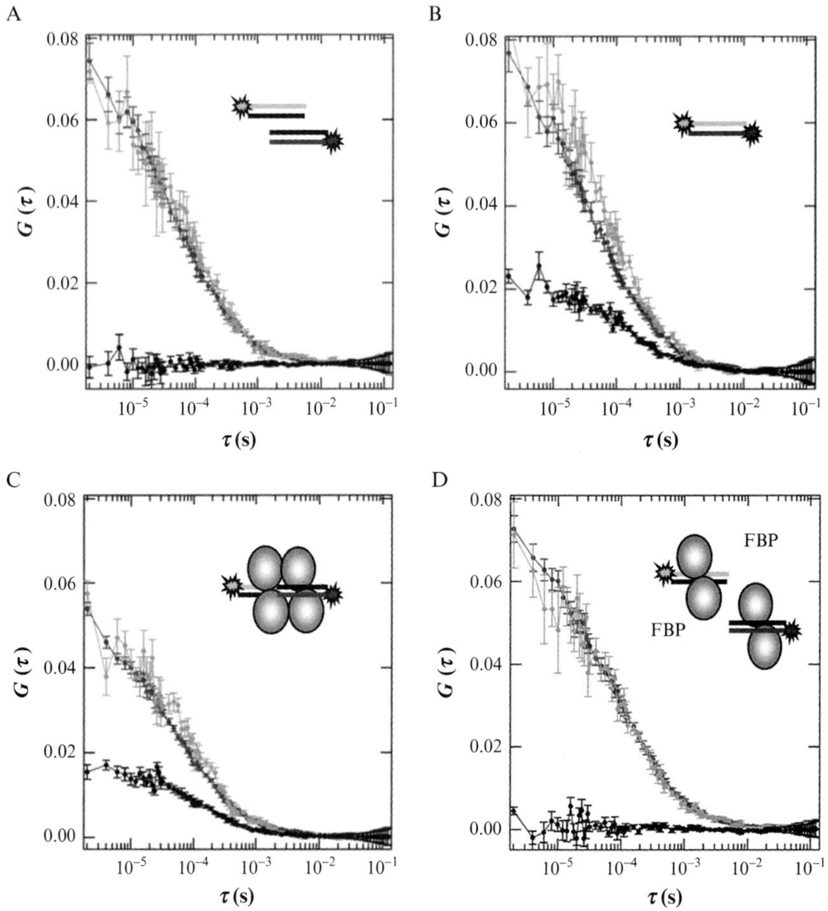


**Figure 7.5** Effect of FBP on CggR-operator interactions. Auto and cross-correlation profiles recovered from FCCS measurements with Atto-647N-labeled or fluorescein-labeled oligonucleotides corresponding to the CggR half-site operator in the presence/absence of the repressor and the inducer metabolite FBP. Schematics of the labeled DNAs and protein present in the sample chambers are shown. Red, green, and black curves correspond to the autocorrelation traces recorded in the red (675 nm) and green (525 nm) detection channels and the cross-correlation curve, respectively. (A) A mixture of the singly labeled dsDNA fragments showing the absence of cross-correlation signal. (B) A doubly labeled DNA hybrid serving as positive control of cross-correlation. The difference in the maximum amplitude of the auto and cross-correlation function ($G(0)$) denotes the partial labeling and hybridization of the two labeled DNA strands. (C) Cross-correlation upon addition of the repressor protein to the singly labeled DNA mixture of (A), demonstrating the CggR-mediated assembly of the DNA fragments. (D) Loss of cross-correlation signal upon addition of the inducer metabolite to the protein/DNA mixture of (C), demonstrating the FBP-induced disruption of the CggR/DNA ternary complex. Concentration was 60 n*M* for the labeled DNA fragments, 300 n*M* for CggR (monomer unit), and 0.5 m*M* for FBP. *Figure taken from Chaix et al. (2010).* (See Color Insert.)

the labeling ratios in such experiments, as this maximizes the dynamic range of the FCCS measurement. Interestingly, we found that in presence of a fivefold excess of CggR (in monomer units), but in absence of FBP, the maximal level of cross-correlation was observed (Fig. 7.5C), indicating that all of the target half-site molecules were involved in tetrameric complexes of CggR, 1 dimer per half-site. This was confirmed by mass spectrometry and SAXS. Upon addition of FBP, all of the cross-correlation disappeared, as the interactions between CggR dimers were abrogated by inducer binding (Fig. 7.5D). These results, along with the SAXS modeling, led us to propose a mechanistic model of repression in which the tetrameric CggR repressor functions as a roadblock to the transcribing RNA polymerase. FBP, by disrupting interactions between CggR dimers, diminishes the overall free energy of binding by eliminating the cooperative free energy. Under these conditions, the probability of CggR dissociation and hence the passage of the roadblock by the polymerase would be significantly enhanced.

Other applications of *in vitro* FFS-based methods on these systems of control of the CCM in *B. subtilis* include the use of FCCS for determining the stoichiometry of interaction of CcpN with its two target operator sites on the *pckA* and *gapB* promoters (Zorrilla, Ortega, et al., 2008). For these studies, we titrated the target oligonucleotides labeled with Atto-647N (red) at a concentration of 5 n$M$ (well below the $K_D$ of the interaction as determined by fluorescence anisotropy) with fluorescein (green)-labeled CcpN repressor over the concentration range indicated in Fig. 7.6. From the amplitudes of the autocorrelation functions at time zero from the green, $G_G(0)$, and red, $G_R(0)$, channels, we calculated the total concentration of labeled repressor and operator, respectively. The time-zero amplitude of the cross-correlation function, $G_x(0)$, is directly proportional to the concentration of complex, such that a plot of the ratio of the amplitudes of the cross-correlation and the autocorrelation from the protein channel, $G_x(0)/G_G(0)$, versus the total concentration of repressor yields the binding isotherm (Zorrilla, Ortega, et al., 2008).

The absolute value of $G_x(0)$ depends upon the labeling ratios of the molecules and the stoichiometry of the final complex. In addition, as CcpN dimerizes over the concentration range of the titration, we also corrected for the self-association state (and hence the lower particle number with respect to monomer) of the protein. Hence, the $G_x(0)/G_G(0)$ ratio at the plateau must be multiplied by this factor of two, yielding an apparent stoichiometry for the CcpN/oligonucleotide complexes of 1.6 for the *gapB* target (which is unphysical) and 3.0 for the *pckA* target. If we further correct the amplitude

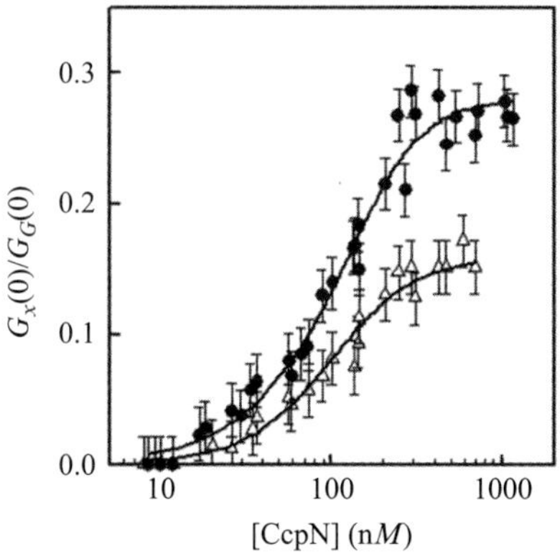


**Figure 7.6** Stoichiometry of protein–DNA interactions by FCCS. FCCS-based titrations of the Atto-647N-labeled *gapB* (triangles) and *pckA* (circles) target oligonucleotides, with the fluorescein-labeled repressor. $G_x(0)$ and $G_G(0)$ are the amplitudes of the cross-correlation and protein autocorrelation functions retrieved from fitting of the profiles. The ratio $G_x(0)/G_G(0)$ represents the fraction of complex with respect to the total concentration of DNA, and the high concentration plateau of the curves is proportional to the stoichiometry of the complexes. Solid lines correspond to the best fits of the data to the proposed energetic models. The concentration of DNA in the titrations was 5 n*M*. *Figure taken from Zorrilla, Ortega, et al. (2008).*

ratios for a 10% contribution from a higher order aggregation (which was present in the CcpN protein solution) and assuming a 12-mer with a diffusion coefficient, $D = 4.8\ \mu m^2/s$, then the stoichiometry is calculated to be 2 CcpN molecules for the *gapB* target and 4 for the *pckA* target. In any case, it is clear from examination of the plateau value for the $G_x(0)/G_G(0)$ ratio that the stoichiometry for the CcpN complex with the *pckA* operator is twice that of the *gapB* operator. We note that this difference in stoichiometry may simply be a result of the length and sequence of the target oligonucleotides used in the study, and may not have a profound biological significance. Indeed, chemical interference foot-printing experiments suggest that the DNA sequence contacted by CcpN in the *gapB* promoter may slightly extend outside the binding sites identified in the original study (Servant et al., 2005) on which the design of our target oligonucleotides was based.

# 3. FFM IN LIVE BACTERIAL CELLS

The results summarized above provided important insight into the physical basis for the regulation of gene expression, in particular in the CCM of *B. subtilis*. However, the link between biophysical measurements *in vitro* and the adaptation to different carbon sources of bacterial populations *in vivo* can be challenging to say the least. Numerous questions arise concerning the extent to which the energetic parameters measured *in vitro* are relevant to the behavior of these systems *in vivo*. For example, what are the real available concentrations of glucose, FBP, and malate in these cells and under different growth conditions? How much do they change upon changing carbon source? How does the crowded environment of the bacterial cytoplasm affect the oligomeric states of the repressors and their interactions with operators, inducer metabolites, or protein partners? What is the intracellular concentration of the regulatory proteins and how does it change upon a metabolic switch? What is the noise associated to the stochastic activity of the regulated promoters in individual bacterial cells and can this biological noise be related to the molecular mechanisms of regulation? To begin to grapple with these fundamental questions, we decided to take the quantitative capacity of two-photon fluctuation fluorescence microscopy and move into measurements in live cells.

## 3.1. Adaptation of fluctuation approaches to imaging bacteria

### *3.1.1 Geometric concerns and the advantages of N&B*

The application of FFS or FFM in live bacterial cells using fluorescent protein (FP) reporter strains presents a number of specific challenges that we had to overcome prior to carrying out *in vivo* studies on *B. subtilis* CCM regulation. The first difficulty stems from the small size of the bacterial cells (Fig. 7.7A), approximately 700 nm in diameter and generally 1–2 μm in length, depending upon growth conditions. Indeed, our two-photon point spread function (PSF) is contained within the cell only when it is centered in the $x$–$y$ direction and extends above and beyond the cell in the $z$ direction. Our PSF in solution is 0.3 fL, but is further diminished by the inherent $z$-section of the bacteria. Indeed, the PSF we determine within bacillus cells is 0.07 fL. We considered that this excitation volume that we termed $Vol_{ex}$ is constant in case of the rode shape cells of *B. subtilis* whose diameter varies very little under normal vegetative growth conditions. Second, the small size of the cell means that fluorescent molecules are highly confined, and hence much more likely to be bleached. Thus, in the time it takes to make a point FCS measurement inside a bacterial cell, all of the fluorescent GFP molecules inside it are bleached (Fig. 7.7B). In order to overcome these limitations due to the geometry of our samples, we turned to two-photon scanning number and brightness (2psN&B) originally developed by the group of Gratton (Digman, Dalal, Horwitz, & Gratton, 2008; Digman,

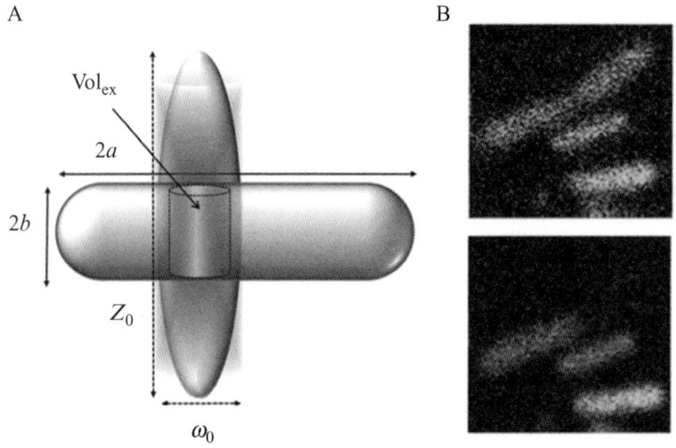


**Figure 7.7** Geometric constraints of FFM in bacteria. Size and geometry of the point spread function (PSF) relative to the bacterial cytoplasm. (A) A schematic showing the size and orientation of the infrared PSF relative to a 1-fL bacterial cell. The vol$_{ex}$ corresponds to the intersection between the two volumes, taking into account the quadratic dependence of excitation with laser power. (B) Fluorescent images of *B. subtilis* cells expressing *gfpmut3*, before (above) and after (below) a typical FCS experiment showing complete photobleaching of the targeted cell. *Figures adapted from Ferguson et al. (2011).* (See Color Insert.)

Wiseman, Choi, Horwitz, & Gratton, 2009). N&B approaches are amply presented in another chapter in this volume and so will not be presented in detail here. Suffice it to say that they are based on the calculation of the variance of the fluorescence intensity at each pixel in multiply raster scanned images. The general expressions for the shot noise corrected molecular brightness, ε, in counts per dwell time per molecule, and absolute number, *n*, in the PSF for a given pixel are given by the expressions:

$$\varepsilon = \frac{\left[(F_i - \langle F \rangle)^2 - \langle F \rangle\right]}{\langle F \rangle} \qquad [7.1]$$

$$n = \frac{\langle F \rangle^2}{\left[(F_i - \langle F \rangle)^2 - \langle F \rangle\right]} \qquad [7.2]$$

The enormous advantage of scanning N&B, as concerns application to bacteria, is that the raster scan, involving very short pixel dwell times (faster than diffusion and in our case 50 μs), means that the beam does not remain in the same area for very long and returns to visited areas with rather low frequency (in our case, every 4 s). Hence, the effects of photobleaching are significantly diminished. With our setup we could increase the laser power to 15 mW when performing N&B (compared to a maximum of 10 mW for point FCS) and noticed no significant photobleaching inside the bacterial cells after a total acquisition time at each pixel of 2.5 ms. Moreover, in this raster scan mode, data can be collected for an entire field of view, containing dozens or even hundreds of individual bacterial cells. This is particularly important for acquiring statistically relevant quantitative data on cell-to-cell variations within bacterial populations. We note that in principle, in raster scanning N&B measurements, the time-dependent information is lost, and therefore, the diffusion rates of the fluorescent particles cannot be determined. However, dynamic information in live cells in the raster scanning mode necessary for measurements in bacteria can be obtained by turning to Raster Scanning Image Correlation Spectroscopy, or RICS (Digman et al., 2005), also presented in this volume.

### *3.1.2 Advantages of two-photon excitation for FFM*

The two-photon excitation mode proved also to be much more compatible with *in vivo* measurements than the one-photon mode using visible lasers. First, the photobleaching was much less intense for laser intensity levels yielding similar signal-to-noise ratios. Second, and perhaps most importantly, the autofluorescent background of the bacterial cytoplasm was drastically reduced with excitation in the infrared (930 nm in the work presented here) and became almost negligible when an appropriate bandpass emission filter is used (e.g., a 525/50 filter (Chroma, Inc.) for GFP). This is particularly important for measuring very low expression levels, especially in case of genes that are strongly repressed or for regulatory proteins that can be present in a few copy numbers per cell. We present below the example of the *gapB* gene, whose expression in cells grown on glucose is too low to be quantified by classical methods, but for which we could successfully perform cell-by-cell measurements by two-photon scanning N&B, revealing a highly heterogeneous basal activity of the $P_{gapB}$ promoter in its repressed state (Ferguson et al., 2011).

## 3.2. Counting up fluorescent molecules in live bacterial cells

The primary advantage of the fluorescence fluctuation approaches compared to methods based on total intensity measurements is that they allow for the direct determination of absolute concentration of diffusing fluorescent particles, an evident interest for quantitative studies in living cells. However, the conditions required for the accurate counting of fluorescent molecules can reduce the range of applicability of the FFS- and FFM-based methods.

First, the microscope must be perfectly aligned, a nontrivial task in case of a two-photon excitation setup, and the size of the PSF has to be precisely determined prior any experiment, typically by point FCS using a 60 n*M* solution of fluorescein (for experiments using green, blue, or yellow FPs) or rhodamine (for red FPs). Second, the amplitude of the light fluctuations must be sufficient in order to be reliably measured. Hence, fluctuation methods such as N&B are particularly well suited in case of very bright proteins that are produced in rather low amount by the bacterial cells, but become problematic when using dimmer proteins. Overproduction of such FPs cannot compensate for their low intensity as this will blur the light variations. One has also to bear in mind that only fluorescent molecules diffusing sufficiently fast in the bacterial cytoplasm can be counted by FCS- and N&B-based methods that are in essence blind to immobile particles. The timescales of single point FCS, scanning FCS and N&B, are very distinct. Single point FCS is only useful for relatively fast diffusing species with diffusion times faster than 10 ms. Scanning FCS can extend this range by a factor of 10. N&B, on the other hand, is most useful for slowly diffusing molecules, as each pixel is sampled every few seconds from one raster scan to the next. Moreover, calculation of absolute molecule numbers by these approaches may be erroneous due to the presence of multiple molecular species with different mobility (e.g., bound or unbound to DNA) and/or brightness (e.g., due to different oligomerization states). As mentioned above, the background fluorescence level is significantly reduced thanks to infrared two-photon excitation, yet it cannot be neglected for the precise quantification of fluorescent reporter proteins of moderate brightness and/or number. For such analysis, it is thus mandatory to perform measurements for the receiver strain (with no FP encoding gene) grown and analyzed as the reporter strains (producing the FPs) in order to determine the background contribution to the calculated averaged numbers.

Following these methodological adaptations and precautions, we have established a flow chart for live-cell quantification of diffusing GFP molecules by two-photon scanning N&B (Ferguson et al., 2011) as summarized in Fig. 7.8. In N&B, in order to obtain absolute numbers of particles, one must correct the measurements for shot noise as noted above (Digman et al., 2008). This is not feasible on an individual pixel level, because in many cases, the average photon counts are smaller than the shot noise, leading to negative true brightness values. Thus, some level of pixel averaging must be carried out. In the case of measuring GFP production, this is quite simple, as the molecular brightness of the GFP is independent of its location in the

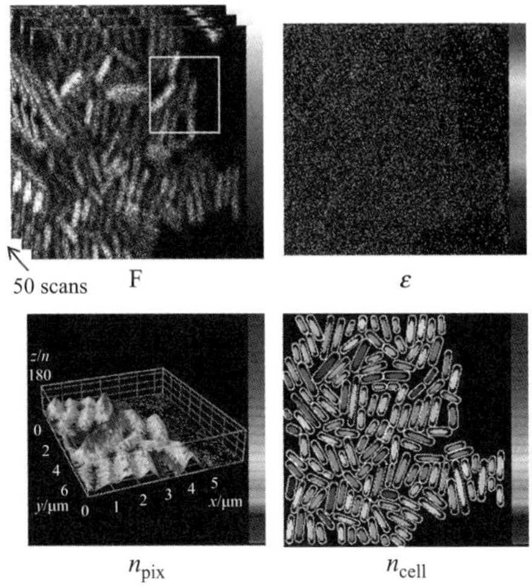


**Figure 7.8** Schematic of 2psN&B experiments in bacteria. A stack of 50 raster scans of agarose-immobilized live cells of *B. subtilis* expressing *gfpmut3* are recorded using infrared (930 nm) laser excitation and a dwell time of 50 μs at each pixel (faster than GFP diffusion); full scale of fluorescence intensity (*F*) is 10 photon counts/pixel/50 μs laser dwell time. The fluorescence fluctuations relative to the mean at each pixel are used to calculate the pixel-based maps of the true (shot noise corrected) molecular brightness (ε, full scale 1 photon/molecule/50 μs dwell time) and the number ($n_{pix}$) of the fluorescent particles detected in the two-photon excitation volume ($vol_{ex}=0.07$ fL inside *B. subtilis*); a 3D surface plot of $n_{pix}$ is shown for the white-delineated area of the above intensity panel. Bottom right panel: cartoon representation of the individual cells autodetected using PaTrack (Espenel et al., 2008) and showing the 50% central pixels used for averaging the particle number in each cell ($n_{cell}$); the full scale for the $n_{pix}$ and $n_{cell}$ maps is 180 molecules/$vol_{ex}$. *Figure taken from Ferguson et al. (2012).* (See Color Insert.)

image. Indeed, there was no spatial bias to the molecular brightness values calculated from 50 rapid raster scans of a field of view containing bacterial cells expressing GFP from a promoter fusion (Fig. 7.8, top panels). Hence, we could average the values of the true molecular brightness and use this average value to calculate the pixel-based true number of molecules in the PSF. Because of the convolution of the PSF with the size of the bacteria, we modified our in-house software (Espenel et al., 2008) to identify bacterial cells in our image and to pick the central 50% of the pixels within each cell. The true number values for these central pixels (Fig. 7.8, bottom left panel) were then averaged to provide a cell-based value of the number of fluorescent molecules in the PSF (Fig. 7.8, bottom right panel). After eventual correction for background contributions, these numbers correspond to the intracellular concentration of GFP in each bacterium in units of number of molecules in the 0.07 fL excitation volume, $Vol_{ex}$. Our software also calculates the volume of each bacterial cell, and hence we can report, as well the absolute number of molecules in each cell, rather than their concentration. We find concentration to be preferable because it is not subject to the halving that occurs with each cell division. We performed a number of control experiments to test our system. For example, we verified that the absolute values for numbers of molecules measured by N&B for solutions of fluorophores under the imaging conditions used for the bacteria were equivalent to those measured by FCS. We also determined that the uncertainty inherent to our measurements of the absolute value of the number of GFP molecules diffusing in solution under the conditions of imaging of the bacteria was 5% (Ferguson et al., 2011).

## 3.3. Biological materials

### 3.3.1 Fluorescent proteins

The choice of the FPs to be used for reporting on the biological phenomenon of interest is particularly crucial for *in vivo* studies by two-photon fluorescence fluctuation approaches (Drobizhev, Makarov, Tillo, Hughes, & Rebane, 2011; Wiednmann, Oswald, & Neinhaus, 2012). As mentioned above, the molecular brightness of the FP is a critical property that determines the dynamic range of intracellular concentrations that can be reliability measured. In our studies, we preferentially use GFP variants (GFPmut2 or mut3) that were originally selected for their exceptionally high brightness and stability in *E. coli* (Cormack, Valdivia, & Falkow, 1996). The use of monomeric FPs (at least in the range of applicability of FF-based methods) is another prerequisite for quantification of absolute molecule numbers, as well as for investigation using FP-tagged proteins to study oligomerization or protein interactions by dual-color approaches. Although a rather wide panoply of monomeric GFP-like proteins is now available (Shaner et al., 2004), only a few of them turned to be useful for *in vivo* studies in bacterial cells. This is in part due to the low expression levels of the FP encoding genes, and in particular, the poor translation rate of the mRNAs because of the differences in the codon usage between bacterial and eukaryotic species (from which the FPs usually originate). To overcome this problem, it is necessary to get a modified version of the *fp* genes in which amino acid codons have been optimized specifically for the receiver species. If not available, the less time-consuming solution is to order synthetic genes. David Rudner at Harvard has graciously provided us with *cfp* and *mCherry* genes with optimized codons for expression in *B. subtilis*. In addition to the translation rate, the folding time, maturation time, and *in vivo* stability of the FPs are other important parameters that may affect the concentration of fluorescent reporters measured at steady state or upon induction. The time and rate-limiting step of the entire synthesis process varies greatly between different FPs or even variants of the same FP. Moreover, chromophore formation and brightness are sensitive to different extent to physiological conditions, in particular, pH and oxygen supply. It is thus recommended to take into account these parameters for the choice of appropriate FP reporters depending on their intended bacterial applications. The GFPmut3 reporter that we used is very fast maturing (under the minute) and highly stable (for over 12 h) in *B. subtilis* (Botella et al., 2010), rendering it particularly well adapted for monitoring transcriptional response to environmental changes, as recently reported in a genome-wide scale study of the reorganization of *B. subtilis* regulatory networks upon nutritional shifts (Buescher et al., 2012). This GFP variant is also known to be rather pH sensitive, its brightness diminishing when exposed to acidic conditions; however, this property (which can be used for monitoring the cytoplasmic pH) is not expected to affect fluorescence measurements in *B. subtilis* cells capable of maintaining their internal pH in a narrow range (around pH 7.5) under a wide variety of growth conditions (Kitko et al., 2009).

### 3.3.2 Bacterial strains

Construction of bacterial strains carrying recombinant genes encoding FPs or FP-tagged proteins is the next, often limiting step of *in vivo* fluorescence studies. In *B. subtilis*, however, incorporation of gene constructs at specific loci by single or double crossover is a rather easy procedure, allowing stable integration and expression of a single copy of the recombinant genes in the context of chromosomal DNA. This is a noticeable advantage compared to the expression from multicopy plasmids, not only for obvious functional reasons but also for technical reasons associated to the above-mentioned limitations of fluctuation-based methods for quantifying overproduced FPs. Shuttle plasmids carrying appropriate *fp* constructs and antibiotic resistance are constructed in *E. coli* and used to transform competent cells of the *B. subtilis* receiver strain. The rather large spectrum of antibiotics that can be used for selection of *B. subtilis* transformants allows for successive gene integration (or deletion) steps. Spectinomycin at 100 μg/mL is nowadays the most popular selection but erythromycin (0.5 μg/mL), chloramphenicol

(5 μg/mL), or phleomycin (0.5 μg/mL) is also commonly used. Except for specific applications, kanamycin should be avoided because the resistance efficiency depends on the gene copy number and it can therefore promote chromosomal amplification of *fp* genes integrated by single crossover. Even when using appropriate antibiotics and double-recombination schemes, multiple integration of the *fp* genes is always possible and it is therefore recommended to test their activity in three independent transformants (usually by bulk fluorescence intensity measurement). Note also that it is also important to verify the functionality of the FP-tagged proteins and if possible compare their activity to that of their wild-type counterparts. In case of CggR- and CcpN-GFP fusions, this could be performed with the conventional lacI reporter system already available for monitoring the activity of the wild-type and mutant repressors (Doan et al., 2008; D. Le Coq, unpublished results).

### 3.3.3 Microscopy samples

The preparation of bacterial cultures for imaging two-photon microscopy requires no specific features. Because *B. subtilis* cells tend to lyse or sporulate after reaching stationary phase, it is important to prepare bacterial samples from exponentially growing cultures. For this purpose, we followed a protocol adapted from that developed by the BaSysBio consortium for high-throughput analysis of *B. subtilis* promoter activities (http://www.basysbio.eu/). Typically, we took cells from glycerol stocks and grew them at 37 °C for 2–3 h in 2 mL LB medium in presence of antibiotic(s), then inoculated serial dilutions in minimal synthetic medium (such as M9) supplemented with appropriate nutrients or inducer metabolite. This was conveniently performed in 24-well plates incubated overnight at 30 °C in a plastic box. The next day we picked one of the cultures that was in exponential phase as indicated by an $OD_{600}$ between 0.2 and 0.6 and used it to inoculate 5 mL of the same M9 medium. Cells were diluted to $OD_{600}$ between 0.1 and 0.2 and grown in culture plates in a shaker at 37 °C. One-hundred microliters of aliquots from the cell cultures were removed to perform OD measurements, as well as recording of the average fluorescence intensity in 96-well microtiter culture plates using a Saphir2 microplate reader (TECAN). This was important to relate bulk fluorescence measurements to single-cell microscopy analysis.

Samples for N&B two-photon microscopy were prepared generally on agarose pads. This allows for immobilization of the bacterial cells, which is an absolute prerequisite for fluctuation analysis, as well as their spreading in a single-cell monolayer on the coverslip. We have also used coverslips or microfluidic devices treated with poly-lysine for immobilization, but the cell density and horizontality could not be easily reproduced. All samples were prepared by centrifuging 0.5–1-mL aliquots of the cell cultures at 10,000 rpm for 2 min and resuspending them in cold M9 medium to an $OD_{600}$ of 10–20. The samples were then left on ice until we were ready for microscopy measurements. The agarose pads were made by depositing 60 μL of 1.5% agarose–M9 medium (eventually supplemented with 0.5% glucose or malate) on a glass coverslip to which a double-sided adhesive silicone spacer ring of 1 × 20 mm (Invitrogen, Carlsbad, CA) was attached. The agarose pad was allowed to dry in a laminar flow hood for 12 min to render the agarose stiff enough to immobilize the bacteria, yet thick enough to touch a second coverslip applied on top. Then, 2–3 μL of the concentrated bacterial samples were deposited in the center of the agarose pad. Bacterial cells were allowed to sediment for 1 min before sealing the second glass coverslip. The sample was then placed in an AttoFluor cell chamber (Molecular Probes, Carlsbad, CA) and imaged quickly (within 15 min) under the microscope. This protocol allowed for a consistent, immobilized monolayer of cells with an intermediate–high cell density.

## 3.4. Measuring promoter activity in live bacterial cells by two-photon scanning N&B

### 3.4.1 Activity of an IPTG-inducible promoter, detection limits, and stochastic noise

Our first interest in applying 2psN&B to live bacterial systems was to measure promoter activities by counting the absolute number of FP molecules produced from transcriptional fusions introduced in the *B. subtilis* chromosome. We tested our approach using an IPTG-inducible promoter, $P_{hyperspank}$, in *B. subtilis* fused to the gene sequence for GFPmut2 (another highly stable and bright variant of GFP (Cormack et al., 1996)) or the codon-optimized CFP by integration via double crossing-over at the *amyE* locus of *B. subtilis* (see (Ferguson et al., 2011) and references therein). We compared the true molecular brightness and true number values obtained for these two fluorescent reporters in single cells as a function of the concentration of IPTG (Fig. 7.9). Expression above the background autofluorescence of the BSB168 receiver strain is detected at similar levels for both the CFP and GFP un-induced strains. The high standard deviation from the mean observed in the absence of inducer reflects the high cell-to-cell variations in the basal activity level of the $P_{hyperspank}$ promoter. Upon addition of IPTG,

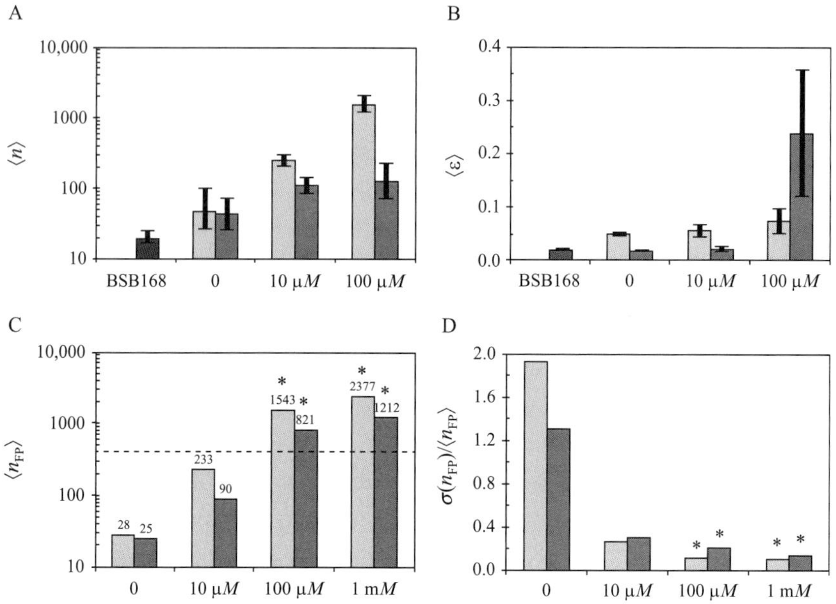


**Figure 7.9** Cell-based N&B analysis of the $P_{hyperspank}$ promoter activity in *B. subtilis*. The fluorescence images obtained by 2p fluctuation microscopy for the GFPmut2 (green or light grey) or CFP (blue or dark grey) producing strains, and the BSB168 background strain was analyzed following the flow chart reported in (Ferguson et al., 2011); $\langle n \rangle$ is the average number of fluorescent molecules detected in the cells (expressed in number of molecules/vol$_{ex}$); $\langle \varepsilon \rangle$ their average molecular brightness (expressed in counts per molecule per 50 μs dwell time); $\langle n_{FP} \rangle$ the average number of GFP or CFP deduced after background subtraction (expressed in number of molecules/vol$_{ex}$); and $\sigma(n_{FP})/\langle n_{FP} \rangle$ is the coefficient of variation of $n_{FP}$ over the mean, reflecting the cell population heterogeneity. Values indicated by an asterisk were calculated from fluorescence intensity measurements and brightness values determined at lower concentration. The dotted line indicates the upper limit for N&B analysis. *Figure adapted from Ferguson et al. (2011).* (For color version of this figure, the reader is referred to the online version of this chapter.)

the expression level increases by nearly 2 orders of magnitude in case of GFP, whereas the true brightness values (ε) remain independent of inducer concentration, as expected, and are about twice the value of the brightness of the CFP. The uncertainty on ε values increases with increasing expression levels because, as explained above, the amplitude of the fluctuations used to calculate the brightness is inversely proportional to the number of particles. At the expression levels afforded by addition of 100 μ*M* IPTG, the true brightness, and hence the true number (deduced after background subtraction), of CFP molecules is very poorly defined, setting the upper limit of FFS measurements under these imaging conditions at about 200 molecules in

$Vol_{ex}$ or around 5 $\mu M$ CFP. Because of the higher molecular brightness of the GFP, this limit is closer to 1000 molecules/$Vol_{ex}$ or 25 $\mu M$ for GFP. Above those intracellular concentrations, the linear relationship between average fluorescence intensity and number of FP molecules can be used as a calibration to estimate promoter activity for very high expression levels. Analysis of the cell-to-cell heterogeneity showed that the high variability in gene expression observed under conditions of repression at steady state is very much reduced after only 2 h of induction at 10 $\mu M$ IPTG. In the fully induced cell population (at 1 m$M$ IPTG), this "biological noise," as measured by the coefficient of variation, tended to a constant value of about 25%, well above the experimental error we determined (5%).

### *3.4.2 Single-cell analysis of gene expression in the CCM of* B. subtilis

Next we applied this approach to measure the carbon source dependence of the transcriptional activities of CCM gene promoters in *B. subtilis* (Ferguson et al., 2012). We focused on the glycolytic *cggR* promoter ($P_{cggR}$, upstream of the *cggR* gene, the first coding sequence of the *gapA* operon, induced on glucose and self-repressed on malate), the gluconeogenic *gapB* ($P_{gapB}$) and *pckA* ($P_{pckA}$) promoters under the control of the CcpN repressor, and the nonregulated $P_{ccpN}$ promoter. Strains bearing nonmutagenic *gfpmut3* transcriptional fusions of these promoters at their original loci were constructed using the pBaSysBioII plasmid designed for Live Cell Arrays (Botella et al., 2010). We measured the true number of molecules produced from these promoters under glucose (glycolytic) and malate (gluconeogenic) (Fig. 7.10). Very strong expression of the $P_{cggR}$ promoter was observed under glucose, while this promoter shows low, but very heterogeneous expression patterns under malate. Expression from $P_{gapB}$, similar to that from $P_{pckA}$ (not shown), is strong under malate but very highly repressed under glucose. Indeed, we were able to measure for these promoters an average of about 3 molecules of GFP in $Vol_{ex}$ above the autofluorescent background. As expected, the expression from the *ccpN* promoter was independent of the carbon source (not shown).

The ability to measure absolute numbers of molecules was crucial to our *in vivo* study of the regulation mechanisms of the CCM repressors, and comparison with the molecular models proposed from *in vitro* studies (Chaix et al., 2010; Licht, Golbik, & Brantl, 2008). Thanks to our 2psN&B analysis, we were able to measure accurately cell-to-cell variations in the activity of CggR- or CcpN-regulated promoters in both their on and off states. From the protein number distributions, we could extract the noise

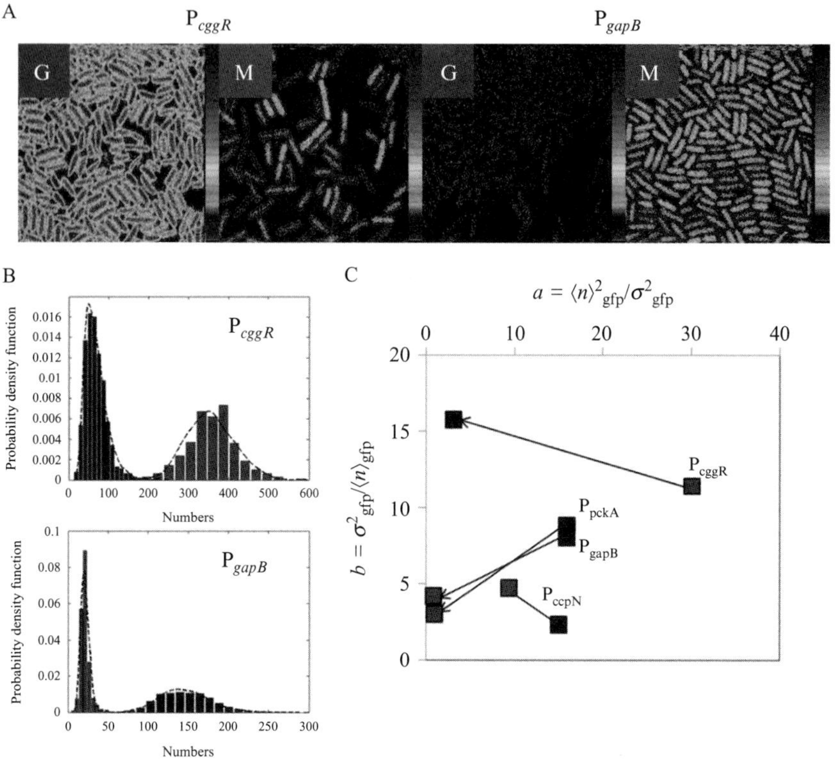


**Figure 7.10** Cell-by-cell quantification and noise in the activity of *B. subtilis* promoters implicated in the main switch between glycolysis and gluconeogenesis. (A) Pixel-based fluorescent particle number maps of *B. subtilis* cells expressing *gfpmut3* transcriptional fusion from $P_{ccgR}$ and $P_{gapB}$. Cells harvested from liquid cultures containing 0.5% glucose (G) or 0.5% malate (M) as the sole carbon source were immobilized on agarose pads and imaged by 2psN&B. The full scale is 360 molecules/vol$_{ex}$. (B) Cell-based particle number ($n_{cell}$) distributions for the indicated promoter–*gfpmut3* fusion strains grown on glucose (red or grey) or malate (blue or black). Models of stochastic gene expression fitting the experimental data (dotted lines) were generated for the repressible promoters based on available knowledge about their control mechanism. (C) Noise patterns obtained by plotting two parameters of stochastic gene expression, computed from the first two moments of the protein number distribution: the apparent frequency of protein production burst per cell cycle ($a = \langle n \rangle^2/\sigma_n^2$) and the average number of protein molecules produced per burst, related to the Fano factor ($b = \sigma_n^2/\langle n \rangle$). Distinct changes in noise patterns were observed upon a switch in carbon source, reflecting distinct underlying molecular mechanisms of repression (see main text). *Figure adapted from Ferguson et al. (2012).* (See Color Insert.)

parameters associated to stochastic gene expression in terms of transcriptional burst size and burst frequency (see the legend of Fig. 7.10C). Interestingly, we found completely different noise patterns for the glycolytic and gluconeogenic promoters. In particular, in their repressed states, the $P_{ccgR}$ and $P_{gapB}$ promoters exhibited very different bursting behavior. Mathematical modeling based on the molecular mechanisms deduced from biological and biophysical data, including the results presented above in this review, allowed interpretation of these noise patterns. The burst size for the $P_{ccgR}$ promoter was very large under repression, consistent with the roadblock mechanism for repression by CggR. Upon transient dissociation of the repressor, multiple RNA polymerase molecules can resume transcription before rebinding of the active tetrameric repressor. In contrast, the CcpN repressor binds to an operator site overlapping the $P_{gapB}$ promoter where it is thought to interact with the RNA polymerase and inhibit promoter escape (Licht & Brantl, 2009; Licht et al., 2008). Upon switching of the promoter from inactive to active states (by dissociation or conformational change of CcpN), only one RNA polymerase molecule can escape the promoter, because the formation of another initiation complex is slower than the switching time of the promoter; hence, the burst size is small.

## 4. CONCLUSIONS AND PERSPECTIVES

In this short review, several examples of the use of fluctuation analysis to understand the molecular mechanisms of gene expression have been presented. The results presented here are not meant to represent the massive number of studies by many groups throughout the world using fluorescence fluctuation techniques to study biomolecular interactions. However, as an ensemble, these studies, particularly focused on the control of the CCM in *B. subtilis*, provide an overview of the application of multiple fluorescence fluctuation strategies to the in-depth characterization of one system. One of the thorniest problems typically encountered in the *in vitro* analysis of such systems is the stoichiometry of the complexes and how it may change as a function of ligand or protein concentration, and salt or other important parameters. Fluorescence fluctuation approaches can be of great help in untangling these issues. While the examples given here are based on cross-correlation results and diffusion coefficients, measurement of the molecular brightness itself is also an obvious approach. Cross-correlation, however, is particularly interesting because it is so definitive. The lack of cross-correlation is absolute proof of the lack of complex formation, and differences in the cross-correlation plateau values (normalized to the amplitude of the autocorrelation) arise from differences in complex stoichiometry.

Beyond the detailed *in vitro* characterizations afforded by fluctuation techniques, the possibilities that fluctuation measurements provide for quantitative characterizations of biomolecular systems *in vivo* at the level of single cells and with single molecule sensitivity appear to be most exciting. We have shown how FFM in 2p-scanning N&B mode can be used to assess absolute expression and biological noise in live *B. subtilis* cells. In unpublished work, we have successfully applied 2psN&B in *E. coli*, yeast cells, and neurons. Adapting such measurements to high-throughput acquisition opens the possibility for true quantitative systems biology. Two color adaptations allow for the simultaneous analysis of expression covariances, and hence correlations in gene expression patterns. Current studies involve as well *in vivo* measurements of the behavior of protein–FP fusions, rather than promoter–FP fusions, and in particular, the characterization of protein complexes. The application of such approaches to any number of key biological networks will provide in the near future key information on their function in the context of fundamental biological processes such as adaptive responses, but also for understanding and modulating such networks in the context of human health and disease.

## REFERENCES

Allemand, F., Haentjens, J., Chiaruttini, C., Royer, C., & Springer, M. (2007). Escherichia coli ribosomal protein L20 binds as a single monomer to its own mRNA bearing two potential binding sites. *Nucleic Acids Research, 35*, 3016–3031.

Ban, N., Nissen, P., Hansen, J., Moore, P. B., & Steitz, T. A. (2000). The complete atomic structure of the large ribosomal subunit at 2.4 A resolution. *Science, 289*, 905–920.

Botella, E., Fogg, M., Jules, M., Piersma, S., Doherty, G., Hansen, A., et al. (2010). pBaSysBioII: An integrative plasmid generating gfp transcriptional fusions for high-throughput analysis of gene expression in Bacillus subtilis. *Microbiology, 156*, 1600–1608.

Buescher, J. M., Liebermeister, W., Jules, M., Uhr, M., Muntel, J., Botella, E., et al. (2012). Global network reorganization during dynamic adaptations of Bacillus subtilis metabolism. *Science, 335*, 1099–1103.

Chaix, D., Ferguson, M. L., Atmanene, C., Van Dorsselaer, A., Sanglier-Cianferani, S., Royer, C. A., et al. (2010). Physical basis of the inducer-dependent cooperativity of the Central glycolytic genes Repressor/DNA complex. *Nucleic Acids Research, 38*, 5944–5957.

Chiantia, S., Ries, J., & Schwille, P. (2009). Fluorescence correlation spectroscopy in membrane structure elucidation. *Biochimica et Biophysica Acta, 1788*, 225–233.

Cormack, B. P., Valdivia, R. H., & Falkow, S. (1996). FACS-optimized mutants of the green fluorescent protein (GFP). *Gene, 173*, 33–38.

Digman, M. A., Brown, C. M., Sengupta, P., Wiseman, P. W., Horwitz, A. R., & Gratton, E. (2005). Measuring fast dynamics in solutions and cells with a laser scanning microscope. *Biophysical Journal, 89*, 1317–1327.

Digman, M. A., Dalal, R., Horwitz, A. F., & Gratton, E. (2008). Mapping the number of molecules and brightness in the laser scanning microscope. *Biophysical Journal, 94*, 2320–2332.

Digman, M. A., & Gratton, E. (2009). Fluorescence correlation spectroscopy and fluorescence cross-correlation spectroscopy. *Wiley Interdisciplinary Reviews. Systems Biology and Medicine, 1*, 273–282.

Digman, M. A., Wiseman, P. W., Choi, C., Horwitz, A. R., & Gratton, E. (2009). Stoichiometry of molecular complexes at adhesions in living cells. *Proceedings of the National Academy of Sciences of the United States of America, 106*, 2170–2175.

Doan, T., & Aymerich, S. (2003). Regulation of the central glycolytic genes in Bacillus subtilis: Binding of the repressor CggR to its single DNA target sequence is modulated by fructose-1,6-bisphosphate. *Molecular Microbiology, 47*, 1709–1721.

Doan, T., Martin, L., Zorrilla, S., Chaix, D., Aymerich, S., Labesse, G., et al. (2008). A phospho-sugar binding domain homologous to NagB enzymes regulates the activity of the central glycolytic genes repressor. *Proteins, 71*, 2038–2050.

Drobizhev, M., Makarov, N. S., Tillo, S. E., Hughes, T. E., & Rebane, A. (2011). Two-photon absorption properties of fluorescent proteins. *Nature Methods, 8*, 393–399.

Espenel, C., Margeat, E., Dosset, P., Arduise, C., Le Grimellec, C., Royer, C. A., et al. (2008). Single-molecule analysis of CD9 dynamics and partitioning reveals multiple modes of interaction in the tetraspanin web. *The Journal of Cell Biology, 182*, 765–776.

Ferguson, M. L., Le Coq, D., Jules, M., Aymerich, S., Declerck, N., & Royer, C. A. (2011). Absolute quantification of gene expression in individual bacterial cells using two-photon fluctuation microscopy. *Analytical Biochemistry, 419*, 250–259.

Ferguson, M. L., Le Coq, D., Jules, M., Aymerich, S., Radulescu, O., Declerck, N., et al. (2012). Reconciling molecular regulatory mechanisms with noise patterns of bacterial metabolic promoters in induced and repressed states. *Proceedings of the National Academy of Sciences of the United States of America, 109*, 155–160.

Fillinger, S., Boschi-Muller, S., Azza, S., Dervyn, E., Branlant, G., & Aymerich, S. (2000). Two glyceraldehyde-3-phosphate dehydrogenases with opposite physiological roles in a nonphotosynthetic bacterium. *Journal of Biological Chemistry, 275*, 14031–14037.

Guillier, M., Allemand, F., Dardel, F., Royer, C. A., Springer, M., & Chiaruttini, C. (2005). Double molecular mimicry in Escherichia coli: Binding of ribosomal protein L20 to its two sites in mRNA is similar to its binding to 23S rRNA. *Molecular Microbiology, 56*, 1441–1456.

Guillier, M., Allemand, F., Raibaud, S., Dardel, F., Springer, M., & Chiaruttini, C. (2002). Translational feedback regulation of the gene for L35 in Escherichia coli requires binding of ribosomal protein L20 to two sites in its leader mRNA: A possible case of ribosomal RNA-messenger RNA molecular mimicry. *RNA, 8*, 878–889.

Kitko, R. D., Cleeton, R. L., Armentrout, E. I., Lee, G. E., Noguchi, K., Berkmen, M. B., et al. (2009). Cytoplasmic acidification and the benzoate transcriptome in Bacillus subtilis. *PLoS One, 4*, e8255.

Lesage, P., Chiaruttini, C., Graffe, M., Dondon, J., Milet, M., & Springer, M. (1992). Messenger RNA secondary structure and translational coupling in the Escherichia coli operon encoding translation initiation factor IF3 and the ribosomal proteins, L35 and L20. *Journal of Molecular Biology, 228*, 366–386.

Licht, A., & Brantl, S. (2006). Transcriptional repressor CcpN from Bacillus subtilis compensates asymmetric contact distribution by cooperative binding. *Journal of Molecular Biology, 364*, 434–448.

Licht, A., & Brantl, S. (2009). The transcriptional repressor CcpN from Bacillus subtilis uses different repression mechanisms at different promoters. *Journal of Biological Chemistry*, *284*, 30032–30038.

Licht, A., Golbik, R., & Brantl, S. (2008). Identification of ligands affecting the activity of the transcriptional repressor CcpN from Bacillus subtilis. *Journal of Molecular Biology*, *380*, 17–30.

Magde, D., Elson, E. L., & Webb, W. W. (1974). Fluorescence correlation spectroscopy. II. An experimental realization. *Biopolymers*, *13*, 29–61.

Sahoo, H., & Schwille, P. (2011). FRET and FCS—Friends or foes? *ChemPhysChem*, *12*, 532–541.

Servant, P., Le Coq, D., & Aymerich, S. (2005). CcpN (YqzB), a novel regulator for CcpA-independent catabolite repression of Bacillus subtilis gluconeogenic genes. *Molecular Microbiology*, *55*, 1435–1451.

Shaner, N. C., Campbell, R. E., Steinbach, P. A., Giepmans, B. N., Palmer, A. E., & Tsien, R. Y. (2004). Improved monomeric red, orange and yellow fluorescent proteins derived from Discosoma sp. red fluorescent protein. *Nature Biotechnology*, *22*, 1567–1572.

Tännler, S., Fischer, E., Le Coq, D., Doan, T., Jamet, E., Sauer, U., et al. (2008). CcpN controls central carbon fluxes in Bacillus subtilis. *Journal of Bacteriology*, *190*, 6178–6187.

Weiss, M. (2008). Probing the interior of living cells with fluorescence correlation spectroscopy. *Annals of the New York Academy of Sciences*, *1130*, 21–27.

Wiednmann, J., Oswald, F., & Neinhaus, G. U. (2012). Fluorescent proteins for live cell imaging: Opportunities, limitations and challenges. *IUBMB Life*, *61*, 1029–1042.

Zorrilla, S., Chaix, D., Ortega, A., Alfonso, C., Doan, T., Margeat, E., et al. (2007). Fructose-1,6-bisphosphate acts both as an inducer and as a structural cofactor of the central glycolytic genes repressor (CggR). *Biochemistry*, *46*, 14996–15008.

Zorrilla, S., Doan, T., Alfonso, C., Margeat, E., Ortega, A., Rivas, G., et al. (2007). Inducer-modulated cooperative binding of the tetrameric CggR repressor to operator DNA. *Biophysical Journal*, *92*, 3215–3227.

Zorrilla, S., Lillo, M. P., Chaix, D., Margeat, E., Royer, C. A., & Declerck, N. (2008). Investigating transcriptional regulation by fluorescence spectroscopy, from traditional methods to state-of-the-art single-molecule approaches. *Annals of the New York Academy of Sciences*, *1130*, 44–51.

Zorrilla, S., Ortega, A., Chaix, D., Alfonso, C., Rivas, G., Aymerich, S., et al. (2008). Characterization of the control catabolite protein of gluconeogenic genes repressor by fluorescence cross-correlation spectroscopy and other biophysical approaches. *Biophysical Journal*, *95*, 4403–4415.

CHAPTER EIGHT

# Studying Ion Exchange in Solution and at Biological Membranes by FCS

**Jerker Widengren**[1]

Experimental Biomolecular Physics, Department of Applied Physics, Royal Institute of Technology (KTH), Albanova University Center, Stockholm, Sweden

[1]Corresponding author: e-mail address: jwideng@kth.se

## Contents



## Abstract

By FCS, a wide range of processes can be studied, covering time ranges from subnanoseconds to seconds. In principle, any process at equilibrium conditions can be measured, which reflects itself by a change in the detected fluorescence intensity. In this review, it is described how FCS and variants thereof can be used to monitor ion exchange, in solution and along biological membranes. Analyzing fluorescence fluctuations of ion-sensitive fluorophores by FCS offers selective advantages over other techniques for measuring local ion concentrations, and, in particular, for studying exchange kinetics of ions on a very local scale. This opens for several areas of application. The FCS approach was used to investigate fundamental aspects of proton exchange at and along biological membranes. The protonation relaxation rate, as measured by FCS for a pH-sensitive dye, can also provide information about local accessibility/interaction of a particular labeling site and conformational states of biomolecules, in a similar fashion as in a fluorescence quenching experiment. The same FCS concept can also be applied to ion exchange studies using other ion-sensitive fluorophores, and by use of dyes sensitive to other ambient conditions the concept can be extended also beyond ion exchange studies.

*Methods in Enzymology*, Volume 519
ISSN 0076-6879
http://dx.doi.org/10.1016/B978-0-12-405539-1.00008-7

## 1. INTRODUCTION

Fluorescent indicators are since long well-established and useful tools to monitor concentrations of physiologically important ions within living cells (Demchenko, 2009; Lakowicz, 2006). These indicators typically respond with changes in fluorescence intensities and/or lifetimes, or with shifts in emission or excitation spectra. Such shifts allow ratioing between different excitation and emission wavelengths (Demchenko, 2010; O'Connor & Silver, 2007), whereby differences in dye concentrations, optical path lengths, or other parameters that affect the fluorescence output are canceled out. In this chapter, an overview will be given of how fluorescence correlation spectroscopy (FCS) can provide an alternative method to these approaches to monitor ion concentrations, offering some specific advantages, in particular, for studying ion concentrations, and exchange kinetics of ions on a very local scale.

FCS is one out of several techniques exploiting the concept of number fluctuation analysis, which have in common that a fluctuating number of particles within a fixed volume is analyzed in terms of its time dependence. This concept was introduced already a long time ago (Chandrasekhar, 1943; Svedberg & Inouye, 1911; von Smoluchowski, 1914). Preceding FCS, several other techniques were developed providing indications of number fluctuations, that is, the dynamic light-scattering technique, exploiting the light-scattering intensity from the particles of interest (Berne & Peccora, 1975; Schaefer, 1973), and the voltage clamp approach (Hodgkin, Huxley, & Katz, 1949), where fluctuation analysis of electrical currents over sections of cellular or artificial membranes is performed (with the later developed patch clamp technique (Neher & Sakmann, 1976) as its single-molecule counterpart).

The theory and first experimental realizations of the FCS technique, utilizing fluorescence intensity as the fluctuating quantity, were originally introduced during the years 1972–1974 (Ehrenberg & Rigler, 1974; Elson & Magde, 1974; Madge, Elson, & Webb, 1972; Magde, Elson, & Webb, 1974). Although showing great potential, the applicability of FCS was strongly reduced at this time due to high background light levels and low detection quantum yields. However, the introduction of small, diffraction-limited observation volumes in FCS measurements, confocal epi-illumination, highly sensitive avalanche photodiodes for fluorescence detection and very selective band-pass filters to discriminate the fluorescence from the background made it possible to improve signal-to-background ratios in FCS measurements by several orders of magnitude (Rigler, Mets, Widengren, & Kask, 1993; Rigler & Widengren, 1990; Rigler, Widengren, & Mets, 1992).

In an FCS measurement, in its most simple realization, fluorescence fluctuations arise from translational diffusion, as the fluorescent molecules are diffusing into and out of the confocal detection volume (Fig. 8.1A). From the fluorescence fluctuations, information can be retrieved about the translational diffusion coefficients and the average number of molecules residing simultaneously in the detection volume. In the absence of any other kinetic process other than translational diffusion affecting the fluorescent molecules, the time-dependent normalized intensity autocorrelation function (ACF) can be written as:

$$G(\tau) = \lim_{T \to \infty} \frac{1}{T} \int_0^T \frac{F(t+\tau)F(t)}{\langle F \rangle^2} dt = \frac{1}{N} \left( \frac{1}{1+4D\tau/\omega_1^2} \right) \left( \frac{1}{1+4D\tau/\omega_2^2} \right)^{1/2} + 1 \qquad [8.1]$$

Here, $N$ denotes the mean number of fluorescent molecules within the detection volume and $D$ the translational diffusion coefficient of the molecules. Brackets denote time average. $\omega_1$ and $\omega_2$ signify the distances in the radial and axial dimensions, respectively, at which the detected fluorescence per unit volume has decreased by a factor of $e^2$, compared to its maximum value in the center of the detection volume.

In a confocal epi-illuminated FCS arrangement, the total detected fluorescence intensity is given from

$$F(t) = Q \int_{-\infty}^{\infty} \mathrm{CEF}(\bar{r}) I_{\mathrm{exc}}(\bar{r}) C(\bar{r}, t) \mathrm{d}V \qquad [8.2]$$

Here, $C(\bar{r}, t)$ is the concentration of fluorescent molecules at position $\bar{r}$ and time $t$. $\mathrm{CEF}(\bar{r})$ is the collection efficiency function of the confocal microscope setup and $I_{\mathrm{exc}}(\bar{r})$ denotes the excitation intensity. $Q = q\sigma_{\mathrm{exc}}\Phi_{\mathrm{f}}$, is a brightness coefficient, where $\sigma_{\mathrm{exc}}$ is the excitation cross section of the fluorescent molecules under study, $\Phi_{\mathrm{f}}$ is their fluorescence quantum yield, and $q$ denotes the efficiency of fluorescence detection when fluorescence is emitted from the center of the laser focus/detection volume. The parameter $q$ includes the solid angle of light collection, the transmission of the microscope optics and the spectral filters, as well as the detection quantum yield of the detector.

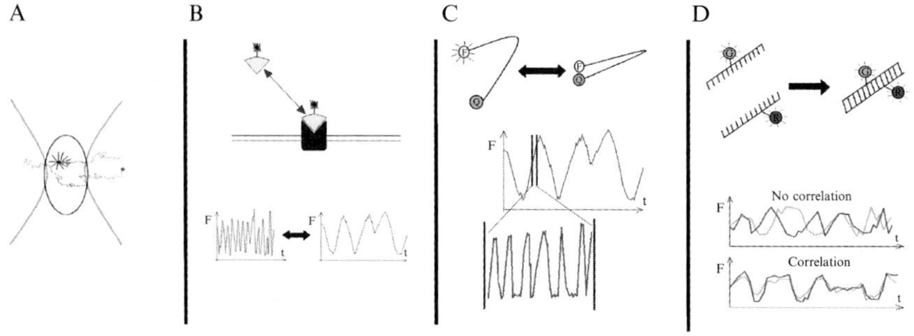


**Figure 8.1** Major categories of molecular dynamic processes or reactions that can be analyzed by an FCS instrument with a single-point, stationary detection volume. (A) Translational diffusion of fluorescent species through the confocal detection volume of an FCS instrument, yielding an ACF given by Eq. (8.1). (B) Reactions that lead to a significant change in the translational diffusion coefficient, *D*, of the fluorescent reaction partner. (C) Molecular dynamic processes or reactions that change the fluorescence brightness, $\Phi_f$, of the studied molecules, yielding an ACF of the form given in Eq. (8.3). (D) FCCS (here, dual-color FCCS), for example, of reaction partners labeled with green (G) and red (R) emitting fluorophores that upon association move in concert and generate correlated fluctuations in the green and red emission range.

Even for a standard FCS instrument with a single-point, stationary detection volume, a wide range of processes can be studied, spanning a time range from subnanoseconds to seconds. In principle, any process at equilibrium conditions, which reflects itself by a change of the detected fluorescence $F(t)$, can be measured. In Fig. 8.1, the three most widely applied FCS-based approaches to extract information from molecules undergoing or participating in dynamic processes or reactions are outlined.

When monitoring fluctuations in the detected fluorescence intensity, $F(t)$, within one emission wavelength range, there are two criteria, of which at least one needs to be fulfilled in order to gain information about a reaction process by FCS:

**A.** The reaction must either lead to a change of the diffusion coefficient, *D*, of the fluorescent species under investigation (Fig. 8.1B). The typical duration of the fluorescence bursts will then change due to the longer passage times of the fluorescent molecules through the detection volume. This is a typical strategy followed for ligand–receptor interaction studies by FCS (Rauer, Neumann, Widengren, & Rigler, 1996).

**B.** The reaction leads to change in the fluorescence quantum yield, $\Phi_f$, or the excitation cross section, $\sigma_{exc}$, of the reactants (Fig. 8.1C). This changes the fluorescence emission rate per molecule, and the fluorescence brightness coefficient $Q$ of the studied molecules. This strategy has been exploited to monitor not only inter- or intramolecular dynamics, influenced by fluorescence quenching (Bonnet, Krichevsky, & Libchaber, 1998; Chattopadhyay, Saffarian, Elson, & Frieden, 2002), a range of photo-physical processes inherent to the fluorophore marker molecules, including triplet state formation (Widengren, Mets, & Rigler, 1995), photoisomerization (Widengren & Schwille, 2000), and photo-induced charge transfer (Widengren et al., 1997; Widengren, Chmyrov, Eggeling, Löfdahl, & Seidel, 2007), but also other processes influencing fluorescence blinking, such as ion exchange involving ion-sensitive fluorophores, as discussed in this review.

If there is no significant change in either diffusion properties or the fluorescence brightness as a consequence of the reaction taking place, a cross-correlation approach can be applied, as an additional alternative (Fig. 8.1D). In fluorescence cross-correlation spectroscopy (FCCS), the fluorescence intensity fluctuations from two spectroscopically (typically spectrally) separate species are correlated (Schwille, Meyer-Almes, & Rigler, 1997). Here, $G(\tau)$ is calculated as in Eq. (8.1), but with $F(t)$ and $F(t+\tau)$ separately detected. The degree of correlation in the fluctuations provides a measure of the extent of interaction (concerted diffusion) between two spectrally separate molecular species (Fig. 8.1D), even if the species display no particular change in the diffusion or brightness properties as a consequence of the interaction.

In this chapter, an overview will be given of how single-point FCS, with a stationary detection volume, can be used to monitor ion exchange in solution, as well as at biological membranes. The concepts are mainly based on the strategy B above (Fig. 8.1C), but it will also be shown how additional specificity may be obtained by incorporating a FCCS approach in the measurements (Fig. 8.1D).

## 2. ION EXCHANGE MONITORING BY FCS—BASIC APPROACH

There exist a range of fluorescent indicators to monitor concentrations of physiologically important ions, which are used not the least for live cell studies (Demchenko, 2010; Lakowicz, 2006; O'Connor & Silver, 2007). These indicators display different responses upon ion binding, including changes in fluorescence intensities and/or lifetimes, or shifts in emission or excitation spectra. FCS can exploit these properties in an alternative fashion not only to monitor ion concentrations but also to extract information about local buffering properties and ion exchange kinetics (Widengren & Rigler, 1997; Widengren, Terry, & Rigler, 1999).

This FCS approach in its basic form is based on the second strategy above (Fig. 8.1C), and the analysis of fluorescence fluctuations arising as a consequence of ions binding to and dissociating from the fluorophores. In general, for a fluorescent species undergoing a reaction influencing the fluorescence, and at the same time diffusing into and out of the detection volume by Brownian diffusion, the normalized ACF of the detected fluorescence fluctuations cannot be expressed in an analytical form. However, for a broad range of reactions, simplifications are possible, and this is also the case for ion exchange reactions influencing the fluorescence of ion-sensitive fluorophores. In general, if diffusion is much slower than the chemical relaxation time(s) and/or the diffusion coefficients of all fluorescent species are equal, then the time-dependent fluorescence correlation function can be separated into two factors (Palmer & Thompson, 1987). In the ACFs, the first factor, $G_D(\tau)$, then depends solely on transport properties (diffusion or flow) and the second, $R(\tau)$, depends only on the reaction rate constants:

$$G(\tau) = G_D(\tau)R(\tau) + 1 \quad [8.3]$$

In its general form, for $M$ species participating in a chemical reaction and with $T_{ij}$ denoting the corresponding matrix of the kinetic rate coefficients, the second factor $R(\tau)$ is given by

$$R(\tau) = \frac{\sum_{i,j=1}^{M} Q_i Q_j X_{ij}(\tau)}{\sum_{i=1}^{M} Q_i^2 \bar{C}_i} \quad [8.4]$$

Here, $Q_i$ is the fluorescence brightness coefficient of state $i$ and $X_{ij}(\tau)$ is the solution to the following set of differential equations and initial conditions:

$$\frac{dX_{ik}(\tau)}{d\tau} = \sum_{j=1}^{M} T_{ij} X_{jk}(\tau) \quad [8.5]$$
$$X_{ik}(0) = \bar{C}_i \delta_{ik}$$

$X_{ij}(\tau)$ describes the probability of finding a molecule in state $j$ at time $\tau$, given that it was in state $i$ at time 0.

As mentioned, a rather broad range of chemical reactions fulfills one or both of the criteria for Eqs. (8.3)–(8.5). Moreover, for a reaction that under standard conditions does not fulfill these criteria, the conditions can often be modified so that the criteria can be fulfilled. Dwell times in the detection volume can be retarded with respect to the chemical relaxation times by expanding the observation volume (Fig. 8.1A), or by speeding up the reactions under study (Fig. 8.1C), for instance, by using higher concentrations of unlabelled reactants.

In its most simple realization, the FCS approach for ion exchange studies exploits the strong contrast in the fluorescence brightness coefficient Q (Eqs. 8.2 and 8.4) that follows from ions binding to or dissociating from an ion-sensitive fluorophore. For the case of pH-sensitive dyes, there are several different dyes commercially available. Taking the pH-sensitive dye fluorescein isothiocyanate (FITC) in a buffered aqueous solution in a close to neutral pH range as an example, the following three equilibria need to be considered:

$$\mathrm{Fl}^{2-} + \mathrm{H}^{+} \underset{k_{-}}{\overset{k_{+}}{\rightleftarrows}} \mathrm{HFl}^{-} \qquad [8.6]$$

$$\mathrm{B}^{-} + \mathrm{H}^{+} \underset{k_{\mathrm{diss}}}{\overset{k_{\mathrm{ass}}}{\rightleftarrows}} \mathrm{BH} \qquad [8.7]$$

$$\mathrm{BH} + \mathrm{Fl}^{2-} \underset{k_{-1}}{\overset{k_{1}}{\rightleftarrows}} \mathrm{B}^{-} + \mathrm{HFl}^{-} \qquad [8.8]$$

In the equations above, $\mathrm{HFl}^{-}$ and $\mathrm{Fl}^{2-}$ denote the protonated anion and the dianion forms of FITC. BH and $\mathrm{B}^{-}$ denote the protonated and the unprotonated forms of the buffer, which are active in the proton exchange with FITC. The ratios $\mathrm{HFl}^{-}/\mathrm{Fl}^{2-}$ and $\mathrm{BH}/\mathrm{B}^{-}$ for a given pH are determined by the $\mathrm{p}K_\mathrm{a}$ of FITC and the buffer, respectively. For FITC, the proton exchange will not notably affect the diffusion properties of the dye. Typically, the proton exchange rate also takes place on a timescale at least an order of magnitude faster than that of translational diffusion of the dye molecules through the observation volume in an FCS experiment. Consequently, in the context of an FCS experiment, one or both of the prerequisites are fulfilled to separate the contribution from proton exchange into a separate factor in the correlation function (Eq. 8.3). This protonation-dependent factor can then be described by

$$R(\tau) = \frac{1}{1-P}\left(1 - P + P\exp\left(-k_\mathrm{p}\tau\right)\right) \qquad [8.9]$$

Assuming that the protonated form of FITC is nonfluorescent, that is, $Q^-_{HFl} = 0, P$ corresponds to the fraction of protonated FITC, that is, $P = [HFl^-]/([Fl^{2-}] + [HFl^-])$. $k_p$ is the protonation rate constant given by

$$k_p = k_+[H^+] + k_- + ([B^-] + [BH])\frac{k_1[H^+]/K_a + k_{-1}}{[H^+]/K_a + 1} \qquad [8.10]$$

Here, $K_a$ is the acidity constant of the buffer.

## 3. MONITORING OF LOCAL ION CONCENTRATIONS AND EXCHANGE IN SOLUTION

The basic features of the FCS-based ion-exchange monitoring approach are illustrated in Fig. 8.2A and B for the pH-sensitive fluorophore FITC. The approach is also applicable to other ions than hydrogen, for instance, for the monitoring of local calcium ion exchange, as previously shown (Widengren & Rigler, 1997). However, in the following, we restrict ourselves to the monitoring of hydrogen ion exchange, and the use of pH-sensitive dyes.

Figure 8.2A shows a series of FCS curves, recorded from FITC in a nonbuffered aqueous solution at different pH. The curves were fitted to Eq. (8.3) using Eq. (8.9), with the addition of a second factor to the correlation function describing singlet–triplet transitions (having a relaxation time in the range of 1–2 μs) (Widengren et al., 1995). In a first approximation, the protonation kinetics of the dye can be considered independent to these singlet–triplet transitions, and later investigations by other groups using the same FCS approach also confirm that the excitation intensity dependence of $k_P$ and $P$, for the dyes pyranine (Wong and Fradin, 2011) and the xanthene dye TG-II (Paredes, Crovetto, Orte, Alvarez-Pez, & Talavera, 2011), is very small. One can note from the correlation curves, and with reference to Eqs. (8.6) and (8.9), that the fraction, $P$, of nonfluorescent, protonated FITC molecules increases and the protonation relaxation time, $1/k_P$, decreases with lower pH and that the fitted $P$ and $k_P$ are well compatible with a one-step protonation with a p$K_a$ of FITC of 6.5 (Fig. 8.2, inset).

From the FCS measurements, the amplitude, $P$, and, with knowledge of the fluorophore p$K_a$, also the pH can be determined. Moreover, from $k_P$, information is obtained about the local exchange kinetics of protons to and from the fluorophores. Figure 8.2B demonstrates the effect of buffer strength on the correlation curves of FITC at a fixed pH. In Fig. 8.2B (inset),

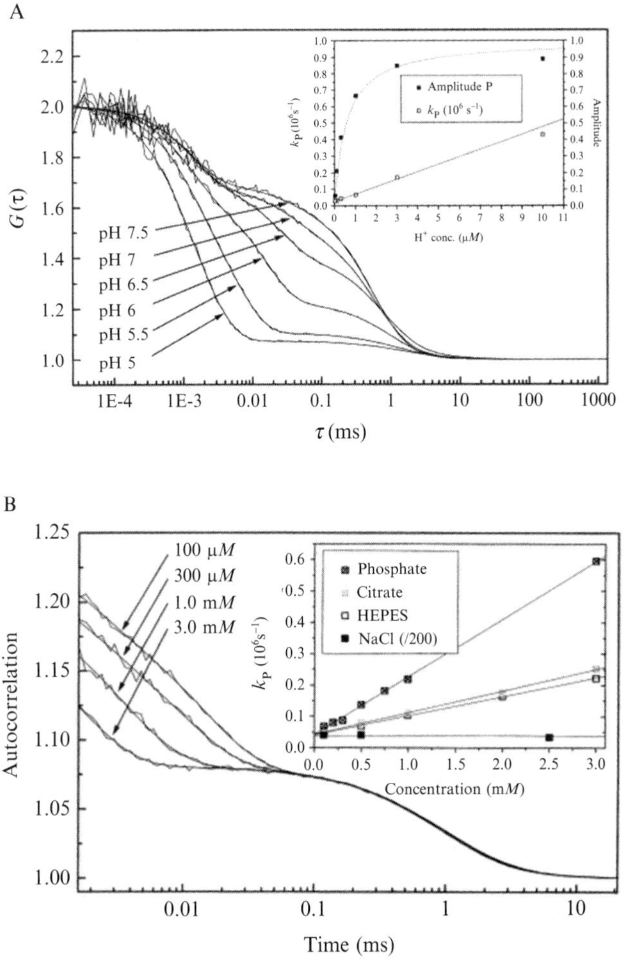


**Figure 8.2** Examples of ion exchange, monitored by FCS. (A) FITC measured at different pH. Decays of the correlation curves in the 1 ms, 2–80 μs, and 1 μs time range are attributed to translational diffusion, proton exchange and single-triplet state transitions of the fluorophores, respectively. Inset: Measured fraction, *P*, of fluorescein molecules in their protonated, anion state (solid squares), and the protonation relaxation rate (open squares) as a function of $[H^+]$. (B) FITC at pH6, with different concentrations of phosphate buffer. Inset: Measured $k_p$ versus concentration of buffer/salt at pH6.

*continued*

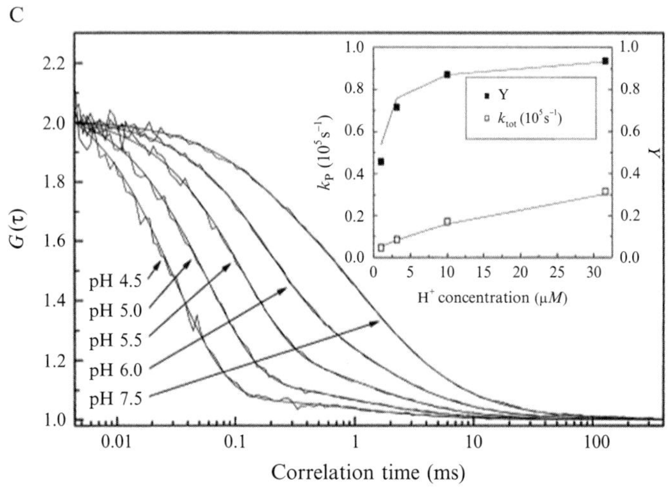


**Figure 8.2—cont'd** (C) FCS curves of GFP (S65T) at different pH. Inset: Fraction of nonfluorescent GFP molecules, $Y$ (solid squares), and the protonation relaxation rate (open squares) as a function of [$H^+$]. *Figure A–C reprinted from Widengren et al. (1999) with permission from Elsevier.*

the relaxation rate of ion exchange, $k_P$, as measured by FCS, can be seen to increase linearly with the concentration of an added phosphate buffer. For comparison, graphs of HEPES and citric acid buffers at the same pH have been added to illustrate the general influence on $k_P$ by an added buffer and the difference observed from one buffer to another. The slope of the graphs in the inset of Fig. 8.2B provides a measure of the buffer strength of the different buffers under the conditions of the measurements. The buffer strength depends on several parameters, including how close the $pK_a$ of the buffer is to the ambient pH and the diffusion coefficient of the buffer molecules, determining the collisional rate between buffer and fluorophore molecules. The slopes can also be influenced by the ionic strength of the solution. Depending on the mutual charges of the buffer molecules and the fluorophores, charge screening through increase of the ionic strength of the solution may either increase (if the buffer molecules and the fluorophores are all equally charged) or decrease the protonation rates (Wong & Fradin, 2011). In analogy with this, similar charge screening effects have been analyzed by FCS for protonation kinetics at biological membranes (Sandén, Salomonsson, Brzezinski, & Widengren, 2010) and for fluorescence quenching of fluorophores close to dielectric interfaces (Blom, Hassler, Chmyrov, & Widengren, 2010). However, since this charge screening affects the interaction between the buffer molecules and the fluorophores, no particular effect of ionic strength on $k_P$ is seen in the absence of a buffer, as illustrated in the inset of Fig. 8.2B, showing the relaxation rates measured at different concentrations of sodium chloride, with no buffer added. Evidently, an effect on $k_P$ can also be expected if the fluorophore is shielded from the buffer molecules with which it can exchange protons. From this point of view, the $k_P$ parameter value, as measured by FCS for a pH-sensitive dye, can provide information about local accessibility/interaction of a particular labeling site and conformational states of biomolecules, in a similar fashion as in a fluorescence quenching experiment. Possible advantages of monitoring local accessibility via $k_P$ rather than by fluorescence quenching are that the rate of quenching can be much lower and can be scaled by the buffer concentration. Thereby, the quenching rate can be adapted for maximum sensitivity to the kinetics of any dynamic process that is to be studied.

As an illustration of how the local accessibility may influence the protonation kinetics is given by green fluorescent proteins (GFPs). A set of correlation curves of the GFP mutant (F64L, S65T) is shown in Fig. 8.2C. This protein displays a similar pH-dependence as that for fluorescein. However, much higher buffer concentrations were required in order to significantly increase the relaxation rate of the proton transfer kinetics, and the relation between the measured $k_P$ rate for GFP and the buffer concentration was nonlinear (inset, Fig. 8.2C). X-ray structures of GFPs (Ormö et al., 1996) have revealed that the fluorescently active part (residues 65–67) of the GFP molecule is surrounded by a very tight barrel, formed by an 11-stranded β-barrel. Most likely, this barrel not only slows the exchange of protons within the microenvironment of the GFP molecule but also physically prevents buffer molecules and protons from directly reaching the fluorescently active residues in the interior of the barrel. Hence, the active residues are only affected indirectly, where changes in fluorescence are a secondary effect mediated intramolecularly, following a proton exchange at some exterior part of the molecule. This two-site model for protonation of the chromophore part of GFPs has subsequently been supported by other experimental and theoretical data (see Bizzarri, Serresi, Luin, & Beltram, 2009 for a review).

## 4. MONITORING OF PROTON EXCHANGE AT BIOLOGICAL MEMBRANES BY FCS

The FCS-based approach for ion exchange studies above does not provide any significant advantages for monitoring ion concentrations over extended three-dimensional volumes. However, it can offer selective advantages over other techniques for measuring local ion concentrations, and, in particular, exchange kinetics of ions on a local scale. Transport of ions across biological membranes is of central importance for a range of cellular processes such as nerve conduction, energy metabolism, and import of nutrients into cells.

Proton gradients across biological membranes act as driving forces for many energy-consuming cellular processes, not the least ATP synthesis by ATP synthase in the mitochondria. To generate the gradients, proton transport at and across membranes is required and involves a series of membrane-spanning proteins in the inner membranes of the mitochondria. The underlying mechanisms for this proton transport have been subject to extensive research but are, nonetheless, not completely understood. One of the key questions concerns the nature of coupling between proton generators, such as cytochrome C oxidase (Cyt*c*O), pumping protons across the membrane, and proton consumers, such as ATP synthase, using the proton gradients across the membrane to drive the ATP synthesis (Medvedev & Stuchebrukhov, 2011). Both the outlet of the generator and inlet of the consumer proteins are located on the same side of the membrane, but the proteins are spatially separated. A major question is how the generated proteins get to the consumers before they are dissociated from the membrane surface. A major experimental technique for these molecular proton exchange studies is the laser-induced proton pulse approach (Gutman, 1986) and other relaxation techniques. Basically, a light flash directed to one site of a membrane releases protons from caged compounds in the membrane, and the response from pH-sensitive fluorophores is then monitored on another site of the membrane. The FCS approach above can offer complementary angles of view to these relaxation techniques. In particular, in FCS measurements, protons associating to and dissociating from pH-sensitive fluorophores are observed at equilibrium conditions. No perturbation into some, often strongly unphysiological, initial condition is required. Moreover, while most other fluorescence-based techniques for ion concentration monitoring require dye concentrations of the order $\mu M$ or more, FCS is typically performed at about 1000-fold lower concentrations. Thereby, the influence of buffering effects due to ions binding to the fluorophores themselves, or to any light absorbing proton emitter, can be avoided. Finally, the proton exchange seen in the FCS measurements reflects the immediate environment around the fluorescent probes and is not influenced by the conditions in the media the protons pass on their way from a light absorbing proton emitter.

Based on these selective advantages of the FCS approach, we have investigated the principal role of biological membranes for proton uptake of membrane-incorporated proteins (Brändén, Sandén, Brzezinski, & Widengren, 2006).

Observations have been made that certain membrane-incorporated proteins can receive protons at a rate faster than that limited by proton diffusion in water (Ädelroth & Brzezinski, 2004). The proton uptake has been proposed to be facilitated by "proton-collecting antennae" composed of surface-accessible negatively charged and protonatable groups close to the entry points of the proton-conducting pathways of the proteins (Checover, Nachliel, Dencher, & Gutman, 1997; Riesle, Oesterhelt, Dencher, & Heberle, 1996). Theoretical studies have also indicated that the membrane itself can contribute to an increased proton uptake by membrane proteins in a similar fashion, that is, yielding a protonation cross section larger than the physical size of the membrane proteins (Georgievskii, Medvedev, & Stuchebrukhov, 2002; Smondyrev & Voth, 2002). However, the occurrence and mechanisms of such proton-collecting antennae or localized proton circuits at the surface of biological membranes had previously not been directly verified experimentally. The FCS approach above was applied to investigate these phenomena, offering the advantages of studying the protonation kinetics at the local level of individual surface proton acceptors/donors on the membrane, at physiologically relevant conditions, and at thermodynamic equilibrium.

On this basis, and in a first round of experiments, we used FCS to investigate the particular role of the biological membrane for proton uptake and transport by monitoring the protonation dynamics at the surface of liposomes with well-defined compositions. In each liposome (~30 nm in diameter), only one of the lipid head groups was covalently labeled with fluorescein (Fig. 8.3A). FCS curves recorded from these fluorescein-labeled liposomes in unbuffered solutions were obtained at different pH. In Fig. 8.3B, a set of FCS curves are shown, measured from liposomes, composed of the lipid DOPG (1,2-dioleoyl-*sn*-glycero-3-[phospho-*rac*-(1-glycerol)]). The corresponding protonation rates, $k_P$, obtained by fitting the curves to Eqs. (8.3), (8.9), and (8.10) are shown in the inset. At low proton concentrations, and from the intercept and the slope of $k_P$ versus the proton concentration, the protonation association and dissociation rates of liposome-associated fluorescein, $k_+$ and $k_-$, could be determined. While the proton dissociation rate of fluorescein in the liposomes was found to be indistinguishable from that of free fluorescein in aqueous

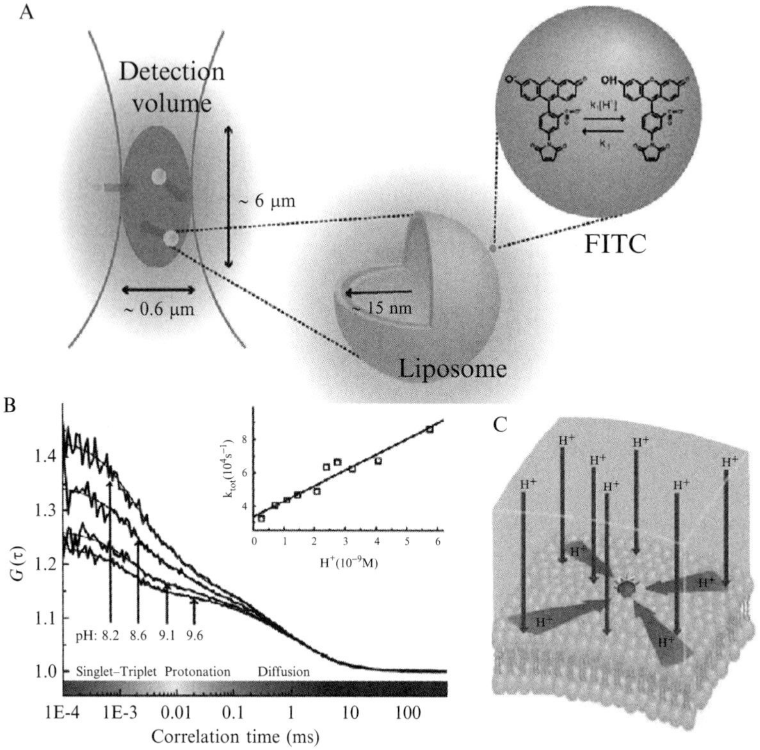


**Figure 8.3** (A) Principal design of FCS experiments to study proton exchange kinetics at biological membranes. Liposomes were labeled with one pH-sensitive fluorophore undergoing fluorescence fluctuations due to protonation/deprotonation. (B) Set of FCS curves recorded from the vesicles at different pH. The FCS curves reflect singlet–triplet transitions in the μs time range, protonation kinetics in the 10–100 μs time range and translational diffusion in the ms time range. Inset: measured protonation relaxation rates versus proton concentration (C) Principle of the proton-collecting antenna effect for a proton acceptor in a biological membrane. *Modified from Brändén et al. (2006). Copyright (2006) National Academy of Sciences, USA.* (See Color Insert.)

solution, the association rate was measured to be two orders of magnitude faster. Changing the lipids of the liposomes from DOPG (negatively charged head group) to DOPC (1,2-dioleoyl-*sn*-glycero-3-phosphocholine), with a zwitterionic head group, slowed down the association rate $k_{+}$, but it still remained about two orders of magnitude larger than that of fluorescein free in solution. The increased association rates can thus not be attributed to electrostatic effects. Instead, the fact that the fluorescein is surrounded by a lipid membrane generates a proton-collecting antenna, dramatically accelerating proton uptake from water to the membrane-anchored proton acceptor of fluorescein. This suggests that proton migration along the surface can be significantly faster than the diffusion of the lipid molecules in the membrane and can efficiently compete with the dissociation rate of protons from the membrane surface to the surrounding water phase. If a specific proton acceptor (in this case, a dye molecule) at a surface is surrounded by protonatable groups (lipid head groups), its proton uptake rate can increase dramatically (Fig. 8.3C). Both DOPG and DOPC are protonatable. However, their p$K_a$ values are in the range of 2, and the lipids are, therefore, very sparsely protonated themselves. On the other hand, addition of the lipid DOPA (1,2-dioleoyl-*sn*-glycero-3-phosphophate), with a p$K_a$ close to that of fluorescein (around 7), into the DOPC/DOPG liposomes results in a further significant increase of the measured protonation relaxation rate, $k_P$. The DOPA lipid head groups can store membrane-localized protons that are in rapid equilibrium with the fluorescein proton acceptor. The increase in $k_P$ is linearly dependent on the added DOPA concentration, in a similar fashion as observed for buffers in solution (inset, Fig. 8.2B), and can be described by Eq. (8.10). In this sense, DOPA acts as a two-dimensional buffer.

As an extension to the experiments above, the proton association and dissociation rates of fluorescein were also studied with the fluorescein labeled to the membrane protein Cyt*c*O, and in the presence or absence of a membrane surrounding the protein (Öjemyr, Sandén, Widengren, & Brzezinski, 2009). As for lipid-labeled fluorescein (Brändén et al., 2006), it was found that the proton association rate, $k_{+1}$, was more than two orders of magnitude larger for fluorescein-Cyt*c*O reconstituted into lipid vesicles (DOPG), than for detergent-solubilized fluorescein-Cyt*c*O, which, in turn, was slightly larger (by a factor of 2) than the $k_{+1}$ rate measured for free fluorescein. The minor increase of $k_{+1}$ for the detergent-solubilized fluorescein-Cyt*c*O, in the absence of a membrane, compared to that of free fluorescein indicates that the particular site of labeling of fluorescein on Cyt*c*O was not surrounded by a proton-collecting antenna on the protein surface. However, the dramatic increase of $k_{+1}$ upon membrane incorporation indicates that the protein surface is in protonic contact with the membrane surface, providing a pathway for proton transfer between proton transporter and proton "consumer" proteins residing within the same membranes in living cells.

Extending the range of pH over which the protonation kinetics of fluorescein-labeled lipids in liposomes were measured, two different proton exchange regimes could be identified (Sandén et al., 2010). At high pH (>8), the $k_{+1}$ rate strongly increases with increasing proton concentrations, whereas at low pH, (<7) the increase of the $k_{+1}$ rate with increasing bulk [$H^+$] is far smaller and comparable to the rate of protonation of the fluorophore expected to take place directly from the bulk solution. The enhancement of the protonation observed for high pH, but not for low pH, can be explained by the liposome membrane acting as a proton-collecting antenna at low proton concentrations, whereas at high proton concentrations, the membrane surface is saturated and only direct flux from the bulk can yield a higher $k_{+1}$ upon increase of bulk $[H^+]$.

Taken together, ion translocation across membranes and between the different membrane protein components is a complex interplay between the proteins and the membrane itself. Our recent studies indicate that the membrane, or the ordered water layer close to the membrane, can act as a proton-conducting link between membrane-spanning proton transporters, provide a proton-collecting antenna effective, in particular, at lower proton concentrations in the bulk, and can be modulated by the ionic strength of the bulk and the lipid composition of the membrane. FCS is well suited to investigate these phenomena and can be applied further to a broad range of investigations of membrane-associated ion exchange, also involving other membrane proteins and the proton exchange between them, as well as exchange of other ions than hydrogen.

## 5. APPROACH FOR ION EXCHANGE MONITORING INCORPORATING DUAL COLOR FLUORESCENCE CROSS CORRELATION SPECTROSCOPY (FCCS)

Although the FCS-based approach to monitor local ion exchange is straightforward and relatively robust, the fluorescence fluctuations due to ion exchange may overlap in time with other fluorescence fluctuation processes, in particular, with singlet–triplet transitions (Widengren et al., 1995). For the case such overlap may be a problem, and as a general means to better discriminate the proton exchange dynamics, we introduced proton exchange monitoring, based on dual-color FCCS applied to ratiometric pH-sensitive dyes (Persson et al., 2009). For the dye studied (NK 138, ATTO-Tec GmbH, Siegen Germany), its excitation, as well as emission spectra, changes as a function of pH. In the FCCS measurements, the fluorescence signal from the predominant emission wavelength range of the protonated form of NK 138 is cross-correlated with that of the deprotonated form. At the same time, by the excitation scheme above, the protonated and the deprotonated forms of the dye are alternatingly predominantly excited. We show that this combination of alternating excitation and FCCS allows a dramatic reduction in cross talk. The principle of the approach and the overall instrumental design is illustrated in Fig. 8.4A.

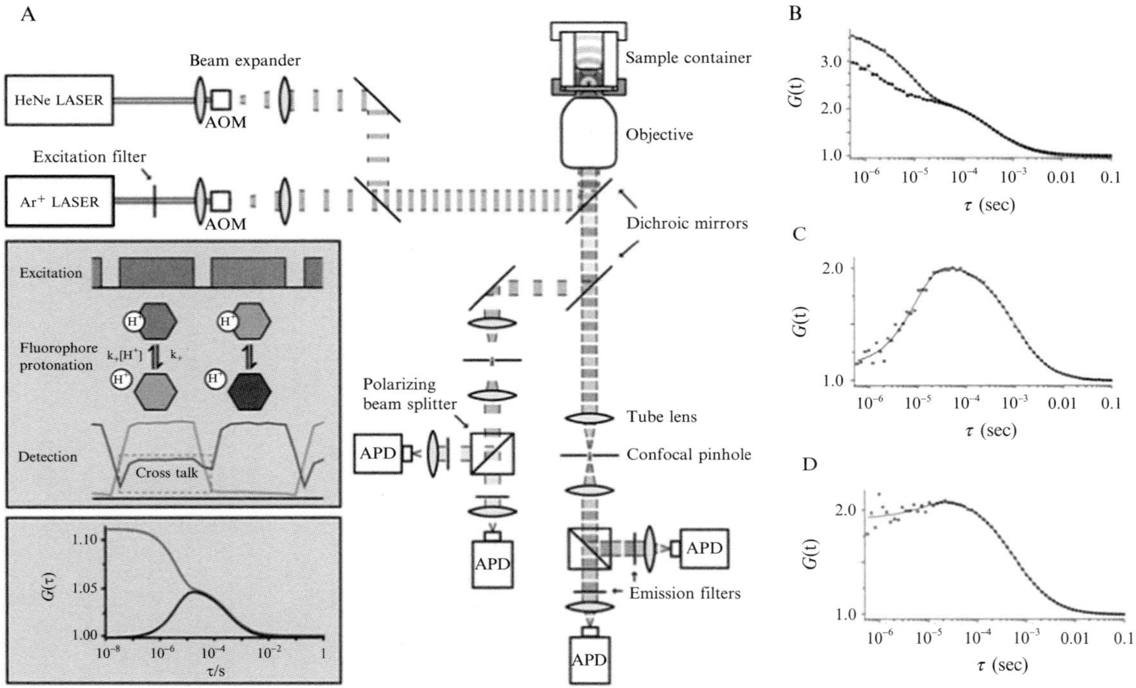


**Figure 8.4** (A) Setup used for FCCS with alternating excitation. Two lasers, emitting at 488 and 594 nm, respectively, are modulated by acousto-optical modulators and serve as excitation sources. The setup has two separate detection pathways. The upper gray box shows how the blue laser preferably excites the protonated species of the fluorophore which is mainly detected in the green detection channels and the yellow laser almost exclusively excites the deprotonated species which is detected in the red detection channels. The lower gray box shows examples of how the switching between the protonated and deprotonated states contributes to the correlation in the case of autocorrelation (upper red curve) and gives rise to an anticorrelation when cross-correlating the signals form the two spectral detection ranges (lower black curve). (B–D) Measurements on the fluorophore NK 138 in 500 μ*M* HEPES buffer at a pH of 7.7. (B) Autocorrelation curves from measurements with CW excitation and detection of either the deprotonated (upper curve, 594 nm excitation) or the protonated species (lower curve, 488 nm excitation). (C) Cross-correlation of the signals from the two spectral ranges in a measurement with alternating excitation. Gating was used to remove cross talk. (D) Cross-correlation without gating. The solid lines in (C) and (D) represents model fits, using Eq. (8.3), with $R(\tau) = 1 - H\exp(-k_p\tau)$, where $H$ is a constant between 0 and 1. $H$ decreases with increasing cross talk. *Figure is modified from Persson et al. (2009), reproduced by permission of the PCCP Owner societies.* (See Color Insert.)

Alternating excitation, introduced some years ago under the acronyms ALEX (Alternating Laser EXcitation) (Kapanidis et al., 2004) and PIE (Pulsed Interleaved Excitation) (Müller, Zaychikov, Bräuchle, & Lamb, 2005), can provide information about which excitation source has generated each detected fluorescence photon. Thereby, spectral cross talk can be strongly suppressed, significantly improving the sensitivity by which the interaction between two spectrally distinct interaction partners can be characterized, on a single-molecule level or by FCCS. Depending on the timescale of the molecular dynamics to be investigated, ALEX and PIE need to be implemented with different rates of alternating excitation to avoid frequency overlaps with the dynamic process to be studied. However, using a recently developed concept to perform FCS with modulated excitation (Persson, Thyberg, & Widengren, 2008) allows retrieval of full timescale correlation curves from FCS measurements with arbitrary timescales of modulation and with arbitrarily low fractions of active excitation. Thereby, we could apply an alternating excitation scheme for FCCS without preknowledge and concern regarding the timescales of the molecular dynamic processes to be studied.

In Fig. 8.4B and C, some correlation curves are shown, illustrating the main features in the autocorrelation and cross-correlation curves recorded for the dye NK 138. Cross-correlating the fluorescence signals originating from the protonated and deprotonated species of the pH-sensitive fluorophore makes the protonation–deprotonation process appear in the fluorescence correlation curve as an anti-correlating component. The prerequisite for this is that both species are fluorescent and that the signals originating from different species can be separated. A shift of both absorption and emission spectra upon protonation makes it possible to both excite and detect the species selectively. Because it is generally not possible to either excite or detect one species exclusively, the combination of alternating excitation and two-color detection provides a significant improvement in contrast.

The results show that alternating excitation and gating allows strong reduction of spectral cross talk in FCCS measurements of ratiometric dyes. A further contrast enhancement may be obtained by separating the different dye species also by their possibly different fluorescence lifetimes, using fluorescence lifetime correlation spectroscopy (FLCS) (Böhmer, Wahl, Rahn, Erdmann, & Enderlein, 2002). FLCS allows discrimination of fluorescence from different emitters with different fluorescence lifetimes by applying temporal filters based on the fluorescence decay. Recently, FLCS was used to discriminate and selectively extract the ACFs of a protonated and non-protonated form of a pH-sensitive dye, regardless of the spectral overlap between the two forms (Paredes et al., 2011). Nonetheless, our generally applicable approach for alternating excitation/detection already drastically improves the contrast, and turning the protonation component of the correlation curve into an anticorrelation by cross-correlation makes it substantially easier to separate from the diffusion component or photo-physical processes in the same time range, which makes the curve fitting more robust. The presented approach is not limited to protonation studies but could also be applied to ratiometric dyes sensitive to other ambient conditions and ions.

## 6. CONCLUSIONS

Based on concepts of experimental fluctuation methods described about one century ago, pioneered for fluorescence measurements 40 years ago, and having its major breakthrough about 20 years ago, the FCS technique and related fluorescence fluctuation methods are today still under strong development. In this review, it is described how FCS and variants thereof can be used to monitor ion exchange, in solution and along biological membranes. FCS, applied to the analysis of fluorescence fluctuations of ion-sensitive fluorophores, offers selective advantages over other techniques for measuring local ion concentrations and, in particular, exchange kinetics of ions on a very local scale. Measurements are performed at equilibrium conditions and at low fluorophore concentrations, which reduces perturbations to an absolute minimum. This offers opportunities to investigate fundamental aspects of proton exchange at and along biological membranes, and new experimental means to investigate the theory of how dimensionality affects molecular transport and reaction processes (Adam & Delbrück, 1968; Berg & Purcell, 1977). The protonation relaxation rate, as measured by FCS for a pH-sensitive dye, can also provide information about local accessibility/interaction of a particular labeling site and conformational states of biomolecules, in a similar fashion as in a fluorescence quenching experiment. By changing the buffer concentration, the quenching rate can be adapted for maximum sensitivity to the kinetics of the particular dynamic process to be studied. The presented approach is not limited to protonation studies but could also be applied to exchange studies of other ions using other ion-sensitive fluorophores, or by use of ratiometric dyes sensitive to other ambient conditions the same concept can also be extended beyond ion exchange studies.

## ACKNOWLEDGMENTS

The recent work by our group (Brändén et al., 2006; Öjemyr et al., 2009; Persson et al., 2009; Sandén et al., 2010) reviewed in this chapter was supported by grants from the Swedish Research Council and the Carl Trygger Foundation. Several persons have contributed to this work and I would like to acknowledge all the authors, and in particular, my former PhD students Dr. Tor Sandén and Dr. Gustav Persson for their work on the proton exchange at biological membranes, and the FCCS approach for proton exchange studies. I would also like to thank Prof. Peter Brzezinski and his research group at Stockholm University for the rewarding collaboration on the work on proton exchange at biological membranes.

## REFERENCES

Adam, G., & Delbrück, M. (1968). Reduction of dimensionality in biological diffusion processes. In A. Rich & N. Davidson (Eds.), *Structural chemistry and molecular biology* (pp. 198–215). San Fransisco: Freeman.

Ädelroth, P., & Brzezinski, P. (2004). Surface-mediated proton-transfer reactions in membrane-bound proteins. *Biochimica et Biophysica Acta, 1655*(1–3), 102–115.

Berg, H. C., & Purcell, E. M. (1977). Physics of chemoreception. *Biophysical Journal, 20,* 193–219.

Berne, B. J., & Peccora, R. (1975). *Dynamic light scattering, with applications to chemistry, biology and physics.* New York: Wiley.

Bizzarri, R., Serresi, M., Luin, S., & Beltram, F. (2009). Green fluorescent protein based pH indicators for in vivo use: A review. *Analytical and Bioanalytical Chemistry, 393,* 1107–1122.

Blom, H., Hassler, K., Chmyrov, A., & Widengren, J. (2010). Electrostatic interactions of fluorescent molecules with dielectric interfaces studied by total internal reflection fluorescence correlation spectroscopy. *International Journal of Molecular Sciences, 11,* 386–406.

Böhmer, M., Wahl, M., Rahn, H. J., Erdmann, R., & Enderlein, J. (2002). Time-resolved fluorescence correlation spectroscopy. *Chemical Physics Letters, 353,* 439–445.

Bonnet, G., Krichevsky, O., & Libchaber, A. (1998). Kinetics of conformational fluctuations in DNA hairpin-locus. *Proceedings of the National Academy of Sciences, 95,* 8602–8606.

Brändén, M., Sandén, T., Brzezinski, P., & Widengren, J. (2006). Localized proton microcircuits at the biological membrane-water interface. *Proceedings of the National Academy of Sciences of the United States of America, 103,* 19766–19770.

Chandrasekhar, S. (1943). Stochastic problems in physics and astronomy. *Reviews of Modern Physics, 15,* 1–89.

Chattopadhyay, K., Saffarian, S., Elson, E. L., & Frieden, C. (2002). Measurement of microsecond dynamic motion in the intestinal fatty acid binding protein by using fluorescence correlation spectroscopy. *Proceedings of the National Academy of Sciences, 99,* 14171–14176.

Checover, S., Nachliel, E., Dencher, N. A., & Gutman, M. (1997). Mechanisms of proton entry into the cytoplasmic section of the proton-conducting channel of bacteriorhodopsin. *Biochemistry, 36*(45), 13919–13928.

Demchenko, A. P. (2009). *Introduction to fluorescence sensing.* Amsterdam: Springer.

Demchenko, A. P. (2010). The concept of λ-ratiometry in fluorescence sensing and imaging. *Journal of Fluorescence, 20,* 1099–1128.

Ehrenberg, M., & Rigler, R. (1974). Rotational Brownian motion and fluorescence intensity fluctuations. *Chemical Physics, 4,* 390–401.

Elson, E. L., & Magde, D. (1974). Fluorescence correlation spectroscopy. 1. Conceptual basis and theory. *Biopolymers, 13,* 1–27.

Georgievskii, Y., Medvedev, E. S., & Stuchebrukhov, A. A. (2002). Proton transport via coupled surface and bulk diffusion. *The Journal of Chemical Physics, 116*(4), 1692–1699.

Gutman, M. (1986). Application of the laser-induced proton pulse for measuring the protonation rate constants of specific sites on proteins and membranes. *Methods in Enzymology, 127,* 522–538.

Hodgkin, A. L., Huxley, A. F., & Katz, B. (1949). Ionic currents underlying activity in the giant axon of the squid. *Archives des Sciences Physiologiques, 3,* 129–150.

Kapanidis, A. N., Lee, N. K., Laurence, T. A., Doose, S., Margeat, E., & Weiss, S. (2004). Fluorescence-aided molecule sorting: Analysis of structure and interactions by alternating-laser excitation of single molecules. *Proceedings of the National Academy of Sciences, 101*, 8936–8941.

Lakowicz, J. R. (2006). *Principles of fluorescence spectroscopy*. New York: Springer.

Madge, D., Elson, E. L., & Webb, W. W. (1972). Thermodynamic fluctuations in a reacting system—Measurement by fluorescence correlation spectroscopy. *Physical Review Letters, 29*, 705–711.

Magde, D., Elson, E. L., & Webb, W. W. (1974). Fluorescence correlation spectrosocpy. 2. An experimental realization. *Biopolymers, 13*, 29–61.

Medvedev, E. S., & Stuchebrukhov, A. A. (2011). Proton diffusion along biological membranes. *Journal of Physics. Condensed Matter, 23*, 1–15.

Müller, B. K., Zaychikov, E., Bräuchle, C., & Lamb, D. C. (2005). Pulsed interleaved excitation. *Biophysical Journal, 89*, 3508–3522.

Neher, E., & Sakmann, B. (1976). Single-channel currents recorded from membrane of denervated frog muscle fibres. *Nature, 260*, 779–802.

O'Connor, N., & Silver, R. B. (2007). Ratio imaging: Practical considerations for measuring intracellular Ca2+ and pH in living cells. *Methods in Cell Biology, 81*, 415–433.

Öjemyr, L., Sandén, T., Widengren, J., & Brzezinski, P. (2009). Lateral proton transfer between the membrane and a membrane protein. *Biochemistry, 48*, 2173–2179.

Ormö, M., Cubitt, A. B., Kallio, K., Gross, L. A., Tsien, R. Y., & Remington, S. J. (1996). Crystal structure of the Aequorea Victoria green fluorescent protein. *Science, 273*, 1392–1395.

Palmer, A. G., & Thompson, N. L. (1987). Theory of sample translation in fluorescence correlation spectroscopy. *Biophysical Journal, 51*, 339–343.

Paredes, J. M., Crovetto, L., Orte, A., Alvarez-Pez, J. M., & Talavera, E. M. (2011). Influence of the solvent on the ground- and excited-state buffer-mediated proton-transfer reactions of a xanthenic dye. *Physical Chemistry Chemical Physics, 13*, 1685–1694.

Persson, G., Thyberg, P., & Widengren, J. (2008). Modulated fluorescence correlation spectroscopy with complete time range information. *Biophysical Journal, 94*, 977–985.

Persson, G., Sandén, T., Sandberg, A. S., & Widengren, J. (2009). Fluorescence cross-correlation spectroscopy of a pH-sensitive ratiometric dye for molecular proton exchange studies. *Physical Chemistry Chemical Physics, 11*, 4410–4418.

Rauer, B., Neumann, E., Widengren, J., & Rigler, R. (1996). Fluorescence correlation spectrometry of the interaction kinetics of tetramethylrhodamine α-bungarotoxin with Torpedo californica acetylcholine receptor. *Biophysical Chemistry, 58*, 3–12.

Riesle, J., Oesterhelt, D., Dencher, N. A., & Heberle, J. (1996). D38 is an essential part of the proton translocation pathway in bacteriorhodopsin. *Biochemistry, 35*(21), 6635–6643.

Rigler, R., Mets, Ü., Widengren, J., & Kask, P. (1993). Fluorescence correlation spectroscopy with high count rate and low background: Analysis of translational diffusion. *European Biophysics Journal, 22*, 169–175.

Rigler, R., & Widengren, J. (1990). Ultrasensitive detection of single molecules by fluorescence correlation spectroscopy. In B. Klinge & C. Owman (Eds.), *Bioscience* (pp. 180–183). Lund, Sweden: Lund University Press.

Rigler, R., Widengren, J., & Mets, Ü. (1992). Interactions and kinetics of single molecules as observed by fluorescence correlation spectroscopy. In O. S. Wolfbeis (Ed.), *Fluorescence spectroscopy* (pp. 13–24). Berlin: Springer-Verlag.

Sandén, T., Salomonsson, L., Brzezinski, P., & Widengren, J. (2010). Surface-coupled proton exchange of a membrane-bound proton acceptor. *Proceedings of the National Academy of Sciences, 107*, 4129–4134.

Schaefer, D. W. (1973). Dynamics of number fluctuations: Motile microorganisms. *Science, 180*, 1293–1295.

Schwille, P., Meyer-Almes, F. J., & Rigler, R. (1997). Dual-color fluorescence cross-correlation spectroscopy for multicomponent diffusional analysis in solution. *Biophysical Journal, 72*, 1878–1886.

Smondyrev, A. M., & Voth, G. A. (2002). Molecular dynamics simulation of proton transport near the surface of a phospholipid membrane. *Biophysical Journal, 82*(3), 1460–1468.

Svedberg, T., & Inouye, K. (1911). Eine neue Methode zur Prüfung der Gültigkeit des Boyle-Gay-Laussacschen Gesetzes f¨r kolloide Lösungen. *Zeitschrift für Physikalische Chemie, 77*, 145–191.

von Smoluchowski, M. (1914). Studien über Molekularstatistik von Emulsionen und dered Zusammenhang mit Brownschen Bewegung. *Wien Berichte, 123*, 2381–2405.

Widengren, J., Chmyrov, A., Eggeling, C., Löfdahl, P. Å., & Seidel, C. A. M. (2007). Strategies to improve photostabilities in ultrasensitive fluorescence spectroscopy. *The Journal of Physical Chemistry. B, 105*, 6851–6866.

Widengren, J., Dapprich, J., & Rigler, R. (1997). Fast interactions between Rh6G and dGTP in water studied by fluorescence correlation spectroscopy. *Chemical Physics, 216*, 417–426.

Widengren, J., Mets, Ü., & Rigler, R. (1995). Fluorescence correlation spectroscopy of triplet states in solution: A theoretical and experimental study. *The Journal of Physical Chemistry, 99*, 13368–13379.

Widengren, J., & Rigler, R. (1997). An alternative way of monitoring ion concentrations and their regulation using fluorescence correlation spectroscopy. *Journal of Fluorescence*, 7(1), 211–213.

Widengren, J., & Schwille, P. (2000). Characterization of photoinduced isomerization and back-isomerization of the cyanine dye Cy5 by fluorescence correlation spectroscopy. *The Journal of Physical Chemistry. A, 104*, 6416–6428.

Widengren, J., Terry, B., & Rigler, R. (1999). Protonation kinetics of GFP and FITC investigated by FCS—Aspects of the use of fluorescent indicators for measuring pH. *Chemical Physics, 249*, 259–271.

Wong, F. H. C., & Fradin, C. (2011). Simulaneous pH and temperature measurements using pyranine as a molecular probe. *Journal of Fluorescence, 21*, 299–312.

CHAPTER NINE

# Fluctuation Analysis of Activity Biosensor Images for the Study of Information Flow in Signaling Pathways

Marco Vilela[*], Nadia Halidi[*], Sebastien Besson[*], Hunter Elliott[*], Klaus Hahn[†], Jessica Tytell[*], Gaudenz Danuser[*,1]

[*]Department of Cell Biology, Harvard Medical School, Boston, Massachusetts, USA
[†]Department of Pharmacology and Lineberger Cancer Center, University of North Carolina at Chapel Hill, Chapel Hill, North Carolina, USA
[1]Corresponding author: e-mail address: gaudenz_danuser@hms.harvard.edu

## Contents



## Abstract

Comprehensive understanding of cellular signal transduction requires accurate measurement of the information flow in molecular pathways. In the past, information flow has been inferred primarily from genetic or protein–protein interactions. Although useful for overall signaling, these approaches are limited in that they typically average over populations of cells. Single-cell data of signaling states are emerging, but these data are

*Methods in Enzymology*, Volume 519
ISSN 0076-6879
http://dx.doi.org/10.1016/B978-0-12-405539-1.00009-9

usually snapshots of a particular time point or limited to averaging over a whole cell. However, many signaling pathways are activated only transiently in specific subcellular regions. Protein activity biosensors allow measurement of the spatiotemporal activation of signaling molecules in living cells. These data contain highly complex, dynamic information that can be parsed out in time and space and compared with other signaling events as well as changes in cell structure and morphology. We describe in this chapter the use of computational tools to correct, extract, and process information from time-lapse images of biosensors. These computational tools allow one to explore the biosensor signals in a multiplexed approach in order to reconstruct the sequence of signaling events and consequently the topology of the underlying pathway. The extraction of this information, dynamics and topology, provides insight into how the inputs of a signaling network are translated into its biochemical or mechanical outputs.

## 1. INTRODUCTION

Optical microscopy has been widely applied to study the dynamics of molecules in biological systems. Accompanied by the development of fluorescently tagged proteins, microscopy can provide insightful information about not only the state of single cells, but also subcellular variation in protein concentration and dynamics (Slavík, 1996). For instance, fluorescence speckle microscopy has been used to track directly the protein motion and aggregation in supramolecular structures (Waterman-Storer, Desai, Bulinski, & Salmon, 1998). Alternatively, fluorescence correlation spectroscopy characterizes protein dynamics by statistical analysis of intensity fluctuations measured within a small volume (Schwille & Haustein, 2009). This method has been used to quantify the motion and interaction of diffusing proteins and organelles (Digman & Gratton, 2011). There are many different microscopy techniques designed to approach a wide variety of biological questions and we refer to Goldman, Swedlow, and Spector (2010) for a complete description.

Although very informative, canonical microscopy techniques are limited to report only local variations of protein concentration. However, the functionality of many proteins depends not only on concentration but also on the protein's activation state. For instance, members of the family of small GTPases are only active when bound to the nucleotide GTP and become inactive when GTP is hydrolyzed to GDP (Raftopoulou & Hall, 2004). A considerable amount of research has been done to include the state of molecular activity as an experimental readout. This effort led to the development of fluorescent constructs that report protein activation in living cells, which are referred to here as "activity biosensors" or simply "biosensors" (Newman, Fosbrink, & Zhang, 2011; VanEngelenburg & Palmer, 2008). The sensitivity of biosensors is often sufficient to resolve subcellular variations in the activation state, even with mild, physiologically relevant stimulation of pathways or with changes due solely to endogenous fluctuations. Therefore, unlike standard fluorescent protein tagging strategies, biosensors can track the spatiotemporal propagation of signals within a cell rather than just the redistribution of protein molecules (Fig. 9.1).

The goal of this chapter is to illustrate the steps necessary to acquire time-lapse image sequences of biosensors and to derive the *information flow* in signaling pathways from the spatiotemporal variations of the sensor's activation. The information flow defines both the topology of signal transduction and the activation kinetics. Importantly, information flow is a generic concept that captures the activation and/or transport of signaling molecules as illustrated in Fig. 9.1, but also morphological events like the assembly and disassembly of supramolecular structures, the motion of a particular subcellular region, or force generation. One of the key strengths of studying

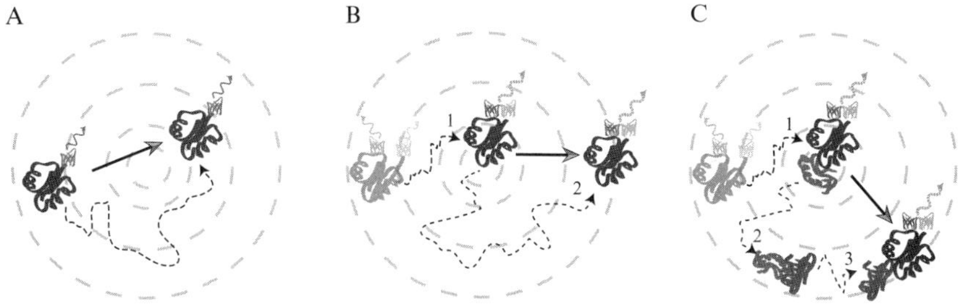


**Figure 9.1** Protein translocation versus spatial propagation of protein signals. Protein translocation is illustrated by dotted arrows and measured signal propagation by solid arrows. In (A), a fluorescently tagged molecule diffuses in space. The fluorescent signal only reports the translocation of the molecule. (B, C) Two different mechanisms for the spatial propagation of signals. In (B), an initially inactive signaling molecule is activated (step 1). The activation state is monitored by a biosensor, in this example, a FRET-based sensor, that reads out conformational changes associated with a state switch of the signaling molecule. In this scenario, the signal is transmitted by physical translocation of the activated molecule by diffusion (step 2). In (C), activation of the signaling molecule (step 1) promotes transient binding of an effector (green), which diffuses and activates a second intermediary molecule (purple, step 2). The latter then binds and activates another signaling molecule of the first kind (step 3). This leads to signal propagation in space which differs from the translocation of the biosensor. (For interpretation of the references to color in this figure legend, the reader is referred to the online version of this chapter.)

information flow is that it does not require a direct link between the observed components. Sampled components may be linked by several unobserved and potentially unknown intermediates, and yet their relationships can still be inferred. Therefore, measuring information flow provides a general means to establish the organization of cellular signal transduction pathways, even when knowledge or observations of the network components are incomplete.

# 2. ACTIVITY BIOSENSORS

## 2.1. Types of activity biosensors

The term "biosensor" has been applied to a wide range of imaging probes that detect localization and/or activation of a particular molecule. Many of them are irreversible in measuring the activation or deactivation of a molecule, making them unsuitable for the analysis of information flows in signaling pathways. To deduce information flows, biosensors must report changes in the activation state of a molecule in both directions, from an inactive to an active state and vice versa. Therefore, for the remainder of the chapter, we focus only on this class of biosensors.

Many activity biosensors share a common design scheme in which an "affinity reagent" that binds only to the active form of the probed signaling molecule is coupled to a "readout module" which changes its optical properties, most often its fluorescence, in response to binding or unbinding of the affinity reagent. Activity biosensors can be divided into two broad categories. The first category, perhaps the most common, uses protein-based affinity reagents and readout modules. These are genetically encoded biosensors in which protein-based fluorophores are incorporated such that binding between the affinity reagent and the target affects fluorescent properties, usually fluorescent resonance energy transfer (FRET) (Periasamy, 2001). FRET is an excellent readout for biosensors because small changes in the distance or orientation between the two fluorophores can cause large changes in FRET efficiency, allowing sensitive detection of protein binding or conformational changes. For example, one member of the FLARE (*fl*uorescence *a*ctivation *re*porter) family of Rho GTPase sensors (Hodgson, Pertz, & Hahn, 2008) consists of the RhoA protein fused to a CFP donor followed by a YFP acceptor fluorophore and finally the RhoA-binding domain (RBD) of the RhoA effector molecule Rhotekin, all in a single protein (Pertz, Hodgson, Klemke, & Hahn, 2006). As RhoA is activated by binding to

GTP, it undergoes a conformational change that increases its affinity for the RBD. RBD binding then folds the sensor so that the two intermediary CFP and YFP fluorophores are brought into close proximity, resulting in a heightened FRET efficiency. Many biosensors of this class with similar design principles have been generated over the past 10 years to monitor the activity of a wide class of molecules. We refer to reviews, such as (Newman et al. (2011) and VanEngelenburg and Palmer (2008), for comprehensive tables and descriptions of these sensors.

The second category of activity biosensor uses a hybrid design where the affinity reagent is a protein but the readout module is an environmentally sensitive dye. For these biosensors, the dye is ligated to the protein domain in a region where binding of the activated molecule of interest alters the local solvent environment near the dye, thereby altering the dye's fluorescent properties. These sensors can be significantly brighter than their fluorescent protein relatives and report activation of endogenous proteins, but they must be mechanically loaded (i.e., via microinjection, electroporation, etc.), which limits the number and type of cells that one can image. One example of this type of biosensor is a Cdc42 biosensor using a domain of WASP, a Cdc42 interacting protein that binds selectively to the activated (GTP-bound) Cdc42 but not to other closely related GTPases (Abdul-Manan et al., 1999; Nalbant, Hodgson, Kraynov, Toutchkine, & Hahn, 2004). This domain was used as the affinity reagent, and an environmentally sensitive merocyanine dye was fused to it (Toutchkine, Kraynov, & Hahn, 2003). When the sensor binds to endogenous Cdc42, the solvent environment of the merocyanine dye changes. This leads to increased fluorescence intensity at a particular wavelength. To distinguish activity-associated changes in intensity from changes in localization the affinity reagent is fused to a second tag, in the case of the Cdc42 biosensor a GFP, serving as a reference signal for ratiometric analyses (see below).

## 2.2. Design of the affinity reagent

One of the most important aspects of biosensor design is the selection of the affinity reagent. The key attribute of the affinity reagent is that it must recognize an inter- or intramolecular change in structure or binding caused by activation of the molecule of interest. Most biosensors are produced using rational design methods where candidate affinity reagents are based on known binding partners. For example, for the Cdc42 and RhoA biosensors, the affinity reagent was based on effector proteins known to specifically bind to the active form of the respective GTPase. As another example, in the Perceval ATP/ADP sensor (Berg, Hung, & Yellen, 2009), a circularly permuted mVenus is connected to a portion of a protein, GlnK1, that changes structure upon ATP binding. In this case, ATP or ADP binding to the GlnK1 domain differentially alters the mVenus structure leading to measurable changes in fluorescence at different wavelengths. Most recently, affinity reagents have been developed by high-throughput screening of fixed biosensor scaffolds, conferring binding affinity for otherwise intractable targets (Gulyani et al., 2011).

## 2.3. Practical considerations

Ideally, a biosensor should have no effect on cellular processes and behavior. However, most biosensors interact with endogenous signaling molecules and, because of this interaction, high levels of biosensor expression can interfere with endogenous signaling through participation in the endogenous signaling process and by sequestering signaling molecules or cofactors. It is therefore important to keep biosensor probe levels as low as possible to minimize these perturbations. The behavior of cells containing biosensor should always be compared with the behavior of cells that are not treated or contain a mock biosensor without interacting domains (e.g., CFP alone). To obtain sufficient signal from cells expressing low levels of the biosensor, light collection must be maximized. However, care must be taken not to increase irradiation to the level where the biosensor bleaches or to where increased phototoxicity becomes significant. There are several approaches to reduce both photobleaching and phototoxicity, including use of neutral density filters and/or long exposure times, rather than short excitation with intense irradiation, as well as the use of enzyme systems that efficiently scavenge free oxygen in the medium to prevent damage from free radical formation (e.g., OxyFluor, Oxyrase Inc.).

When imaging the spatiotemporal dynamics of a FRET biosensor, one has to consider the low dynamic range of activation. FRET-based sensors generally measure binary changes (inactive vs. active) between a low and a high FRET state. The difference between the two states varies widely between sensors and can be small. Thus, it is important to determine the differences in the acceptor-to-donor emission ratios between the active and inactive states in order to establish the relevant activation range of a biosensor. To accomplish this, the biosensor construct should be mutated (creating dominant-negative or dominant-positive mutants) in order to determine the minimum and maximum FRET signals in a native cellular environment.

## 2.4. Image acquisition and data processing

While many methods exist for measuring FRET efficiency in FRET-based biosensors, the most common involves acquiring raw localization and activation images. These images are then processed into a ratiometric image that indicates the local fraction of active and total amount of signaling protein. This method is referred to as "sensitized FRET" (Periasamy, 2001). When using a CFP as donor and YFP as an acceptor fluorophore, images from three channels are recorded: CFP excitation with CFP emission (donor localization image), CFP excitation with YFP emission (FRET; activation image), and YFP excitation with YFP emission (acceptor localization image). Ideally, images should be captured simultaneously to avoid artifacts caused by cell movement in-between frames. However, depending on the rate of change of the activity being measured and the morphodynamic activity of the cell, they can be captured sequentially.

Measurement of FRET efficiency via ratiometric analysis relies on the differences between the localization and activation images, which are frequently subtle. This requires that any other potential differences between these images be removed prior to calculating the ratio image. Therefore, several corrections are required, and they are specific to the imaging system used to collect the raw data. The first two corrections are termed dark current and shade corrections, and they ensure that the measured spatial variations in image intensities are accurate within each image and comparable across the different image channels. Dark current noise refers to activation of the image sensor independent of incident light, which can show significant spatial variation depending on the camera. The shade correction compensates for the nonhomogeneous illumination of the sample, which typically declines in a smooth gradient from the center to the edge of the illuminated field. Background subtraction and photobleach corrections ensure that the measured intensities are comparable over time and across experiments at the whole-image level. Background subtraction corrects for differences in spatially uniform, nonbiosensor-derived image intensities such as media autofluorescence, over which the biosensor image intensities are superimposed. Photobleach correction adjusts for the changes in fluorescence intensities over time associated with the bleaching of either donor or acceptor. Finally, spectral overlap and imperfect spectral filters cause "bleed-through" between the donor localization image and the acceptor localization and activation images, respectively. Bleed-through corrections therefore produce fully independent activation and localization images. These are typically not used for biosensors in which all components are combined in a single chain.

Our lab provides a software package that implements these corrections (download from lccb.hms.harvard.edu). The workflow of the software is shown in Fig. 9.2. It is also possible in this package to correct for image misalignments associated with chromatic aberration and/or mechanical shifts between different cameras (transformation step in Fig. 9.2). Further details can be found in the online documentation and in Hodgson et al. (2008) and Machacek et al. (2009).

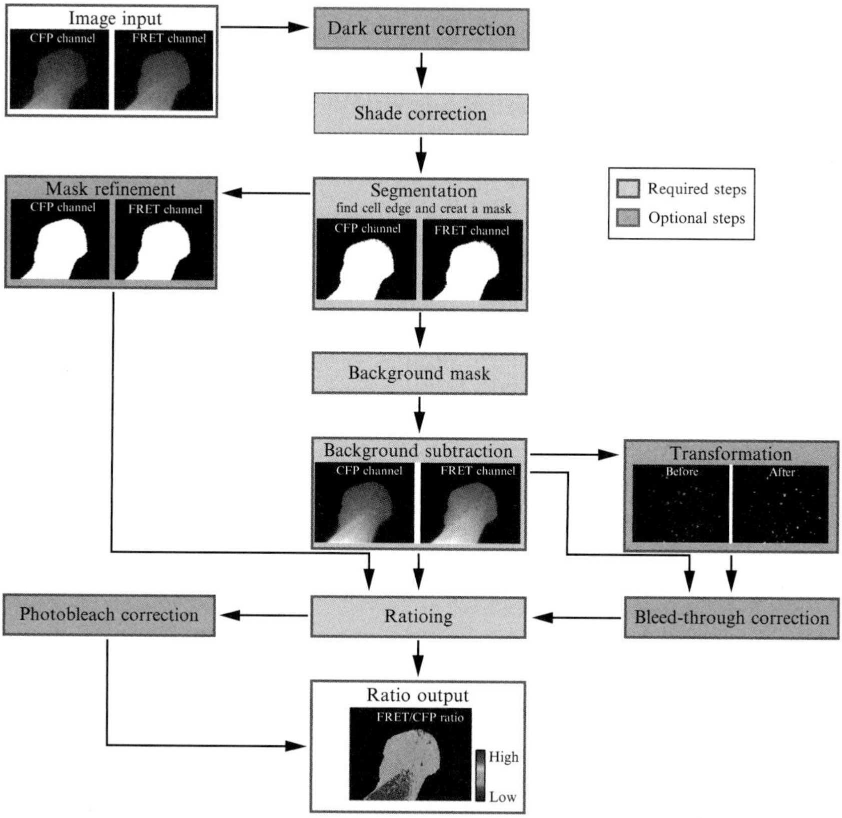


**Figure 9.2** Image corrections and processing required for FRET-based biosensor readouts of signaling activities. The end product of the workflow is a ratio image that indicates the spatial biosensor activity at each frame of the movie. (See Color Insert.)

# 3. EXTRACTING ACTIVITY FLUCTUATIONS IN A CELL SHAPE INVARIANT SPACE

Many signaling pathways are highly regulated and compartmentalized. Moreover, the same signaling protein can be involved in different pathways at different cellular locations. For instance, the small GTPase Rac1 promotes actin polymerization through the recruitment of actin nucleators in cell lamellipodia, while it also regulates focal adhesion maturation just few microns away from the actin nucleation sites (Burridge & Wennerberg, 2004). In order to understand such differences in regulation, signaling events need to be probed with a resolution that matches the spatial variability.

To locally probe signals in living cells, we propose an *in silico* compartmentalization of the cell area that is adaptive to cell shape changes (Lim, Sabouri-Ghomi, Machacek, Waterman, & Danuser, 2010; Machacek & Danuser, 2006; Machacek et al., 2009). Using time-lapse image sequences of cells containing activity biosensors, the cell perimeter is segmented into sampling windows (see Fig. 9.3A) in each of which the local signaling activity is determined by averaging the biosensor readout over all its pixels. The segmentation is performed in all frames of the sequence. Therefore, each window gives rise to a time series that represents the *local fluctuation in biosensor signal.*

A major challenge in implementing the windowing strategy is to match corresponding windows from one frame to the next. This is an important requirement because the time series extracted from one window should represent signal fluctuations of a unique cellular region. This prerequisite becomes difficult to satisfy when the cell undergoes significant changes in

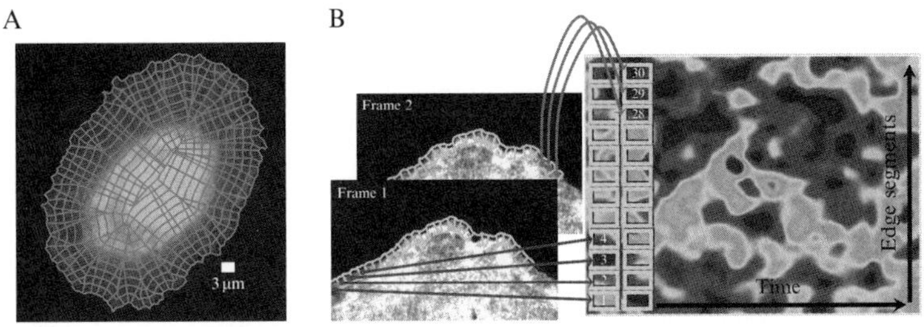


**Figure 9.3** Windowing process. (A) Segmentation of a cell into sampling windows. (B) Sampling of the fluorescence signal and construction of the spatiotemporal activity map. *Figure is reproduced, with permission, from references Lim et al. (2010) and Welch, Elliott, Danuser, and Hahn (2011).* (See Color Insert.)

morphology, either by changing the cell edge shape or the total area. Different solutions to this problem have been proposed (Bosgraaf, van Haastert, & Bretschneider, 2009; Tyson, Epstein, Anderson, & Bretschneider, 2010). Our lab has focused on studies of the connection between the spatiotemporal organization of signaling activities and cell morphological outputs like protrusion, retraction, and migration. Therefore, we developed a strategy for the definition of a cell shape invariant window mesh—that is, an *in silico* compartmentalization that can be applied irrespective of cell shape or shape changes. After identifying and tracking the local motion of the cell edge, the sampling windows at the cell border follow the frame-to-frame edge displacement. The sampling windows in the cell interior are then constructed relative to these windows in a manner that maintains a fixed relationship to the cell edge. For subsequent processing of the signaling fluctuations, the sampled image values are mapped window by window, time point by time point into an activity map (Fig. 9.3B). Importantly, this mathematical representation of image variables is independent of cell shape—hence it is *cell shape invariant*—allowing comparison of signaling patterns between cells with distinct morphologies. Moreover, in experiments where multiple image variables are acquired, such as simultaneous imaging of multiple biosensors, this mapping enables the analysis of the spatial and temporal relations between variables by correlation methods (described below). Using this approach we have recently explored the relationships between cell morphodynamics and the underlying forces, cytoskeleton dynamics, and regulatory signaling (Ji, Lim, & Danuser, 2008; Lim et al., 2010; Machacek et al., 2009).

Knowledge of the spatial scale of signaling and morphodynamic events is crucial for a meaningful definition of sampling window size. If the window size is too large relative to the spatial variation of the sampled signals, significant fluctuations will be averaged out. If the window size is too small, the readout may be too noisy and neighboring windows may measure the same signaling event. Both issues prevent meaningful analysis of signaling dynamics via fluctuation series. A practical tool to define the window size is the spatial autocorrelation of the activity map (Welch et al., 2011). The autocorrelation can be interpreted as a measure of self-similarity and is discussed in detail below. By choosing the full width at the half maximum of the spatial autocorrelation as the window size, the windows offer a practical compromise between spatial resolution, noise, and self-similarity.

# 4. CORRELATION ANALYSIS OF ACTIVITY FLUCTUATIONS FOR PATHWAY RECONSTRUCTION

This section describes a set of statistical techniques that can be applied to time series data generated from a biosensor movie that has been processed and sampled by the methods described above. The goal of this analysis is to determine correlations, time delays, and spatiotemporal scales of the sampled signals with the ultimate goal of piecing together the sequence of signaling events in a pathway.

## 4.1. Defining the spatiotemporal scale of events

The length and time scales at which signaling events occur are not only biologically meaningful but are important factors in defining the parameters of data acquisition and data analysis. As discussed above, the spatial scale of signal variations determines the appropriate window size to be used for the sampling of activity maps. Analogously, the temporal scale of signaling variations dictates the frame rate at which biosensor movies must be acquired. Both the spatial and temporal scales are *a priori* unknown properties of the studied pathway. Here, we introduce autocorrelation and power spectrum as two methods for determining these scales and for ensuring compliance of the experimental setup and data analysis with the Nyquist theorem. The Nyquist theorem asserts that a continuous, noise-free signal has to be sampled with a rate greater than twice the fastest frequency present in the signal in order to fully reconstruct the original signal (Brigham, 1988). Although conceptually simple, the theorem has important practical implications for experimental design. For instance, PtK1 cells exhibit a protrusion/retraction cycle with a period of ~130 s (Tkachenko et al., 2011). Converted into a frequency, this yields 0.008 cycles per second or 8 mHz. However, these long cycles may be superimposed by faster switches between protrusion and retraction that occur every ~40 s (25 mHz). According to the Nyquist theorem, one would therefore need to acquire an image faster than every 20 s (50 mHz) to capture the processes that produce both slow/long and fast/short edge movements. In practice, sampling at the Nyquist frequency will not be sufficient for a meaningful analysis because of the measurement noise present in the signal. As a rule of thumb, the sampling should be at least twice the Nyquist frequency. Thus, in the example of PtK1 cell protrusions, movies have to be acquired with frame rates of 10 s or faster.

### 4.1.1 Autocorrelation

The autocorrelation function (ACF) defines how data points in a time series are related, on average, to the preceding data points (Box, Jenkins, & Reinsel, 1994). In other words, it measures the self-similarity of the signal over different delay times. Accordingly, the ACF is a function of the delay or lag $\tau$, which determines the time shift taken into the past to estimate the similarity between data points. For instance, in a structured process where nearby measurements have similar values but distant points have no relation, the autocorrelation decreases as the lag $\tau$ increases. Conversely, the autocorrelation of an unstructured processes like white noise is, in theory, equal to zero for all values of $\tau > 0$ because there is no effect from one time point on another. This fact is exploited to determine the significance of the autocorrelation values. This significance can be estimated by comparing the autocorrelation of a given time series $X$ with the standard error of the autocorrelation of a white noise series with the same variance and number of points as in $X$. A value is considered significant if its magnitude exceeds the standard error of the white noise (Box et al., 1994). A positive autocorrelation value for a particular lag $\tau$ can be interpreted as a measure of persistence of data points separated by this lag to stay above and/or below the mean value of the signal. A negative autocorrelation indicates that data points separated by this lag tend to alternate about the mean value. An important piece of information provided by the ACF is the maximum lag $\tau_{max}$ that still has a significant value. This lag indicates the "memory" or temporal persistence of the fluctuation series. Data points separated by time lags greater than $\tau_{max}$ are uncoupled. The ACF is often redundantly plotted for positive and negative values of $\tau$, although by definition it is symmetric about $\tau = 0$. Of note, the ACF can also be computed in space. In spatial autocorrelation, the lag $\tau$ is then interpreted as a distance between data points. In either case, the characteristics of the temporal and spatial autocorrelation of a signaling process help us to understand the scale at which the pathway operates. These scales help us to define appropriate sampling and provide information on the spatiotemporal characteristics of the associated signal transduction network.

### 4.1.2 Power spectrum

The spatial and temporal scales at which cellular signaling operates can be further dissected by analyzing the power spectrum of extracted time series. The power spectrum measures how the variance of a time series is distributed over different frequencies (Box et al., 1994). The interpretation of the power spectrum is linked to the definition of Fourier series, which describe a signal as a sum of sine and cosine waves with different frequencies and amplitudes (Brigham, 1988). In this sum, each pair of sine and cosine waves with a given frequency $\omega$ has a specific amplitude. The power spectrum delineates the amplitudes for all sampled frequencies $\omega$, giving a measure of the contribution of each particular frequency to the net temporal behavior of the signaling system. In practice, the power spectrum is calculated from an averaging process. The signal is split into $N$ overlapping windows and Fourier-transformed, and the amplitude values in each frequency are averaged over all windows to create a global power spectrum density. This averaging process corrects for the fact that the variance of the spectrum increases with the number of points if the entire signal is used as one window. Additionally, it also provides the confidence interval based on the standard deviation calculated from all the overlapping windows (Brillinger & Krishnaiah, 1983). The power spectrum is closely related to the ACF and in fact can be mathematically defined as the Fourier transform of the ACF. Like the ACF, the power spectrum is symmetric about the $y$-axis. We discuss below the relationship between the ACF, power spectrum, and temporal resolution, but the very same considerations apply to data sequences sampled in space. Whether analyzing spatial or temporal behaviors, the power spectrum allows us to identify specific scales or ranges of scales that dominate the spatiotemporal behavior of the signaling network being observed.

### 4.1.3 Optimizing the spatiotemporal sampling of activity fluctuations

As mentioned above, the accurate measurement of the topology and kinetics of information flow in signaling networks requires sampling of the associated activities at appropriate spatiotemporal scales. These scales are rarely known prior to the experimental process, and it is therefore necessary to estimate them from measurements of the signaling system of interest. We describe here how the ACF and power spectrum support this scale selection. The former is a time domain method that estimates the overall memory of the system that generated the time series whereas the power spectrum shows the combination of frequencies or frequency bands that compose the signal. For activity biosensor movies, it is generally easier to consistently estimate the ACF rather than the power spectrum. This is because a reliable estimation of the power spectrum requires the acquisition of longer time series (Box et al., 1994). Yet, both techniques can assist in identifying the sampling rate required for the reconstruction of signaling events. In general, this involves iterating between experiment and estimation of the ACF and power spectrum until certain conditions are met. For instance, starting with an image acquisition rate $F_0$, one can estimate the autocorrelation of the sampled signals and record the maximum significant lag $\tau_{max}$. To test whether $F_0$ is sufficient, one can estimate the autocorrelation using a down-sampled version of the signals, where, for example, every other frame is excluded from the analysis. If the maximum lag $\tau_{max}^{down}$ of the down-sampled signal has the same value as $\tau_{max}$, then the current sampling rate $F_0$ is more than sufficient and can be decreased to reduce image acquisition artifacts such as phototoxicity or photobleaching. However, if the new maximum lag $\tau_{max}^{down}$ is smaller than $\tau_{max}$, no conclusions can be drawn about the sufficiency of $F_0$. A new experiment with a faster frame rate $F_1$ needs to be performed. Once again, the ACF and the maximum lag $\tau'_{max}$ associated with the new frame rate $F_1$ need to be estimated and compared with $\tau_{max}$. Similarly to the previous comparison, $F_1$ oversamples the signals if $\tau'_{max} = \tau_{max}$ but no conclusions can be drawn if $\tau'_{max} > \tau_{max}$. New experiments with faster frame rates are needed until the condition $\tau'_{max} = \tau_{max}$ is satisfied. The satisfaction of this oversampling condition implicitly translates into compliance with the Nyquist theorem. Similarly, the power spectrum can also be used to elucidate the necessary spatiotemporal sampling scales. Starting with an undersampled signal, gradual increases in the frame rate should result in increasing amplitudes in higher frequency bands of the power spectrum. This is because higher acquisition rates allow measurement of fluctuations associated with high-frequency signaling behaviors. Oversample conditions are reached when an increase in the sampling rate does not result in additional significant amplitudes in the power spectrum. This indicates that the highest-frequency signaling behaviors have already been captured, and faster imaging will provide little additional information.

The same procedures described above can be applied to the spatial component of the sampled signals. Here, the analysis needs to determine first whether the image pixel size is sufficiently small to capture the spatial variation of the observed signaling activity. If this is not the case, then the imaging setup must be modified by either an increase in magnification and/or a decrease in camera pixel size. If, however, the pixel size is sufficiently small, the spatial ACF or power spectrum can be used to determine the allowable spatial binning of the signal, that is, the size of the sampling windows. For FRET-based biosensors, utilization of immersion objectives is usually necessary to collect the weak fluorescence signal these probes emit. Immersion objectives have a magnification of 40 × and more, which implies submicron pixel sizes (depending on the camera). Considering the range of diffusion rates of signaling molecules in cells, this is generally sufficient for the sampling of signaling events. Hence, the spatial scale analysis is generally limited to defining the appropriate binning of an inherently oversampled signal into sampling windows.

Figure 9.4 shows an example of the effects of the chosen sampling rate on the reconstruction of a theoretical signal. The simulated signal used in this example has two frequency bands [0.009–0.01] Hz and [0.04–0.05] Hz, with lower amplitude values for the second band. In Fig. 9.4, both autocorrelation and power spectrum were calculated by sampling the original signal every 5, 10, and 20 s (or 0.2, 0.1, and 0.05 Hz). The immediate decay of the ACF to an insignificant value in Fig. 9.4B would suggest a short memory in this time series. However, the power spectrum in Fig. 9.4C clearly shows information in the 0.009–0.01-Hz frequency band. This example illustrates two key properties of the time scale analysis via ACF and power spectrum. First, per the Nyquist theorem, at a sampling rate of 0.05 Hz, no signal faster than 0.025 Hz can be reconstructed. Therefore, the sampling in this example

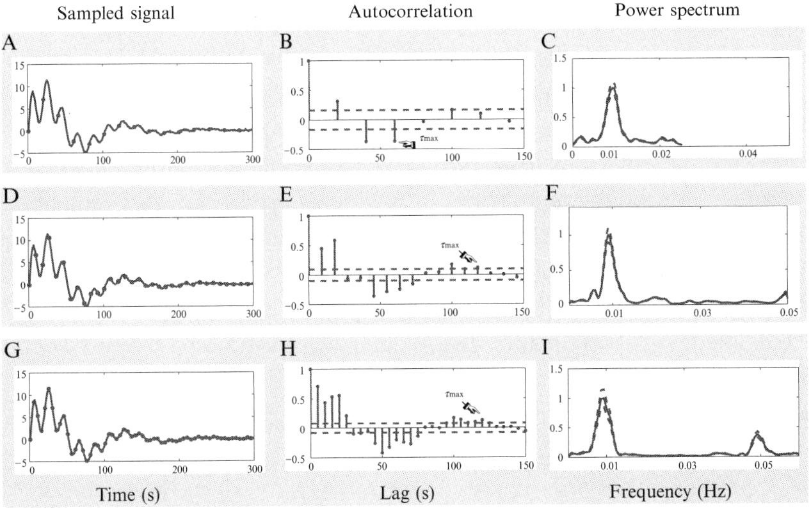


**Figure 9.4** Sampling effects in the autocorrelation and power spectrum. The first column (A, D, and G) shows the continuous signal (in blue) and the signal samples (in red) used to calculate the autocorrelation and power spectrum. The second column (B, E, and H) shows their autocorrelation functions. The red dashed lines indicate the 95% confidence level of autocorrelation values. The third column (C, F, and I) illustrates the power spectrum. The red dashed lines indicate the confidence interval with $p$ value of 0.05. The confidence interval in this case indicates the precision of the power spectrum estimation. (See Color Insert.)

is insufficient for a complete recovery of the full information contained in the signal. Second, while the computation of the ACF is more robust for short time series, the power spectrum can recover partial information about the signal (only the first frequency band of the signal was recovered in Fig. 9.4C). Following the logic introduced above for optimizing the time sampling, increasing the sampling frequency to 0.1 Hz results in both a more informative ACF and power spectrum (Fig. 9.4E and F), although the power spectrum still cannot fully resolve the entire range of frequencies in the signal. A further increase to 0.2 Hz does not change the maximum lag in the autocorrelation (Fig. 9.4H), indicating 0.2 Hz as a reliable frequency for reconstruction of the original signal, and allowing complete reconstruction of the signal's frequency components in the power spectrum. This illustrates how, even without *a priori* knowledge of the spatiotemporal scales, iteration between experiment and analysis needs to be implemented for selection of the appropriate sampling scales.

## 4.2. Establishing relationships between pathway events

The ACF and power spectrum are valuable tools for understanding the dynamics of a signal and therefore a single component of a signaling network. However, much of the functionality of a biological system relies on the interactions among their constituents. We introduce here two statistical tools that can be used for uncovering relationships between measured signals and thereby allow inference of the nature of interactions between the measured signaling components: the cross-correlation and coherence. Analogous to how the ACF and power spectrum measure the relationship between a signal and itself at different time delays or frequencies, the cross-correlation function (CCF) and coherence quantify linear relationships between two different signals in the time and frequency domain, respectively. Combining these with spatially localized sampling, the relationship among signal events can be probed for different cellular regions.

### *4.2.1 Cross-correlation*

Analogous to the ACF, the CCF determines the strength of any linear relationship between *two* sampled time series representing two different signaling activities as a function of a given lag $\tau$ (Box et al., 1994). One can think of the lag in the following way: a positive lag means that one time series is fixed as the reference and the second time series is shifted into the past, that is, the events in the second time series happen after the potentially corresponding events in the reference series. With a negative lag, the second time series is shifted into the future, that is, the events in the second time series happen before the potentially corresponding events in the reference series. The cross-correlation value for a particular $\tau$ indicates how strong the similarity of the two time series is at that particular lag. Unless the two time series are identical or symmetric, the CCF is not symmetric about $\tau = 0$. Once the CCF is computed, the key question is whether the magnitude of the function maximum is statistically significant. The cross-correlation between two signals $X$ and $Y$ is considered significant if it exceeds for at least one time lag $\tau$, the CCF of two uncorrelated random signals with the same variance, and number of points as in $X$ and $Y$. Among several mechanisms, a likely explanation for a significant positive cross-correlation could be that the events of one time series partially activate the events of the second time series. Conversely, a significant negative magnitude likely indicates that the events of one time series inhibit events of the second time series. Although cross-correlation is not a strictly causative measure (Vilela & Danuser, 2011), the time lag associated with the CCF maximum defines which of the two time series happens, on average, first, suggesting upstream–downstream relations between the activities. Thus, the CCF provides insight not only of the strength and nature of the relationship between two signaling activities but also predicts the temporal organization and kinetics of this relationship.

### 4.2.2 Coherence

Complementary to the cross-correlation, the coherence is a measure of the relationship of two signals in the frequency domain (Brillinger & Krishnaiah, 1983). Mathematically, it is defined as the Fourier transform of the cross-correlation. The coherence quantifies the overall linear coupling of two time series as a function of the specific frequencies or frequency bands shared between them. Because of this selectivity of shared frequencies, the coherence can resolve situations where one signaling activity relates to multiple other signaling activities, but at different frequency bands (Brillinger & Krishnaiah, 1983).

Figure 9.5 illustrates the use of cross-correlation and coherence for characterizing the relationship between two hypothetical activities $X$ and $Y$. Figure 9.5A shows the two time series and how their information is transmitted through a *communication channel* (Feinstein, 1958). The cross-correlation and coherence analyses serve the purpose of identifying whether there is any linear information flow between $X$ and $Y$ through the channel. In a cellular context, this communication channel conceptualizes the cascade of physicochemical events that link the activation/deactivation of one particular signal to the activation/deactivation of another signal. Dependent on the kinetics and the

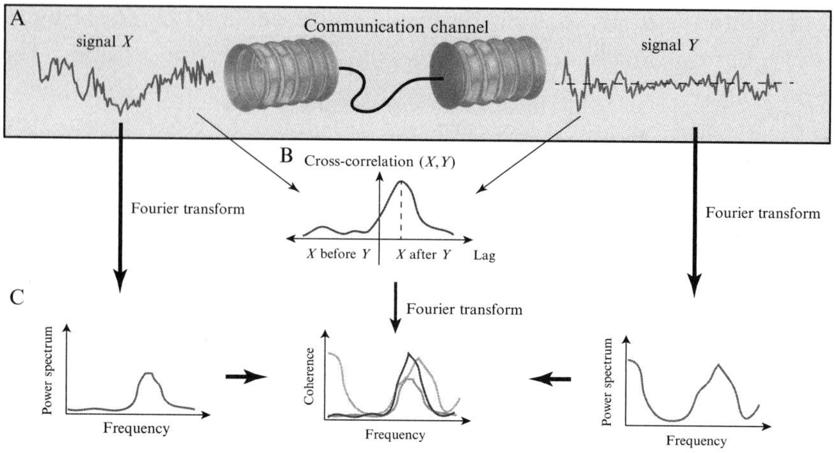


**Figure 9.5** Characterization of information flow between two activities *X* and *Y* through a communication channel. (A) The communication channel conceptualizes the cascade of molecular events that is triggered by one of the activities and contributes to the modulation of the other activity. (B) Cross-correlation between the activities. Here, activity *Y* is used as the reference. Accordingly, the positive time lag of the peak correlation value suggests that the fluctuations in activity *Y* lag those of activity *X*, leading to prediction that *Y* may be upstream of *X*. (C) Coherence analysis. The left and right panels show the power spectra of the two activities. The center panel illustrates that the coherence (in red) represents the overlap of the two spectra. (For interpretation of the references to color in this figure legend, the reader is referred to the online version of this chapter.)

complexity of this event cascade, the information transfer between the signals may lead to more or less delay, which is decoded by the time lag $\tau$ of the dominating cross-correlation maximum or minimum. Also, in the absence of strong feedback, the sign of the time lag indicates the directionality of information flow. As illustrated in Fig. 9.5C, the coherence informs us about the frequencies that are transmitted through the channel. Importantly, frequency and time delay are not equivalent. Two particular signals may be coupled through distinct frequency bands but both bands may have the same time lag because the molecular processes underlying the information flow obey the same overall kinetics. On the other hand, one particular signal may communicate with two other signals in the same frequency band but with different time lags.

## 4.3. Integrating results: Averaging over multiple windows and cells

We have described in the previous sections statistical tools that allow the analysis of a single time series or a pair of time series extracted from one local sampling window of a biosensor data set. However, the data from an individual sampling window are very noisy. Therefore, correlation, power spectra, and coherence measurements must be averaged over multiple windows and over multiple cells. Averaging these metrics requires some caution as simple mean values may be biased due to a relatively small number of potentially nonnormally distributed data points. Here, we illustrate the use of the bootstrap technique to allow accurate averaging. This technique generates a large number of samples by randomly resampling the existing data with replacement (Zoubir & Iskander, 2004). For more robust results, variance stabilization methods can be added (Zoubir & Iskander, 2004). Figure 9.6 shows a mean ACF bootstrapped from the time series of different sampling windows in a mouse embryonic fibroblast expressing a FRET-based activity biosensor of the small GTPase Rac1 (Machacek et al., 2009). First, the autocorrelation for time series extracted at individual windows is calculated. Then the bootstrap algorithm samples with replacement the autocorrelation values from all windows for a given lag to estimate one final value with a confidence interval. This process is repeated for all lags resulting in a global ACF for the entire cell.

The same approach can be taken to compute an average CCF between two activities. Importantly, the data entering the bootstrap can originate from windows sampled in a single cell or sampled over multiple cells. The fundamental assumption underlying the analysis is that although each

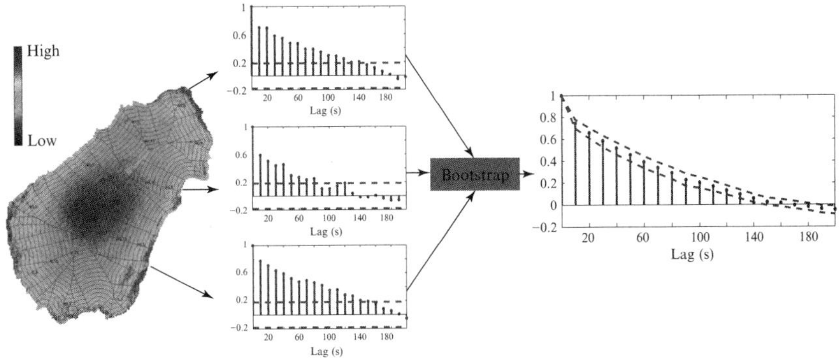


**Figure 9.6** Bootstrap method to extract an average autocorrelation function of a molecular activity (in this example, Rac1 activation) sampled in all windows along the cell edge. The autocorrelation is first calculated for time series in individual windows. In the sampling process, values of the autocorrelation that fall inside the confidence bounds (red dashed lines) are set to zero. A 95% confidence interval is estimated for each value of the bootstrapped autocorrelation based on the empirical distribution built by the algorithm. (See Color Insert.)

of these windows generates a random fluctuation series, their statistical properties are conserved between windows and between cells. Practically, this means that data from windows with similar properties are integrated, for example, from all windows at the boundary of moving cell edges, or from all windows at the boundary of quiescent cell edges, or from all windows 5 μm from the cell edge. How these windows are categorized varies with the specific application and research question. Given these assumptions, the bootstrap allows accurate aggregation of results across cells and cell regions, increasing the statistical power of these results and the generality of their biological implications.

### 4.4. Integrating results: Multiplexing of different activities using a common fiduciary

Current biosensor designs and imaging technology do not allow the simultaneous observation of more than two, or maximally three, molecular activities in living cells at sufficient spatiotemporal resolution (Hodgson et al., 2008; Welch et al., 2011). However, the goal of these live cell fluctuation studies is to reconstruct the flow of information in pathways with tens of components. To achieve this goal, fluctuation data of different biosensors imaged separately in different experiments must be integrated *in silico*. We refer to this approach as computational multiplexing (Welch et al., 2011). To allow computational multiplexing, two important requirements need to be fulfilled. First, identical experimental conditions must be maintained across all experiments. Second, each experiment must measure one activity which is common to at least one other experiment. This common activity shared between experiments provides a reference or "fiduciary," allowing the time series from different experiments to be linked (Machacek et al., 2009; Welch et al., 2011). The simplest strategy for computational multiplexing is to relate all experimental data to a single common fiduciary across all experiments. This strategy was established for the first time by Machacek et al. (2009) where the cell edge velocity was exploited to characterize the coordination of the small GTPases Rac1, RhoA, and Cdc42 during cell protrusion. Basal fluctuations of these signaling molecules were measured over time in the context of cells undergoing directed migration. Each experiment imaged the activity of one GTPase at the time. Based on the cross-correlation analysis between biosensor activity and cell edge velocity, the timing of each one of the GTPases relative to the onset of protrusion was identified. This alignment of GTPase activity and cell edge motion indirectly made predictions as to how the GTPases would be timed (and spatially shifted) relative to one another. These predictions were then confirmed in experiments where two spectrally orthogonal biosensors were imaged concurrently (Machacek et al., 2009). Thus, by exploiting a fiduciary common to several experiments, computational multiplexing allows us to infer the flow of information in signaling networks with many more components than can be observed in one experiment.

### 4.5. Integrating results: Comparing correlation and coherence data between different subcellular locations

The propagation of signaling events is organized not only in time but probably also in space. Here, we give a glimpse of how local sampling of biosensor activity fluctuations in small windows can be exploited to test this notion. We demonstrate the variation in the relation between the activity of the small GTPase Rac1 and cell edge motion at various distances from the cell boundary. Rac1 is thought to activate the formation of protruding lamellipodia (Raftopoulou & Hall, 2004). Thus, it would be expected that signaling information would flow from Rac1 activation to cell edge protrusion. Furthermore, this relationship would be expected to taper off rapidly with increasing distance from the protruding edge. Figure 9.7 shows Rac1 activity sampled in 45 windows at the cell boundary (A) and in 45 windows 2 μm away from the cell edge (B). For the windows at the cell edge velocity values of the local cell edge motion are sampled as well (C). Both cross-correlation and coherence reveal a stronger interaction between the cell edge velocity and Rac1 activity sampled 2 μm away from the cell edge. The time lag of the cross-correlation peak indicates that Rac1 is activated, on average, ~40 s after the increase in cell edge velocity. The cross-correlation peak for windows at the cell edge is weaker than for those at 2 μm distance, and the time shift between edge motion and Rac1 activation increases. These fundamentally distinct behaviors of Rac1 at the cell edge versus further away from it are corroborated by distinct bands of significant coherence. At 2 μm from the cell edge, the coherence peaks at 0.01 Hz or in a cycle of 100 s. This cycle time coincides with the ~100-s period of the protrusion/retraction cycles in these cells, suggesting that Rac1 activity at 2 μm from the edge is part of a feedback mechanism that links edge motion to the reactivation of GTPase signals away from the cell edge, probably in maturing adhesions. The coherence function at the cell edge covers a wider range of frequencies. This indicates that activation of Rac1 at these distances is more random and not directly related to the protrusion/retraction cycle. Current work in our

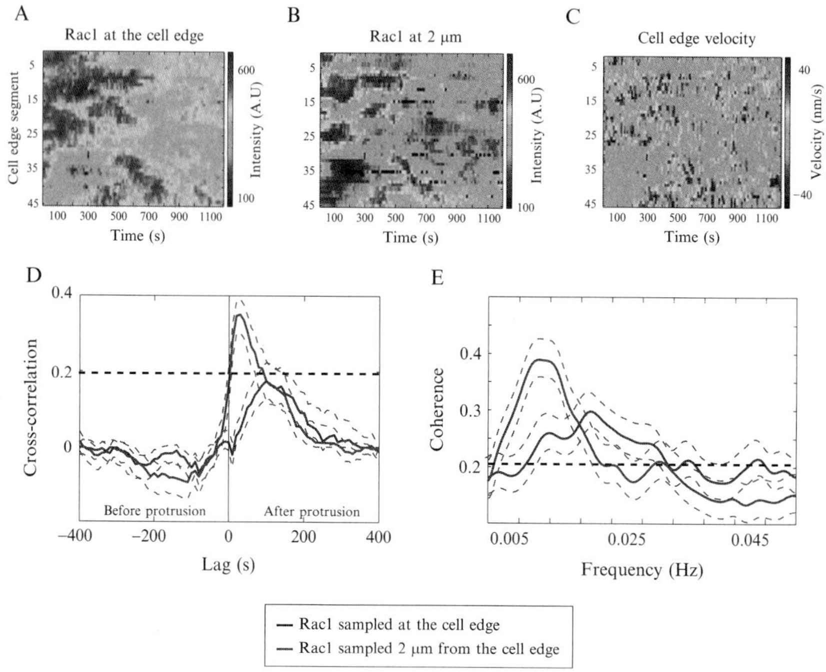


**Figure 9.7** Spatial variation of the relationship between cell edge velocity and Rac1 signaling sampled at different distances from the cell edge. (A, B) Spatiotemporal activity maps of Rac1 signaling sampled at the cell edge and 2 μm inward, respectively. (C) Cell edge velocity map. (D, E) Cross-correlation (with edge velocity as reference) and coherence between the cell edge velocity and Rac1 activation sampled at the edge and 2 μm inward. (See Color Insert.)

labs is focused on investigating the molecular differences between these distinct regimes of Rac1 regulation. This example highlights how the combination of approaches described in this chapter can provide unprecedented understanding of the dynamics and variability of signal transduction with subcellular resolution.

## 5. OUTLOOK

We present in this chapter the basic concepts of using fluctuations in signaling activity as measured by biosensors for the reconstruction of information flows in signaling networks. Autocorrelation and power spectral analyses can characterize the spatiotemporal properties of individual signaling components, and coherence and cross-correlation provide a measure of the relationships between different signaling components. Furthermore, in combination with an experimental fiduciary, methods like cross-correlation and coherence can be used to computationally multiplex data from different experiments in pathway models that consider many more components than can be observed directly in a single experiment. Although informative, these basic, linear statistical methods are unable to uncover more complex relationships among signaling components such as feedback loops. In order to clarify such interaction, we foresee the use of more sophisticated tools that can further decompose the link between two signals and probe the possibility of bi-directional information flow. Some tools from the fields of economics and neuroscience possess this capability; however, a substantial effort is still necessary to adopt these tools to biosensor fluctuation data.

## ACKNOWLEDGMENT

This chapter builds on work in the Danuser and Hahn labs funded by a collaborative T-R01 GM090317.

## REFERENCES

Abdul-Manan, N., Aghazadeh, B., Liu, G., Majumdar, A., Ouerfelli, O., Siminovitch, K., et al. (1999). Structure of Cdc42 in complex with the GTPase-binding domain of the 'Wiskott-Aldrich syndrome' protein. *Nature, 399*, 379–462.

Berg, J., Hung, Y., & Yellen, G. (2009). A genetically encoded fluorescent reporter of ATP: ADP ratio. *Nature Methods, 6*, 161–167.

Bosgraaf, L., van Haastert, P. J., & Bretschneider, T. (2009). Analysis of cell movement by simultaneous quantification of local membrane displacement and fluorescent intensities using Quimp2. *Cell Motility and the Cytoskeleton, 66*, 156–165.

Box, G. E. P., Jenkins, G. M., & Reinsel, G. C. (1994). *Time series analysis: Forecasting and control* (3rd ed.). Upper Saddle River, NJ: Prentice Hall.

Brigham, E. O. (1988). *The fast Fourier transform and its applications*. Englewood Cliffs, NJ: Prentice Hall.

Brillinger, D. R., & Krishnaiah, P. R. (1983). *Time series in the frequency domain*. Amsterdam: North-Holland.

Burridge, K., & Wennerberg, K. (2004). Rho and Rac take center stage. *Cell, 116*, 167–179.

Digman, M., & Gratton, E. (2011). Lessons in fluctuation correlation spectroscopy. *Annual Review of Physical Chemistry, 62*, 645–713.

Feinstein, A. (1958). *Foundations of information theory*. New York: McGraw-Hill.

Goldman, R. D., Swedlow, J., & Spector, D. L. (2010). *Live cell imaging: A laboratory manual* (2nd ed.). Cold Spring Harbor, NY: Cold Spring Harbor Laboratory Press.

Gulyani, A., Vitriol, E., Allen, R., Wu, J., Gremyachinskiy, D., Lewis, S., Dewar, B., Graves, L. M., Kay, B. K., Kuhlman, B., Elston, T., & Hahn, K. M. (2011). A biosensor generated via high-throughput screening quantifies cell edge Src dynamics. *Nat Chem Biol.*, 7(7), 437–444.

Hodgson, L., Pertz, O., & Hahn, K. M. (2008). Design and optimization of genetically encoded fluorescent biosensors: GTPase biosensors. *Methods in Cell Biology, 85*, 63–81.

Ji, L., Lim, J., & Danuser, G. (2008). Fluctuations of intracellular forces during cell protrusion. *Nature Cell Biology*, *10*, 1393–1400 U1338.

Lim, J. I., Sabouri-Ghomi, M., Machacek, M., Waterman, C. M., & Danuser, G. (2010). Protrusion and actin assembly are coupled to the organization of lamellar contractile structures. *Experimental Cell Research*, *316*, 2027–2041.

Machacek, M., & Danuser, G. (2006). Morphodynamic profiling of protrusion phenotypes. *Biophysical Journal*, *90*, 1439–1452.

Machacek, M., Hodgson, L., Welch, C., Elliott, H., Pertz, O., Nalbant, P., et al. (2009). Coordination of Rho GTPase activities during cell protrusion. *Nature*, *461*, 99–103.

Nalbant, P., Hodgson, L., Kraynov, V., Toutchkine, A., & Hahn, K. (2004). Activation of endogenous Cdc42 visualized in living cells. *Science*, *305*, 1615–1624.

Newman, R. H., Fosbrink, M. D., & Zhang, J. (2011). Genetically encodable fluorescent biosensors for tracking signaling dynamics in living cells. *Chemical Reviews*, *111*, 3614–3666.

Periasamy, A. (2001). Fluorescence resonance energy transfer microscopy: A mini review. *Journal of Biomedical Optics*, *6*, 287–378.

Pertz, O., Hodgson, L., Klemke, R., & Hahn, K. (2006). Spatiotemporal dynamics of RhoA activity in migrating cells. *Nature*, *440*, 1069–1141.

Raftopoulou, M., & Hall, A. (2004). Cell migration: Rho GTPases lead the way. *Developmental Biology*, *265*, 23–32.

Schwille, P., & Haustein, E. (2009). Fluorescence correlation spectroscopy: An introduction to its concepts and applications. *Spectroscopy*, *94*, 1–34.

Slavík, J. (1996). *Fluorescence microscopy and fluorescent probes*. New York: Plenum Press.

Tkachenko, E., Sabouri-Ghomi, M., Pertz, O., Kim, C., Gutierrez, E., Machacek, M., et al. (2011). Protein kinase A governs a RhoA-RhoGDI protrusion-retraction pacemaker in migrating cells. *Nature Cell Biology*, *13*, 660–667.

Toutchkine, A., Kraynov, V., & Hahn, K. (2003). Solvent-sensitive dyes to report protein conformational changes in living cells. *Journal of the American Chemical Society*, *125*(14), 4132–4145.

Tyson, R., Epstein, D., Anderson, K., & Bretschneider, T. (2010). High resolution tracking of cell membrane dynamics in moving cells: An electrifying approach. *Mathematical Modelling of Natural Phenomena*, *5*, 34–89.

VanEngelenburg, S. B., & Palmer, A. E. (2008). Fluorescent biosensors of protein function. *Current Opinion in Chemical Biology*, *12*, 60–65.

Vilela, M., & Danuser, G. (2011). What's wrong with correlative experiments? *Nature Cell Biology*, *13*, 1011.

Waterman-Storer, C., Desai, A., Bulinski, J., & Salmon, E. (1998). Fluorescent speckle microscopy, a method to visualize the dynamics of protein assemblies in living cells. *Current Biology*, *8*, 1227–1257.

Welch, C. M., Elliott, H., Danuser, G., & Hahn, K. M. (2011). Imaging the coordination of multiple signalling activities in living cells. *Nature Reviews Molecular Cell Biology*, *12*, 749–756.

Zoubir, A. M., & Iskander, D. R. (2004). *Bootstrap techniques for signal processing*. Cambridge University Press.

CHAPTER TEN

# Probing the Plasma Membrane Organization in Living Cells by Spot Variation Fluorescence Correlation Spectroscopy

**Cyrille Billaudeau**[*,†,‡,1], **Sébastien Mailfert**[*,†,‡,1], **Tomasz Trombik**[*,†,‡,1], **Nicolas Bertaux**[§,¶], **Vincent Rouger**[*,†,‡], **Yannick Hamon**[*,†,‡], **Hai-Tao He**[*,†,‡], **Didier Marguet**[*,†,‡,2]

[*]Centre d'Immunologie de Marseille-Luminy (CIML), Aix-Marseille University, UM2, Marseille, France
[†]Institut National de la Santé et de la Recherche Médicale (Inserm), U1104, Marseille, France
[‡]Centre National de la Recherche Scientifique (CNRS) UMR7280, Marseille, France
[§]Institut Fresnel, Centre National de la Recherche Scientifique (CNRS) UMR7249, Marseille, France
[¶]École Centrale Marseille, Technopôle de Château-Gombert, Marseille, France
[1]These authors contributed equally to this work.
[2]Corresponding author. e-mail address: marguet@ciml.univ-mrs.fr

## Contents



*Methods in Enzymology*, Volume 519
ISSN 0076-6879
http://dx.doi.org/10.1016/B978-0-12-405539-1.00010-5

## Abstract

While intrinsic Brownian agitation within a lipid bilayer does homogenize the molecular distribution, the extremely diverse composition of the plasma membrane, in contrast, favors the development of inhomogeneity due to the propensity of such a system to minimize its total free energy. Precisely, deciphering such inhomogeneous organization with appropriate spatiotemporal resolution remains, however, a challenge. In accordance with its ability to accurately measure diffusion parameters, fluorescence correlation spectroscopy (FCS) has been developed in association with innovative experimental strategies to monitor modes of molecular lateral confinement within the plasma membrane of living cells.

Here, we describe a method, namely spot variation FCS (svFCS), to decipher the dynamics of the plasma membrane organization. The method is based on questioning the relationship between the diffusion time $\tau_d$ and the squared waist of observation $w^2$. Theoretical models have been developed to predict how geometrical constraints such as the presence of adjacent or isolated domains affect the svFCS observations. These investigations have allowed significant progress in the characterization of cell membrane lateral organization at the suboptical level, and have provided, for instance, compelling evidence for the *in vivo* existence of raft nanodomains.

## ABBREVIATIONS

**FCS** fluorescence correlation spectroscopy
**FRAP** fluorescence recovery after photobleaching
**SPT** single-particle tracking
**STED** stimulated emission depletion
**svFCS** spot variation fluorescence correlation spectroscopy

## 1. INTRODUCTION

The fluorescence correlation spectroscopy (FCS)-based approaches have gained renewed interest for cell biology investigations (Bacia, Kim, & Schwille, 2006), in particular due to their sensitivity for studying plasma membrane organization in living cells (see, for review, Chiantia, Ries, & Schwille, 2009; He & Marguet, 2011). These methods are minimally invasive (low light intensity) and require low probe concentrations ($\sim$n$M$ range), thus preventing experimental bias due to the overexpression of fluorescently tagged membrane components. Moreover, FCS methods provide a suitably high temporal resolution with which to probe the translational diffusion within cell membranes. However, the heterogeneity of biological membrane makes it difficult to separate by classical FCS the multiple subpopulations characterized by non-Brownian diffusion. Indeed, two subsets of molecular components are barely distinguishable when their respective diffusion coefficients *D* differ by a factor below ~1.6 (Meseth, Wohland, Rigler, & Vogel, 1999). This constraint becomes more important when considering the need to describe the inhomogeneous organization of the plasma membrane in which an overlapping distribution of the *D* values is commonly reported for the same molecule. Such behavior is usually described by an anomalous diffusion model in which molecules diffuse with a nonlinear relationship in time (Bouchaud & Georges, 1990; Saxton, 2007, 2008). This is usually attributed to the combination of cytoskeleton interactions (Kusumi et al., 2005; Kusumi, Shirai, Koyama-Honda, Suzuki, & Fujiwara, 2010), molecular crowding, confinement in lipid-dependent domains, or any other cause.

To overcome the major limitation of assessing the origin of anomalous subdiffusion, a powerful strategy has been proposed named spot variation FCS (svFCS) in which the focal volume of observation is varied allowing to measure the diffusion time $\tau_d$ at different spatial scales (Lenne et al., 2006; Wawrezinieck, Lenne, Marguet, & Rigneault, 2004; Wawrezinieck, Rigneault, Marguet, & Lenne, 2005). By this method, the plot of $\tau_d$ versus the square of the beam waist $w^2$ yields two key observables: $D_{eff}$, the effective diffusion coefficient represented by the inverse of the slope, and $t_0$, the intercept on the diffusion time axis indicative of potential confinement of the studied molecule. svFCS is thus a perfect analog to single-particle tracking (SPT) in the time domain (Saxton, 2005; Wieser & Schutz, 2008).

The idea of developing the svFCS method in which appropriate temporal resolution could be maintained without any loss of spatial information was mainly inspired by the seminal work of Yechiel and Edidin (1987) in which fluorescence recovery after photobleaching (FRAP) experiments were performed by varying the size of the bleached area. The authors modulated the illuminated area by using microscope objectives of different magnifications, numerical apertures (NAs), and immersion media allowing the variation of the beam radius between 0.35 and 5 μm (Yechiel & Edidin, 1987). By analyzing the dependence of the mobile fraction on the square of the beam radius for both lipids and proteins, the authors quoted the concept of micrometer-scale domains as an organizing principle of the plasma membrane. In the original study, the beam radius was not varied continuously and different objectives led to different optical aberrations that changed the spot shape.

svFCS has been applied to monitor the modes of molecular lateral diffusion in biomembranes. The generalization of its application among different cellular models is expected to yield significant insight into particular cell membrane organization at the suboptical level.

## 2. OPTICAL SETUPS FOR SIZING THE EXCITATION VOLUME

svFCS method is mainly based on a classical microscope and does not require a sophisticated optical setup. In our laboratory, the measurements are performed on a homemade confocal setup built around an inverted Axiovert 200M microscope (Zeiss, Germany) equipped with a Zeiss C-Apochromat 40 × objective with high NA (1.2) (see Fig. 10.1). For green fluorescent protein (GFP) and related fluorescent probes, a 488-nm excitation laser wavelength illuminates the back aperture of the objective with a power of ~3 μW. The emitted fluorescence is collected by the same objective, separated from the

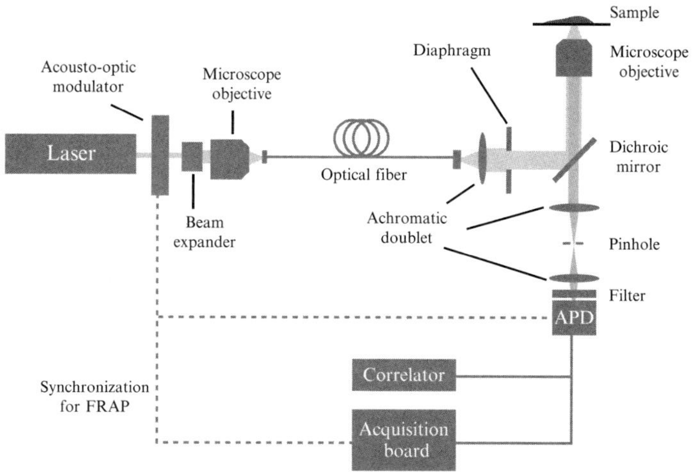


**Figure 10.1** Schematic representation of the svFCS–FRAP experimental setup. A 488-nm laser is focused onto an optical fiber by the use of a beam expander and a microscope objective. The output signal of the fiber is then collimated and sized by a diaphragm. The fluorescence emitted by the sample is collected through a confocal light path and recorded on an APD. A multiple-tau hardware correlator is connected directly to the detector. An AOM allows us to precisely control the power of the laser that we use for svFCS (few μW) and for FRAP (few mW) with high temporal accuracy. An acquisition board is used to control all the devices and to record signals.

excitation line by a 488-nm dichroic mirror (Chroma Technology Corp, USA) and directed to the detector. After rejecting the out-of-focus fluorescence through the use of a confocal pinhole, the light is collected through a 545/20-nm bandpass filter (Chroma Technology Corp, USA) by an avalanche photodiode (APD) (PerkinElmer Optoelectronics, Canada) which has high photon-counting detection efficiency and low dark count rate. The fluorescence intensity signal is processed by a multiple-tau hardware correlator (Correlator.com, USA) which has multiple sampling and delay times. It measures the autocorrelation function (ACF) with high precision for delays ranging from less than a microsecond to a few minutes. An *xyz* piezo nanopositioning stage (Physik Instrument, Germany) is used to locate the excitation beam onto the appropriate area of the biological sample.

Different experimental strategies have been used to spatially size the excitation volume illuminating a defined area of the plasma membrane. These are based on three main principles: (1) the svFCS method in which the back aperture of the objective is under-filled by the laser beam (Masuda, Ushida, & Okamoto, 2005; Wawrezinieck et al., 2004), (2) the *z*-scan FCS in which the focal volume is moved in the *z*-direction to intercept plasma membrane area of variable sizes (Humpolickova et al., 2006), and (3) the STED–FCS, in which FCS is coupled with stimulated emission depletion (STED) microscopy (Eggeling et al., 2009). In the following text, we refer indifferently to the beam radius or the waist as the decay to $1/e^2$ of the 3D Gaussian beam in the plane of symmetry *xy* of the confocal volume.

## 2.1. Spot variation FCS

The easiest way to analyze the diffusion times through observation volumes of different sizes is to introduce a diaphragm between the laser beam expander and the dichroic mirror, in order to select the extension of the laser beam falling on the back aperture of the microscope objective. Because the resulting point spread function (PSF) is a direct function of the diaphragm aperture, the more under-filled the objective is, the larger the PSF. Thereby, the size of the observation volume is easily and continuously tuned between 0.2 and 0.5 μm in radius (Wawrezinieck et al., 2004) (Fig. 10.2A).

Alternatively, Masuda et al. (2005) sized the waist by varying the extension of the excitation beam and thus changing the NA of the objective. The optical setup is very simple and needs only a motorized variable beam expander in the illumination light path to size the incoming laser, ensuring that the beam remains collimated but with a variable radius. This motorized variable beam

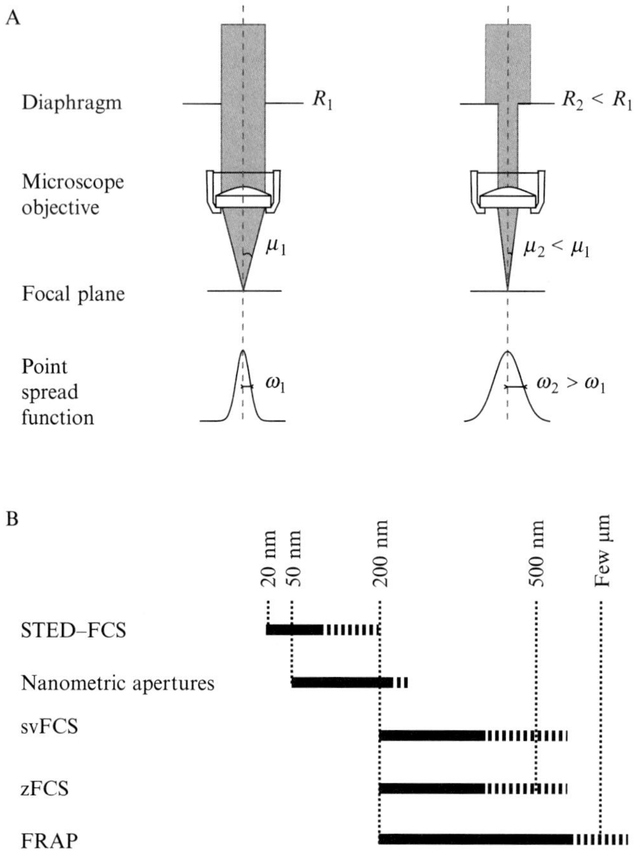


**Figure 10.2** Optical principle for sizing the volume of observation. (A) Principle of extending the beam waist using a diaphragm. A reduced diaphragm size before the objective increases the beam waist after the objective. (B) Spatial domains of use covered by the different techniques. Two of the techniques achieve a subwavelength resolution by physical (nanometric apertures) or optical (STED–FCS) reduction of the beam waist.

expander consists of a lens and a motorized zoom lens. The Gaussian distribution of the laser intensity is conserved as is the incoming laser power allowing constant power density of illumination. With this setup, the effective waist can also vary continuously between 0.2 and 0.5 μm (Masuda et al., 2005).

The two strategies require prior calibration of the spot of illumination (see Section 3.1).

## 2.2. *z*-scan FCS

The "calibration-free" technique called *z*-scan FCS is based on intercepting the plasma membrane at different *z*-altitudes of the PSF. This is easily done by moving the excitation volume along the *z*-axis (Humpolickova et al., 2006).

The radius of the Gaussian beam is then a function of the distance between the sample and the minimum laser beam diameter ($z$ dependence, namely $\Delta z$). This technique does not require a specific optical setup and can be performed on commercial confocal microscopes. In order to analyze $\tau_d$ dependency as a function of $\Delta z$, the ACFs are acquired along the $z$-axis at different positions, typically of 1 μm above and below a focal plane initially set at the plasma membrane. As a result, the diffusion time is a parabolic function of $\Delta z$. At the minimum waist, the diffusion time $\tau_d$ and the number of molecules $N$ have a minimum value, whereas the photon detected per molecule per second (i.e., count rate) is maximal. As such it is possible to analyze $N(\Delta z)$ and $\tau_d(\Delta z)$ without any calibration of the spot size. However, it is compulsory to take into account that, for an optical setup free of optical aberrations, both variables $N(\Delta z)$ and $\tau_d(\Delta z)$ must follow similar mathematical functions.

## 2.3. Super-resolution svFCS by single nanometric apertures and STED microscopy

Both the svFCS and $z$-scan FCS methods are limited by the classical optical diffraction limit (i.e., $\sim$200 nm) (Fig. 10.2B). To overcome this weakness and to allow nanoscale observations, alternative approaches have been developed providing a higher spatial resolution.

The first approach illuminates the biological sample through single nanometric apertures of radii varying between 75 and 250 nm (Wenger et al., 2007). A focused ion beam technique is used to mill circular nanometric apertures in an aluminum film of 220 nm thickness deposited on standard microscope glass coverslips. The svFCS method can thus be significantly extended to document diffusion at small waists. Compared to conventional svFCS, this method reaches the spatial resolution necessary to quantify the size of membrane heterogeneities at the submicron scale. Nevertheless, this approach requires a mechanical contact of the sample with the nanometric apertures.

Alternatively, it has been proposed to combine FCS with the emerging STED microscopy (Eggeling et al., 2009). The concept of STED enables the sizing of an effective fluorescent spot of observation below the optical diffraction barrier by creating a doughnut-shape depletion spot around the spot of excitation (Hell & Wichmann, 1994; Willig, Rizzoli, Westphal, Jahn, & Hell, 2006). The effective excitation volume is tuned down to 30 nm in radius by modulating the power of the STED depletion laser. The calibration of the effective fluorescence spot established by STED needs to be tightly evaluated. This approach is rather more sophisticated than the others described above. When combined with FCS, this strategy has, however, provided compelling observations that have not only reinforced those made by regular svFCS but have also provided new insights allowing an enhanced understanding of plasma membrane organization (see, for review, He & Marguet, 2011).

### 2.4. Combining FRAP and svFCS

FRAP and FCS are alternative methods to measure lateral diffusion in biomembranes (Niv, Gutman, Kloog, & Henis, 2002). The quantification of the immobile fraction of molecules is only possible by FRAP. This parameter is of great interest to avoid epiphenomenal observations by svFCS. While the two methods are inversely sensitive to the concentration of fluorescent probes, there exists a range for which measurements are compatible. Implementing FRAP and FCS on the same setup is easy to do with an acousto-optic modulator (AOM) to precisely adjust the laser power (Fig. 10.1). For FRAP experiments, thanks to an acquisition board (National Instruments) to drive the AOM and APD, the fluorescence is recorded at $\sim 3\ \mu W$ before and after a bleaching period of 1 ms at $\sim 3$ mW during which the APD is shutdown.

## 3. GENERAL CONSIDERATIONS FOR svFCS ACQUISITION

The svFCS technique relies on the measurement of mean diffusion time of fluorescent molecules within the confocal spot of varying size. In practice, we collect the fluorescence signal coming from four to five sizes of spot with waist $w$ ranging usually from 200 up to 500 nm (Wawrezinieck et al., 2005). A systematic determination of the confocal volume is, therefore, obligatory before recording experimental data on living cells.

Accurate measurements by svFCS depend on multiple parameters that must be defined before data acquisition on biological samples. Below, we discuss the points regarding more specifically this method, bearing in mind that more general considerations for standard FCS that are described in this book obviously apply to svFCS acquisitions.

### 3.1. Spot size calibration

Three different methods have been proposed so far (see, for review, Ruttinger et al., 2008). The most commonly used method relies on calculating the average diffusion time $\tau_d$ of fluorescent molecules used as a standard within the effective volume. The $\tau_d$ value is extracted from the fit of the

ACF. From the conventional diffusion coefficient established for a standard dye (i.e., Rhodamine-6G (Rh6G)), the waist *w* of the PSF is calculated. The accuracy of *w* determination will depend on the dye concentration normally within the pico- to nanomolar scale range.

Alternatively, it is possible to measure a dilution series of a fluorescent dye of known concentration. The average number *N* of fluorescent molecules within the confocal volume is calculated from the inverse of the amplitude $G(0)$ of the autocorrelation curve. The waist *w* is then extracted from the slope of the linear relationship between *N* and the dye concentration (ideally between 50 p*M* and 100 n*M* (Ruttinger et al., 2008)). The accuracy of the determination of *w* relies on the quality of the dilution series which, at low concentrations, might be biased by adsorption onto the support or by solvent evaporation. It is also important to point out that the triplet state effect has a direct impact on the estimation of *N*.

Finally, a third method exists that is based on the scanning of a subresolution-sized fluorescent bead ($\sim$100 nm in diameter) in all axes. The extraction of *w* is then achieved from the 2D Gaussian fitting.

We routinely use the first method to calibrate the spot following each modification of the diaphragm, sizing the lateral extension of the laser beam. In practice, $\sim$200 μL of a solution of Rh6G at 10 p*M* is placed on a glass coverslip laid down directly onto the water-immersion objective. The 488-nm laser illumination is powered at 300 μW as measured at the back aperture of the objective. The fluorescence signal is recorded over 10 runs of 20 s. The resulting averaged ACF is fitted to the 3D diffusion model with triplet state (see Section 5.1) from which the $\tau_d$ value is extracted. We calculate the $w^2$ value from the known value of *D* for the Rh6G in water ($2.8 \times 10^{-6}$ cm$^2$ s$^{-1}$; Elson & Magde, 1974; Lee, Sato, Ushida, & Mochida, 2011). This calibration is thoroughly performed after each diaphragm adjustment.

## 3.2. Setting optimal laser illumination

While FCS is considered a minimally invasive method on living cells, adjusting the laser power to optimal illumination levels secures the accuracy of the observations. This is governed by the need to efficiently excite the fluorescent molecules and achieve good signal-to-noise ratio (SNR) on the one hand but to limit triplet state generation and photodamage due to photobleaching or phototoxicity on the other. This is particularly important for slow diffusion processes such as those occurring within the cell membrane as compared to molecules in solution (Garcia-Saez & Schwille, 2008).

Synthetic fluorophores and fluorescent proteins (FPs) are classically used in FCS experiments but they have different physicochemical properties. It is, therefore, mandatory to adapt the experimental conditions to each biological system. In our experimental conditions, we use a 488 nm laser illumination ranging from 1 to 4 μW, and work with cells expressing eGFP-tagged proteins or proteins labeled with Alexa Fluor 488 antibodies. This laser power usually gives a signal ranging between 20 and 100 kHz depending on the level of fluorescence labeling. The photobleaching is monitored as the stability of the fluorescence intensity over the acquisition time of a single run.

## 3.3. Labeling strategies

Numerous synthetic fluorophores are available for labeling proteins: among them, the Alexa-Fluor or ATTO series display appropriate brightness and stability for FCS applications. They are available with different reactive groups (NHS ester, iodoacetamide) for efficient covalent coupling onto the protein of interest (monoclonal antibodies, Fab fragments, ligands, etc.). However, it is difficult to achieve the addition of a single fluorescent label per molecule, which is a common concern when quantitative studies based on the brightness of the molecules are anticipated. Recent developments in the protein tagging technologies have overcome this issue: a peptide sequence is genetically grafted onto a recombinant protein; the addition of a single fluorescent molecule is ensured by the specificity of the enzymatic reaction coupling the fluorescent substrate to the tag (Gautier et al., 2008; Griffin, Adams, & Tsien, 1998; Miller, Cai, Sheetz, & Cornish, 2005).

The genetic tagging of FPs onto the protein of interest is also possible. Nowadays, good intrinsically FPs have been engineered with photophysical properties comparable to those of synthetic probes. These FPs cover a large excitation spectrum (for review, see Chudakov, Matz, Lukyanov, & Lukyanov, 2010). Nevertheless, eGFP, mRFP, and mCherry remain the most frequently used FPs in FCS. Moreover, the use of FP-tagged recombinant proteins allows the experimentalist to probe dynamic processes that occur within intracellular compartments. However, particular attention must be given to the genetic construction itself. The choice of monomeric forms of FP (Zacharias, Violin, Newton, & Tsien, 2002) should be favored, although we have not found any experimental difference in terms of diffusion rates between eGFP and monomeric eGFP (with the mutation $A_{206}K$)-tagged constructs (Lassere R, Hamon Y, Mailfert S, Marguet D, He H.T.). Other factors such as the linker length or possible unspecific interaction of

FPs with endogenous proteins should also be considered. In any case, the coupling reaction or the creation of recombinant proteins may modify the structure of the labeled proteins and consequently alter their affinity for the proteins of interest or their functional properties (Baens et al., 2006), respectively.

The mode and time of cell transfection (i.e., Exgen 500, Jet-PEI, Lipofectamine, Amaxa, etc.), as well as the quantity and quality of DNA, are intrinsic factors of each particular experiment and have to be established independently. It should be noted that where expression of chimeric proteins in lymphoid cells is needed, the transfection is generally performed by nucleofection (Amaxa GmbH, Germany). Lipofection in this case is inefficient. In some cases, it is useful to establish stably transfected cell lines by classical transfection and selection methods. However, FCS observations should be performed once the cells have been functionally validated.

For the immunolabeling of surface molecules, it is more appropriate to use monovalent Fab fragments or recombinant single-chain variable fragments rather than bivalent antibodies that could aggregate the molecules, change their diffusion properties, or trigger nonspecific signaling. Special care must be taken when bovine serum albumin (BSA) is used in buffers. The BSA alone can modify the lipid composition of the plasma membrane and alter its organization. The antibody (Fab) concentration for labeling must be determined for each experiment. A range of labeled molecules should match the FCS experimental constraints (i.e., low concentration of fluorescently labeled molecules). It is also important to ensure that the labeling conditions do not allow nonspecific cellular activation. Moreover, lowering the temperature of the staining steps to 10 °C helps to minimize endocytosis or surface molecule recycling.

The intrinsic nature of the cell membranes might require probing directly the diffusion behavior of lipid components. Although the use of fluorescent lipid analogs does have its drawbacks, the FCS-based observations do not require high probe concentrations contrary to the classical FRAP experiments. This should limit potential artifacts due to the membrane incorporation of lipid analogs.

## 4. MEASUREMENTS ON LIVING CELLS

Performing FCS on living cells requires that the cells are maintained in suitable physiological conditions throughout the observation period. In our laboratory, such experiments are performed at 37 °C by warming the whole microscope within an incubator box (Life Imaging Services, Basel, Switzerland). The measurements do not impose long periods of observation and, therefore, do not necessitate adding a $CO_2$ controller. The cells are simply kept in buffered medium during the acquisition (Hank's Balanced Salt Solution (HBSS) supplemented with 10 m*M* HEPES; cell culture medium without phenol red supplemented with Nutridoma-SP (Roche)).

## 4.1. Procedure for measurement on adherent cells

The svFCS method was first established on Cos7 cells (Lenne et al., 2006; Wawrezinieck et al., 2005) before being applied to different adherent cell lines (HeLa, HEK293, MDCK, etc.). Generally, cells are grown on coverslips of good optical quality before being observed on the microscope. It is easier, however, to grow the cells directly within 8-well chambered coverglasses for inverted microscopes (NUNC A/S, Denmark). For weakly adherent cells or if experimental conditions disturb the adherence, a precoating with poly-L-lysine or with culture medium supplemented with 10% serum is performed before seeding the cells. Typically, cells are seeded at ~25,000 cells/cm$^2$, 24 h prior to the FCS acquisitions, in the DMEM (Gibco, preferentially without phenol red) supplemented with 10% fetal bovine serum (FBS) (Sigma-Aldrich) and 1 m*M* sodium pyruvate (Gibco). Confluence should be kept at around 30–50% to optimize the number of isolated cells.

For FP-tagged proteins, cells are transfected directly in the chambered coverglass the day before the measurements. The lag time before starting FCS acquisitions has to be adjusted according to the expression kinetics of the recombinant protein: small proteins are generally efficiently targeted to the plasma membrane within less than 24 h, whereas polytopic membrane proteins need more time to mature and reach the plasma membrane. Before acquisitions, the cells are washed three times and maintained in acquisition medium. pH buffering is not necessary when acquisitions are performed in a $CO_2$-controlled environment. For sensitive cells, it is necessary to keep them in a rich medium (see Section 4.2).

The fluorescently labeled cells are chosen using wide-field fluorescence. An *xy* coarse image is recorded to check the fluorescence distribution followed by a *z*-scan over the nucleus (Fig. 10.3A). This allows localizing a fluorescent signal at the plasma membrane under and above the nucleus. The observation volume is then located at the maximum of fluorescence intensity in the *z*-axis (Fig. 10.3B). This procedure minimizes the fluorescence

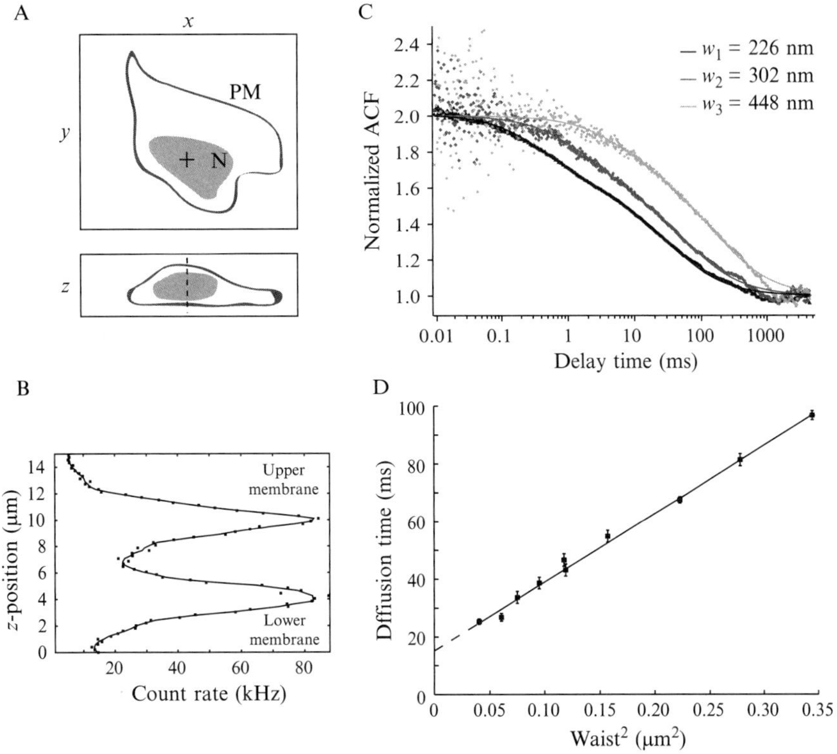


**Figure 10.3** svFCS on living cells. (A) Schematic representation of an *xy*-axes scan of the adherent cell and the *z*-axis plane. The cell is scanned along the *z*-axis (dashed line) after positioning the confocal spot in the nucleus area (N) (plus). (B) Example of the *z*-axis scan on a Cos7 cell expressing GPI-anchored eGFP showing the upper and lower plasma membrane. The confocal spot is positioned on one of the two peaks before recording fluorescence signal fluctuations. (C) Experimental normalized ACFs for three different waists as measured on Cos7 cells expressing the GPI-anchored Thy1 tagged with eGFP. Dots represent experimental values and solid lines the corresponding ACF fits. (D) Plot of the mean value of $\tau_d$ versus the squared waist $w^2$ for Thy1-eGFP proteins.

coming from intracellular compartments. In most cases, we have been unable to record distinct behaviors between the "lower" and "upper" membranes either in terms of diffusion measurement (FCS, spot FRAP) or mobile fraction (spot FRAP); however, the membrane in contact with the coverslip does have a more planar geometry and stability. Picking one rather the other will depend on biological considerations.

The number and time of acquisitions have to take into account some biological aspects such as the diffusion behavior, the recycling, or the internalization of the molecule of interest. The acquisition time of individual measurements varies between 5 and 20 s/run and depends on the diffusion of molecules. Indeed, slowly diffusing molecules need longer acquisition times, whereas for rapidly internalized molecules a short acquisition time is more appropriate. We typically collect series of 10–20 runs each lasting 5–10 s. For one svFCS point (i.e., corresponding to a given spot area), we usually perform a series of 10 acquisitions on 10–20 different cells. In any case, we privilege longer acquisition times in order to record a better-constructed ACF; however, this directly depends on signal stationarity. The collected ACFs are fitted with an appropriate model and the average diffusion time $\tau_d$ is calculated (see Section 5).

### 4.2. Special considerations for measurements on nonadherent cells

FCS method has been adapted to nonadherent cells such as T cell lines (Jurkat, T cell hybridoma) or primary cells (T lymphocytes (Lasserre et al., 2008) and natural killer cells (Guia et al., 2011)). As a prerequisite, the cell culture conditions have to be properly set. Typically, the T cell lines are usually grown in RPMI-1640 (Gibco) medium supplemented with 5–10% FBS (Sigma-Aldrich), 1 m*M* sodium pyruvate (Gibco), and 2 m*M* L-glutamine (Gibco). Primary T cells are maintained in DMEM-F12 medium (Gibco) supplemented with 1% Nutridoma-SP (Roche). For specific biological experiments, the cells are usually starved overnight in DMEM-F12 medium without phenol red (Gibco) supplemented with 1% Nutridoma-SP (Roche) before the measurements. In such a case, the use of 10 m*M* HEPES-buffered HBSS significantly modifies the cell morphology. Moreover, the cells should be systematically harvested by soft centrifugation (reduced gravitational acceleration at $\sim 100 \times g$). Considering the need to work at low temperatures for immunostaining, prewarming the cells ($\sim 10^6$ cells/cm$^2$) at 37 °C for 5–10 min is required prior to FCS acquisitions. This also allows the nonadherent cells to attach on precoated chambered coverglass. For primary T cells or T cell lines, we lower the laser power to between 1.5 and 2 μW. We generally record 5–8 different waists ranging from 200 to 330 nm.

More generally, svFCS becomes meaningful when it is part of a dialogue approach with other techniques. The plot of the diffusion time $\tau_d$ as a function of the squared waist $w^2$ can provide a molecular explanation for signaling events analyzed at the level of a cell population. Conversely, changes in cell behavior or responses can be explained by a modulation of the diffusion mode of the molecules of interest. Similarly, the plot of $\tau_d$ versus $w^2$ is not necessarily sufficient by itself and, when applicable, should be supported by complementary approaches. For instance, FRAP is useful to assess the mobile fraction of molecules, the one which only contribute to the fluctuations recorded by FCS.

# 5. svFCS DATA ANALYSIS AND CURVE FITTING

## 5.1. From fluorescent fluctuations to diffusion constant

In FCS, the fluorescent molecules constantly diffuse into and out of the observation volume, yielding fluctuations in concentration (both in space and time domains) and, therefore, also in the measured fluorescence intensity. These fluctuations are stochastic and differ one from another over time and position thus preventing the accurate determination of a diffusion coefficient by a single fluctuation measurement. However, the diffusion process has a direct impact on the rate at which these fluctuations dissipate within a temporal measurement, defined as a degree of correlation. The statistical tool dedicated to the analysis of the temporal fluctuations of the fluorescence intensity $F(t)$ is the ACF:

$$g^{(2)}(\tau) = \frac{\langle F(t)F(t+\tau)\rangle}{\langle F(t)\rangle^2} \qquad [10.1]$$

where $\langle F(t)\rangle$ is the temporal average of the signal $F(t)$ and $\tau$, the lag time. Special attention has to be given to the experimental conditions to ensure that fluctuations only originate from the passage of molecules within the observation area, meaning that the volume of illumination in FCS should be small and with a reasonable number of fluorescent markers.

The resulting ACF will finally depend on the kind of phenomenon occurring in the studied systems (e.g., diffusion, convection, chemical reaction) and has already been studied by many authors (Bacia & Schwille, 2003; Elson, 2011; Rigler & Elson, 2001). To ensure a proper analysis of the ACF, a specific model should be correctly chosen considering the observed system. In the following, we focus on the case of a pure diffusion experiment involving fluorescent markers in cell, which are excited by a laser with a Gaussian profile:

$$I(x,y,z) = I_0 \exp\left[-2\left(\frac{x^2+y^2+s^2z^2}{w^2}\right)\right] \qquad [10.2]$$

where $w$ denotes the waist and $s$ is the structural parameter of the excitation laser defined as the ratio between the lateral waist and the axial waist. If we consider that the probes diffuse partly at the plasma membrane (lateral diffusion in 2D) and within its vicinity (3D diffusion), then the general model for the autocorrelation is

$$G(\tau) = 1 + f_{\mathrm{triplet}}(\tau)\frac{1}{N}\left(\frac{A}{1+\frac{\tau}{\tau_{d1}}} + \frac{1-A}{\left(1+\frac{\tau}{\tau_{d2}}\right)\sqrt{1+s^2\frac{\tau}{\tau_{d2}}}}\right) \quad [10.3]$$

where $N$ is the total number of particles in the effective volume, $A$ is the ratio of particles diffusing only in 2D (as compared to $N$), and $\tau_{d1}$ and $\tau_{d2}$ are, respectively, the mean diffusion times associated to the particles diffusing either in 2D (at the membrane) or in 3D (in the vicinity of the membrane) in the effective volume $V_{\mathrm{eff}} = \pi^{3/2} s w^3$. In an ideal situation, the fluctuations detected by the ACF are related to the displacement of the probes within the excitation volume. However, a percentage of the dyes can be excited in a triplet state and as such do not emit photons for a characteristic time $\tau_T$, which is typically in the microsecond range. In the autocorrelation model, the triplet state is then considered by the term $f_{\mathrm{triplet}}(\tau)$ defined as

$$f_{\mathrm{triplet}}(\tau) = 1 + \frac{T}{1-T}\exp\left(-\frac{\tau}{\tau_T}\right) \quad [10.4]$$

where $T$ is the fraction of nonfluorescent dyes in triplet state. In the case of a pure free diffusion process, the diffusion time $\tau_d$ depends on $D$ and $w^2$ as $\tau_d = w^2/4D$. The diffusion time measurement at different spatial scales increases the accuracy of the $D$ estimation (Fig. 10.3C). We plot the diffusion time $\tau_d$ as a function of the squared waist $w^2$ which, as we explain in Section 6, we use to describe the mechanisms of confinement during the diffusion process.

## 5.2. Noise in FCS

Even when special care is taken to prepare and observe the biological samples, physical phenomena will generate fluctuations contributing to the noise and restricting the accuracy of the measurement. These fluctuations originate mainly from the quantum nature of light (shot noise) and the stochastic nature of the fluorescent fluctuations itself. Overall, this is modeled by an additive Gaussian perturbation on the measurements (see Section 5.3). Under classical FCS conditions (Gaussian molecular detection, 3D Brownian diffusion, negligible background, and small sampling time $\Delta\tau \rightarrow 0$), it has been demonstrated (Koppel, 1974; Wenger, Gerard, Aouani, & Rigneault, 2009; Wohland, Rigler, & Vogel, 2001) that the SNR could be approximated by

$$\mathrm{SNR}_{\tau \rightarrow 0} \approx \mathrm{CRM}\sqrt{\frac{T_{\mathrm{tot}}\Delta\tau}{1+1/\langle N\rangle}} \qquad [10.5]$$

where CRM is the count rate per molecule, $T_{\mathrm{tot}}$ the total time, and $\Delta\tau$ the correlator channel width. Accuracy in FCS can be partially improved by increasing the laser power to modify the CRM, but is limited by the saturation or photobleaching of the dyes. Theoretically, the increase of $T_{\mathrm{tot}}$ should have only a weak impact on the SNR. However, the FCS analysis applies for a system in a stationary state on which phenomena such as membrane fluctuations, dye aggregations, or endocytosis preclude extending the acquisition time. Therefore, in order to increase the accuracy of the estimated parameters ($\tau_{\mathrm{d}}$, $w^2$, $D$, etc.), it is more appropriate to repeat measurements and perform proper procedures based on statistical estimation taking into consideration such noise contributions.

## 5.3. Estimation of the svFCS observables and their errors

svFCS relies on the analysis of the relationship between the diffusion time $\tau_{\mathrm{d}}$ and the waist $w^2$. The accuracy of the result is assessed by a linear regression analysis accounting for the estimation errors of $\tau_{\mathrm{d}}$ and $w^2$.

Let $\left\{\left(\tilde{\tau}_{\mathrm{d}i}, \widetilde{w_i^2}\right)\right\}_{i\in[1,P]}$ be the set of $P$ measures that are perturbed by simultaneous errors in the diffusion time and the waist such that $\tilde{\tau}_{\mathrm{d}i} = \tau_{\mathrm{d}i} + n_{\tau_i}$ and $\widetilde{w_i^2} = w_i^2 + n_{w_i^2}$ $\forall i \in [1,P]$. The errors $n_{\tau_i}$ and $n_{w_i^2}$ would be expected to independently follow a Gaussian distribution; the standard deviations of these errors are experimentally known and defined for each $i^{\mathrm{th}}$ measurement by $\left\{\left(\sigma_{\tau_{\mathrm{d}i}}, \sigma_{w_i^2}\right)\right\}_{i\in[1,P]}$. The linear model is given by $\tau_{\mathrm{d}i} = t_0 + (1/4D)w_i^2$ such that for each $i$, the measures can be written as

$$\begin{cases} \widetilde{w_i^2} = w_i^2 + n_{w_i^2} \\ \tilde{\tau}_{\mathrm{d}i} = t_0 + \dfrac{1}{4D}w_i^2 + n_{\tau_i} \end{cases} \quad \forall i \in [1,P] \qquad [10.6]$$

In this equation, $t_0$ and $a = 1/4D$ are the unknown parameters of interest, and $\{w_i^2\}_{i\in[1,P]}$ is the set of unknown nuisance parameters. The maximum likelihood estimator provides the estimations of $\hat{t}_0$ and $\hat{a}$ by minimization of likelihood criterion:

$$\hat{t}_0, \hat{a} = \arg\min_{t_0, a} \mathrm{LC}(t_0, a) \qquad [10.7]$$

where

$$\mathrm{LC}(t_0, a) = \sum_{i=1}^{P} \left[ \left( \frac{\widetilde{w}_i^2 - \hat{w}_i^2(t_0, a)}{\sigma_{w_i^2}} \right)^2 + \left( \frac{\widetilde{\tau}_{di} - t_0 - a\hat{w}_i^2(t_0, a)}{\sigma_{\tau_{di}}} \right)^2 \right] \qquad [10.8]$$

with

$$\hat{w}_i^2(t_0, a) = \frac{\dfrac{\widetilde{w}_i^2}{\sigma_{w_i^2}^2} + a\dfrac{\widetilde{\tau}_{di} - t_0}{\sigma_{\tau_{di}}^2}}{\dfrac{1}{\sigma_{w_i^2}^2} + \dfrac{a^2}{\sigma_{\tau_{di}}^2}} \quad \forall i \in [1, P] \qquad [10.9]$$

The Cramer–Rao bound (CRB) provides a lower bound on the variance of any unbiased estimator. Given classical assumptions, the maximum likelihood estimator is asymptotically efficient and the variance of the maximum likelihood estimator is equal to the CRB, which is given by $\mathbf{CRB} = \mathbf{J}^{-1}$, where $\mathbf{J}$ is the Fisher information $(P+2) \times (P+2)$ matrix and $()^{-1}$ defines the matrix inversion. The likelihood of the parameters $t_0, a$ and $\{w_i^2\}_{i \in [1,P]}$ when $\left\{ \left( \widetilde{\tau}_{di}, \widetilde{w}_i^2 \right) \right\}_{i \in [1,P]}$ are measured is

$$L\left( t_0, a, \{w_i^2\}_{i \in [1,P]} \right) = cte - \frac{1}{2} \sum_{i=1}^{P} \left[ \left( \frac{\widetilde{w}_i^2 - w_i^2}{\sigma_{w_i^2}} \right)^2 + \left( \frac{\widetilde{\tau}_{di} - t_0 - a w_i^2}{\sigma_{\tau_{di}}} \right)^2 \right] \qquad [10.10]$$

The calculation of the Fisher matrix leads to the result:

$$\mathbf{J}\left( t_0, a, \{w_i^2\}_{i \in [1,P]} \right) = \begin{bmatrix} J_{a,a} & J_{a,t_0} & \begin{bmatrix} J_{a,w_1^2} & J_{a,w_2^2} & \cdots & J_{a,w_P^2} \end{bmatrix} \\ J_{a,t_0} & J_{t_0,t_0} & \begin{bmatrix} J_{t_0,w_1^2} & J_{t_0,w_2^2} & \cdots & J_{t_0,w_P^2} \end{bmatrix} \\ \begin{bmatrix} J_{a,w_1^2} \\ J_{a,w_2^2} \\ \vdots \\ J_{a,w_P^2} \end{bmatrix} & \begin{bmatrix} J_{t_0,w_1^2} \\ J_{t_0,w_2^2} \\ \vdots \\ J_{t_0,w_P^2} \end{bmatrix} & \begin{bmatrix} J_{w_1^2,w_1^2} & 0 & \dots & 0 \\ 0 & J_{w_2^2,w_2^2} & \cdots & 0 \\ \vdots & \vdots & \ddots & \vdots \\ 0 & 0 & \dots & J_{w_P^2,w_P^2} \end{bmatrix} \end{bmatrix} \qquad [10.11]$$

with

$$J_{a,a} = \sum_{i=1}^{P} \left( \frac{w_i^2}{\sigma_{\tau_{di}}} \right)^2, \quad J_{t_0,t_0} = \sum_{i=1}^{P} \frac{1}{\sigma_{\tau_{di}}^2}, \quad J_{a,t_0} = \sum_{i=1}^{P} \frac{w_i^2}{\sigma_{\tau_{di}}^2},$$

$$J_{a,w_i^2} = a \frac{w_i^2}{\sigma_{\tau_{di}}^2}, \quad J_{t_0,w_i^2} = \frac{a}{\sigma_{\tau_{di}}^2}, \quad \text{and } J_{w_i^2,w_i^2} = \frac{1}{\sigma_{w_i^2}^2} + \frac{a^2}{\sigma_{\tau_{di}}^2} \quad \forall i \in [1,P]$$

Finally, we obtain an approximation of the variance of estimation of $t_0$ and $a = 1/4D$ by

$$\mathrm{var}(\hat{a}) = \left[ \mathbf{J}\left( \hat{t}_0, \hat{a}, \{\hat{w}_i^2\}_{i\in[1,P]} \right)^{-1} \right]_{1,1} \qquad [10.12]$$

$$\text{and} \quad \mathrm{var}(\hat{t}_0) = \left[ \mathbf{J}\left( \hat{t}_0, \hat{a}, \{\hat{w}_i^2\}_{i\in[1,P]} \right)^{-1} \right]_{2,2}$$

where $[\mathbf{B}]_{l,c}$ defines the scalar at the $l^{\text{th}}$ line and the $c^{\text{th}}$ column in the matrix **B**.

For example, these parameters have been estimated for Thy-1 (where the plot $\tau_d(w^2)$ is shown in Fig. 10.3D): $t_0 = 15.2 \pm 0.8$ ms and $D = 1.05 \pm 0.02\ \mu\text{m}^2\,\text{s}^{-1}$.

# 6. THE NATURE OF THE MOLECULAR CONSTRAINTS ON LATERAL DIFFUSION

## 6.1. Modeling lateral diffusion in inhomogeneous membranes

Although the intrinsic Brownian agitation within a lipid bilayer does homogenize the molecular distribution, the extreme diversity of the plasma membrane composition favors the development of inhomogeneity due to the propensity of such a system to minimize its total free energy. As a consequence, a mosaic pattern of molecular distribution at the cell membrane emerges that is somewhat different from the one which would result from strict Brownian diffusion. The svFCS method has the appropriate spatio-temporal resolution to reveal the dynamics of such molecular arrangements. The rationale sustaining the svFCS approach is based on the strict proportionality between the diffusion time and the probed area that respects the classical Fick's second diffusion law.

Here, we sum-up the simulations performed to decipher the possible effect of different geometrical membrane architectures on svFCS observations

(Wawrezinieck et al., 2005). In the simplest case, that is, in the absence of any constraint, Brownian motion would occur which is characterized by a constant diffusion coefficient $D$ corresponding to the mean surface explored by a molecule per unit of time: $\tau_d = w^2/4D$. By plotting $\tau_d$ as a function of $w^2$, which relates directly to the observed area, one would expect a straight line with a null value for $t_0$, the intercept with the time axis, and a slope inversely proportional to $D$ (Fig. 10.4A).

We next introduced in the model the adjacent permeable domains to mimic physical barriers such as those imposed by the actin-based cytoskeleton (Kusumi et al., 2005). The molecules diffuse freely inside the domains (with a microscopic diffusion coefficient $D_\mu$) before reaching the barriers which can be crossed with a given probability relating to the confinement strength. As long as the observation area is smaller than $l$, the size of the meshwork, a linear regime is observed very similar to the one previously observed for free diffusion. At larger scales of observation, the barriers are perceived as obstacles; they contribute to increase the diffusion time as compared to a free diffusion process (Fig. 10.4A). The plot of $\tau_d$ versus $w^2$ conserves the straight linearity, but the slope is now proportional to an effective

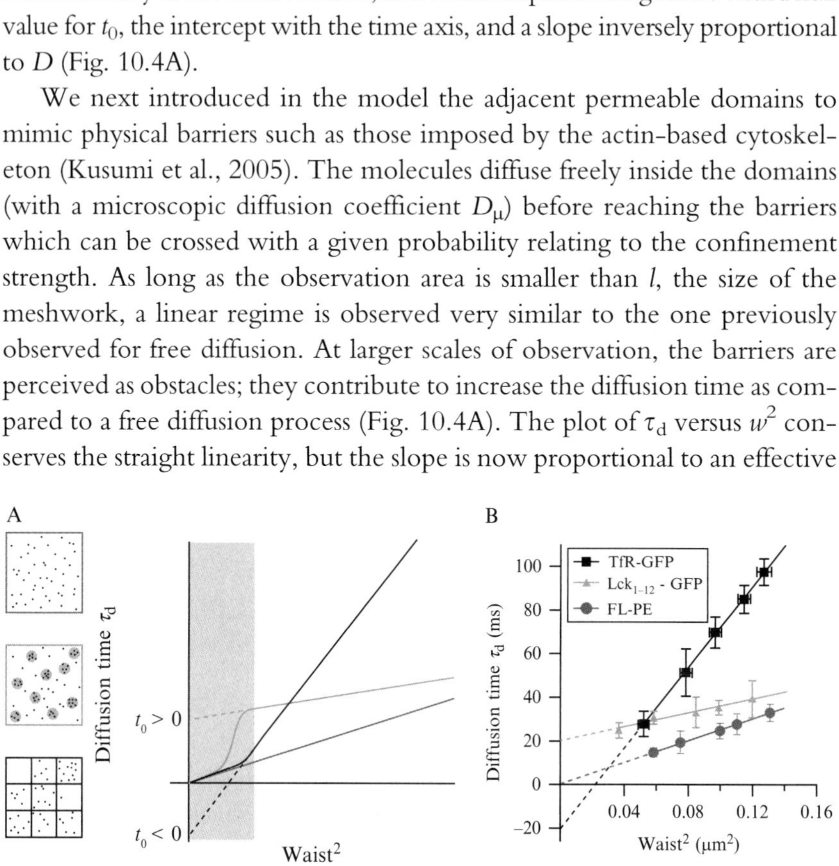


**Figure 10.4** (A) Diffusion models for membrane organization. In the free diffusion model (top panel), molecules (black dots) exhibit a pure Brownian motion within the plane of the membrane. In the presence of isolated domains (center panel), the molecules can diffuse in and out of the domains (gray spot) where they are transiently trapped. In the adjacent domain model, the presence of permeable barriers prevents the free diffusion within the plane of the membrane. The diffusion time $\tau_d$ is plotted as a function of the squared waist of observation $w^2$ for the different models. The free diffusion (dark gray curve) model shows a straight linearity of $\tau_d$ with $w^2$ and a null intercept with the time axis ($t_0 = 0$). In the presence of isolated domains (light gray curve), $t_0$ is positive, whereas with adjacent domains (black curve), $t_0$ is negative. (B) Experimental svFCS data for different membrane components (see Section 6.2). Error bars on the $x$- and $y$-axes give the standard deviation of the means.

diffusion coefficient $D_{\text{eff}}$ and the extrapolated $t_0$ intercept takes a negative value. As compared to the free diffusion model, $D_{\text{eff}}$ is lower: it depends on $D_{\mu}$ and the ratio $\tau_{\text{conf}}/\tau_{\text{dom}}$, where $\tau_{\text{conf}}$ and $\tau_{\text{dom}}$ are the confinement time and the diffusion time into the domain, respectively. The negative $t_0$ intercept is mainly dominated by $\tau_{\text{conf}}$ when $\tau_{\text{dom}}$ becomes negligible. The offset of $t_0$ was derived as a function of the size parameter $l$ and the effective diffusion coefficient as $t_0 \approx l^2/(36D_{\text{eff}})$ (Destainville, 2008).

Alternatively, permeable isolated domains (Fig. 10.4A) have been introduced mimicking, for instance, lipid-dependent nanodomains (see, for review, He & Marguet, 2011). In such a situation, particles diffuse into and out of the domains with a transit time that depends on the confinement strength. The $\tau_d$ value reported by svFCS is the sum of the time the molecules are freely diffusing and the time spent within domains. $\tau_d$ depends on the domain density, confinement strength, and diffusion coefficients inside and outside the domains. Again, when the observation area is smaller than the size of the domains, a linear regime is very similar to the one previously observed for the free diffusion model. At larger scales of observation, the plot of $\tau_d$ versus $w^2$ conserves a straight linearity, but the slope is proportional to an effective diffusion coefficient $D_{\text{eff}}$ and the extrapolated $t_0$ intercept takes a positive value. In this model, $t_0$ depends on the fraction of molecules $\alpha$ confined within the domains and the confinement time: $t_0 \approx 2\alpha(\tau_{\text{conf}} - \tau_{\text{dom}})$, and $D_{\text{eff}}$ is sensitive to the free diffusion constant and $\alpha$ by the following equation $D_{\text{eff}} \approx (1-\alpha)D_{\text{free}}$. Recently, a more detailed description for these parameters has been proposed (Ruprecht, Wieser, Marguet, & Schutz, 2011).The positive $t_0$ offset value gives information on the confinement strength and is derived from general analytical expressions:

$$t_0 = \beta\tau_{\text{trap}} \exp\left(-\frac{4}{\sqrt{1-\beta}}\frac{D_{\text{trap}}}{D_{\text{free}}}\right) \qquad [10.13]$$

$$D_{\text{eff}} = (1-\beta)D_{\text{free}} + \beta D_{\text{trap}} \qquad [10.14]$$

where $\beta = \tau_{\text{trap}}/(\tau_{\text{trap}} + \tau_{\text{free}})$ denotes the trapped time fraction, $D_{\text{free}}$ and $D_{\text{trap}}$ the diffusion coefficient of free and trapped molecules, and $\tau_{\text{trap}}$ and $\tau_{\text{free}}$ the duration of trapped and free status, respectively.

## 6.2. Interpretation of svFCS experimental data

The svFCS strategy has been applied to analyze the plasma membrane organization of different components and in different cell types (Cahuzac et al., 2006; Chakrabandhu et al., 2007; Guia et al., 2011; Lasserre et al., 2008;

Lenne et al., 2006). Figure 10.4 exemplifies the observations obtained for molecules exhibiting distinctive $t_0$ and $D_{eff}$ values. These molecules are also known to interact differently with their environment and to diffuse either freely for the BODIPY-phosphatidylethanolamine lipid analog (FL-PE) or with constraints due either to the actin-based cytoskeleton for the transmembrane transferrin receptor (TfR-GFP) or to the lipid-dependent nanodomains for the GFP tagged to the inner leaflet of the plasma membrane by the N-terminal 12-amino acid sequence of Lck ($Lck_{1-12}$-GFP) (Fig. 10.4B).

There exists a strong correlation between the $t_0$ values and the behavior of these molecules. As expected, the FL-PE lipid analog freely diffuses within the plasma membrane with an almost null $t_0$ and a $D_{eff}$ value usually measured with classical methods. The negative deviation of the $t_0$ reported for the TfR-GFP suggests a confinement due to physical obstacles (cytoskeleton fences). This was assessed by performing svFCS after pharmacological treatments that destabilize the actin meshwork (Lenne et al., 2006). In contrast, the positive deviation of the $t_0$ reported for the $Lck_{1-12}$-GFP would suggest that confinement does occur but in isolated domains, the nature of which were identified when the observations were performed following treatments with drugs reducing the sphingomyelin and cholesterol levels (Lasserre et al., 2008).

Altogether and in accordance with the predictive models, these data demonstrate that the svFCS method offers the spatiotemporal resolution necessary to characterize in living cells the inhomogeneous lateral organization of the plasma membrane. It is important to stress that the geometrical modes of confinement described in this chapter are not mutually exclusive: a deviation of $t_0$ in one direction or the other solely reveals a dominant mode of confinement (Lenne et al., 2006).

## 7. SUMMARY AND FUTURE OUTLOOK

The svFCS method in which the measurements are performed by varying the size of the focal volume of observation has demonstrated its potency at characterizing different geometrical constraints that hinder the lateral diffusion of membrane components (Lenne et al., 2006; Wawrezinieck et al., 2004, 2005).

The svFCS method is based on plotting the diffusion time $\tau_d$ as a function of the squared waist of observation $w^2$. From models with constrained lateral diffusion, three regimes can be revealed depending on the observable area: (i) for small observation spots, diffusion appears as free with a straight linear plot; (ii) complex transient regimes are observed when the size of observation spots is comparable to that of domains; (iii) at larger scales of observation, the plot of $\tau_d$ versus $w^2$ conserves a straight linearity but the slope is proportional to an effective diffusion coefficient $D_{eff}$ and the extrapolated $t_0$ intercept correlates with the mode of confinement ($t_0 < 0$ for adjacent domains, $t_0 > 0$ for isolated domains).As the size of the domains that hinder lateral diffusion is below the optical diffraction limit (~200 nm), the experimental observations made by svFCS are mainly intended to subsequently evaluate the effective diffusion coefficient $D_{eff}$ and extrapolated $t_0$ intercept.

Recently, the svFCS method has been further improved by increasing the resolution far below the optical diffraction limit using innovative experimental strategies (Fig. 10.2B). By using metallic films drilled with nanoapertures, it has been possible to illuminate very small membrane areas with $w$ varying between 75 and 250 nm. The transition regime predicted for isolated domain organization was thus observable allowing a refinement of the characterization of domain size (Wenger et al., 2007). Elsewhere, the feasibility of FCS based on the nanometric illumination of near-field scanning optical microscopy has been achieved to probe lateral diffusion on living cells (Manzo, van Zanten, & Garcia-Parajo, 2011). Moreover, by combining STED illumination (Hell & Wichmann, 1994) with FCS, the spot of observation was reduced in waist to between 30 and 250 nm. The STED–FCS observations have confirmed the transient confinement of specific lipids and proteins in cholesterol-isolated domains within living cells (Eggeling et al., 2009; Mueller et al., 2011). Data from both approaches have supported the existence of the diffusion behavior predicted by the simulations.

## ACKNOWLEDGMENTS

This research project was supported by institutional grants from Inserm and CNRS, and by specific grants from ANR, ARC, FRM, INCa, and CNRS. V. R. was awarded fellowships from the LNFCC. We thank Emilie Witty (AngloScribe) for editing the English.

## REFERENCES

Bacia, K., Kim, S. A., & Schwille, P. (2006). Fluorescence cross-correlation spectroscopy in living cells. *Nature Methods*, *3*, 83–89.

Bacia, K., & Schwille, P. (2003). A dynamic view of cellular processes by in vivo fluorescence auto- and cross-correlation spectroscopy. *Methods*, *29*, 74–85.

Baens, M., Noels, H., Broeckx, V., Hagens, S., Fevery, S., Billiau, A. D., et al. (2006). The dark side of EGFP: Defective polyubiquitination. *PLoS One*, *1*, e54.

Bouchaud, J.-P., & Georges, A. (1990). Anomalous diffusion in disordered media: Statistical mechanisms, models and physical applications. *Physics Reports*, *195*, 127–293.

Cahuzac, N., Baum, W., Kirkin, V., Conchonaud, F., Wawrezinieck, L., Marguet, D., et al. (2006). Fas ligand is localized to membrane rafts, where it displays increased cell death-inducing activity. *Blood, 107*, 2384–2391.

Chakrabandhu, K., Herincs, Z., Huault, S., Dost, B., Peng, L., Conchonaud, F., et al. (2007). Palmitoylation is required for efficient Fas cell death signaling. *The EMBO Journal, 26*, 209–220.

Chiantia, S., Ries, J., & Schwille, P. (2009). Fluorescence correlation spectroscopy in membrane structure elucidation. *Biochimica et Biophysica Acta, 1788*, 225–233.

Chudakov, D. M., Matz, M. V., Lukyanov, S., & Lukyanov, K. A. (2010). Fluorescent proteins and their applications in imaging living cells and tissues. *Physiological Reviews, 90*, 1103–1163.

Destainville, N. (2008). Theory of fluorescence correlation spectroscopy at variable observation area for two-dimensional diffusion on a meshgrid. *Soft Matter, 4*, 1288–1301.

Eggeling, C., Ringemann, C., Medda, R., Schwarzmann, G., Sandhoff, K., Polyakova, S., et al. (2009). Direct observation of the nanoscale dynamics of membrane lipids in a living cell. *Nature, 457*, 1159–1162.

Elson, E. L. (2011). Fluorescence correlation spectroscopy: Past, present, future. *Biophysical Journal, 101*, 2855–2870.

Elson, E. L., & Magde, D. (1974). Fluorescence correlation spectroscopy. I. Conceptual basis and theory. *Biopolymers, 13*, 1–27.

Garcia-Saez, A. J., & Schwille, P. (2008). Fluorescence correlation spectroscopy for the study of membrane dynamics and protein/lipid interactions. *Methods, 46*, 116–122.

Gautier, A., Juillerat, A., Heinis, C., Correa, I. R., Jr., Kindermann, M., Beaufils, F., et al. (2008). An engineered protein tag for multiprotein labeling in living cells. *Chemistry & Biology, 15*, 128–136.

Griffin, B. A., Adams, S. R., & Tsien, R. Y. (1998). Specific covalent labeling of recombinant protein molecules inside live cells. *Science, 281*, 269–272.

Guia, S., Jaeger, B. N., Piatek, S., Mailfert, S., Trombik, T., Fenis, A., et al. (2011). Confinement of activating receptors at the plasma membrane controls natural killer cell tolerance. *Science Signaling, 4*, ra21.

He, H. T., & Marguet, D. (2011). Detecting nanodomains in living cell membrane by fluorescence correlation spectroscopy. *Annual Review of Physical Chemistry, 62*, 417–436.

Hell, S. W., & Wichmann, J. (1994). Breaking the diffraction resolution limit by stimulated emission: Stimulated-emission-depletion fluorescence microscopy. *Optics Letters, 19*, 780–782.

Humpolickova, J., Gielen, E., Benda, A., Fagulova, V., Vercammen, J., Vandeven, M., et al. (2006). Probing diffusion laws within cellular membranes by Z-scan fluorescence correlation spectroscopy. *Biophysical Journal, 91*, L23–L25.

Koppel, D. E. (1974). Statistical accuracy in fluorescence correlation spectroscopy. *Physical Review, A10*, 1938–1945.

Kusumi, A., Nakada, C., Ritchie, K., Murase, K., Suzuki, K., Murakoshi, H., et al. (2005). Paradigm shift of the plasma membrane concept from the two-dimensional continuum fluid to the partitioned fluid: High-speed single-molecule tracking of membrane molecules. *Annual Review of Biophysics and Biomolecular Structure, 34*, 351–378.

Kusumi, A., Shirai, Y. M., Koyama-Honda, I., Suzuki, K. G., & Fujiwara, T. K. (2010). Hierarchical organization of the plasma membrane: Investigations by single-molecule tracking vs. fluorescence correlation spectroscopy. *FEBS Letters, 584*, 1814–1823.

Lasserre, R., Guo, X. J., Conchonaud, F., Hamon, Y., Hawchar, O., Bernard, A. M., et al. (2008). Raft nanodomains contribute to Akt/PKB plasma membrane recruitment and activation. *Nature Chemical Biology, 4*, 538–547.

Lee, J. I., Sato, M., Ushida, K., & Mochida, J. (2011). Measurement of diffusion in articular cartilage using fluorescence correlation spectroscopy. *BMC Biotechnology, 11*, 19.

Lenne, P. F., Wawrezinieck, L., Conchonaud, F., Wurtz, O., Boned, A., Guo, X. J., et al. (2006). Dynamic molecular confinement in the plasma membrane by microdomains and the cytoskeleton meshwork. *The EMBO Journal, 25*, 3245–3256.

Manzo, C., van Zanten, T. S., & Garcia-Parajo, M. F. (2011). Nanoscale fluorescence correlation spectroscopy on intact living cell membranes with NSOM probes. *Biophysical Journal, 100*, L8–L10.

Masuda, A., Ushida, K., & Okamoto, T. (2005). New fluorescence correlation spectroscopy enabling direct observation of spatiotemporal dependence of diffusion constants as an evidence of anomalous transport in extracellular matrices. *Biophysical Journal, 88*, 3584–3591.

Meseth, U., Wohland, T., Rigler, R., & Vogel, H. (1999). Resolution of fluorescence correlation measurements. *Biophysical Journal, 76*, 1619–1631.

Miller, L. W., Cai, Y., Sheetz, M. P., & Cornish, V. W. (2005). In vivo protein labeling with trimethoprim conjugates: A flexible chemical tag. *Nature Methods, 2*, 255–257.

Mueller, V., Ringemann, C., Honigmann, A., Schwarzmann, G., Medda, R., Leutenegger, M., et al. (2011). STED nanoscopy reveals molecular details of cholesterol- and cytoskeleton-modulated lipid interactions in living cells. *Biophysical Journal, 101*, 1651–1660.

Niv, H., Gutman, O., Kloog, Y., & Henis, Y. I. (2002). Activated K-Ras and H-Ras display different interactions with saturable nonraft sites at the surface of live cells. *The Journal of Cell Biology, 157*, 865–872.

Rigler, R., & Elson, E. (2001). *Fluorescence correlation spectroscopy: Theory and applications.* Berlin, Heidelberg, New York: Springer-Verlag.

Ruprecht, V., Wieser, S., Marguet, D., & Schutz, G. J. (2011). Spot variation fluorescence correlation spectroscopy allows for superresolution chronoscopy of confinement times in membranes. *Biophysical Journal, 100*, 2839–2845.

Ruttinger, S., Buschmann, V., Kramer, B., Erdmann, R., Macdonald, R., & Koberling, F. (2008). Comparison and accuracy of methods to determine the confocal volume for quantitative fluorescence correlation spectroscopy. *Journal of Microscopy, 232*, 343–352.

Saxton, M. J. (2005). New and notable: Fluorescence correlation spectroscopy. *Biophysical Journal, 89*, 3678–3679.

Saxton, M. J. (2007). A biological interpretation of transient anomalous subdiffusion. I. Qualitative model. *Biophysical Journal, 92*, 1178–1191.

Saxton, M. J. (2008). A biological interpretation of transient anomalous subdiffusion. II. Reaction kinetics. *Biophysical Journal, 94*, 760–771.

Wawrezinieck, L., Lenne, P.-F., Marguet, D., & Rigneault, H. (2004). Fluorescence correlation spectroscopy to determine diffusion laws: Application to live cell membranes. *Proceedings of SPIE, 5462*, 92–102.

Wawrezinieck, L., Rigneault, H., Marguet, D., & Lenne, P. F. (2005). Fluorescence correlation spectroscopy diffusion laws to probe the submicron cell membrane organization. *Biophysical Journal, 89*, 4029–4042.

Wenger, J., Conchonaud, F., Dintinger, J., Wawrezinieck, L., Ebbesen, T. W., Rigneault, H., et al. (2007). Diffusion analysis within single nanometric apertures reveals the ultrafine cell membrane organization. *Biophysical Journal, 92*, 913–919.

Wenger, J., Gerard, D., Aouani, H., & Rigneault, H. (2009). Nanoaperture-enhanced signal-to-noise ratio in fluorescence correlation spectroscopy. *Analytical Chemistry, 81*, 834–839.

Wieser, S., & Schutz, G. J. (2008). Tracking single molecules in the live cell plasma membrane—Do's and don't's. *Methods, 46*, 131–140.

Willig, K. I., Rizzoli, S. O., Westphal, V., Jahn, R., & Hell, S. W. (2006). STED microscopy reveals that synaptotagmin remains clustered after synaptic vesicle exocytosis. *Nature, 440*, 935–939.

Wohland, T., Rigler, R., & Vogel, H. (2001). The standard deviation in fluorescence correlation spectroscopy. *Biophysical Journal, 80*, 2987–2999.
Yechiel, E., & Edidin, M. (1987). Micrometer-scale domains in fibroblast plasma membranes. *The Journal of Cell Biology, 105*, 755–760.
Zacharias, D. A., Violin, J. D., Newton, A. C., & Tsien, R. Y. (2002). Partitioning of lipid-modified monomeric GFPs into membrane microdomains of live cells. *Science, 296*, 913–916.

# AUTHOR INDEX

Note: Page numbers followed by "*f*" indicate figures, "*t*" indicate tables, and "*np*" indicates footnotes.

## C

## D

## I

## J



## K

## L

## S

## T

## X



## Y



## Z

# SUBJECT INDEX

Note: Page numbers followed by "*f*" indicate figures, and "*t*" indicate tables.

## G

## I



## L



## M



## N



## P

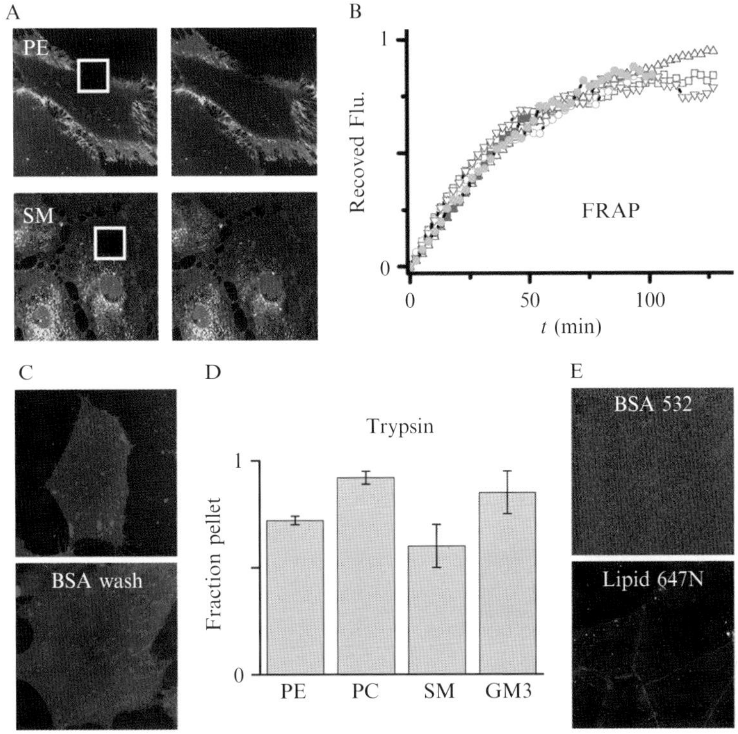


**Veronika Mueller *et al.*, Figure 1.2** Cellular labeling with fluorescent lipid analogs. (A) Confocal scanning images ($70 \times 70\ \mu m^2$) of living PtK2 cells labeled with an Atto647N-labeled phosphoethanolamine (PE, upper panel) and sphingomyelin (SM, lower panel) after (left panels) photobleaching a $15 \times 15\ \mu m^2$ large area (white box) and after recovery of signal (right panels). (B) FRAP curves of Atto647N-labeled PE and SM on different cells or on different parts of the cell. (C) BSA washing: Confocal scanning images ($80 \times 80\ \mu m^2$) of the plasma membrane of living PtK2 cells incorporating Atto647N-labeled SM before (upper) and after (lower) washing with BSA. (D) Trypsin treatment: Fraction of signal in the cell pellet, that is, fraction of fluorescent lipid analogs (PE: Atto647N-labeled phosphoethanolamine, PC: Atto647N-labeled phosphocholine, SM: Atto647N-labeled sphingomyelin, GM3: Atto647N-labeled GM3) still incorporated after trypsin treatment. (E) Confocal scanning images ($80 \times 80\ \mu m^2$) of living PtK2 cells labeled with Atto647N-labeled PE incorporated by a complex with Atto532-labeled BSA: Atto532 (upper image) and Atto647N fluorescence (lower image).

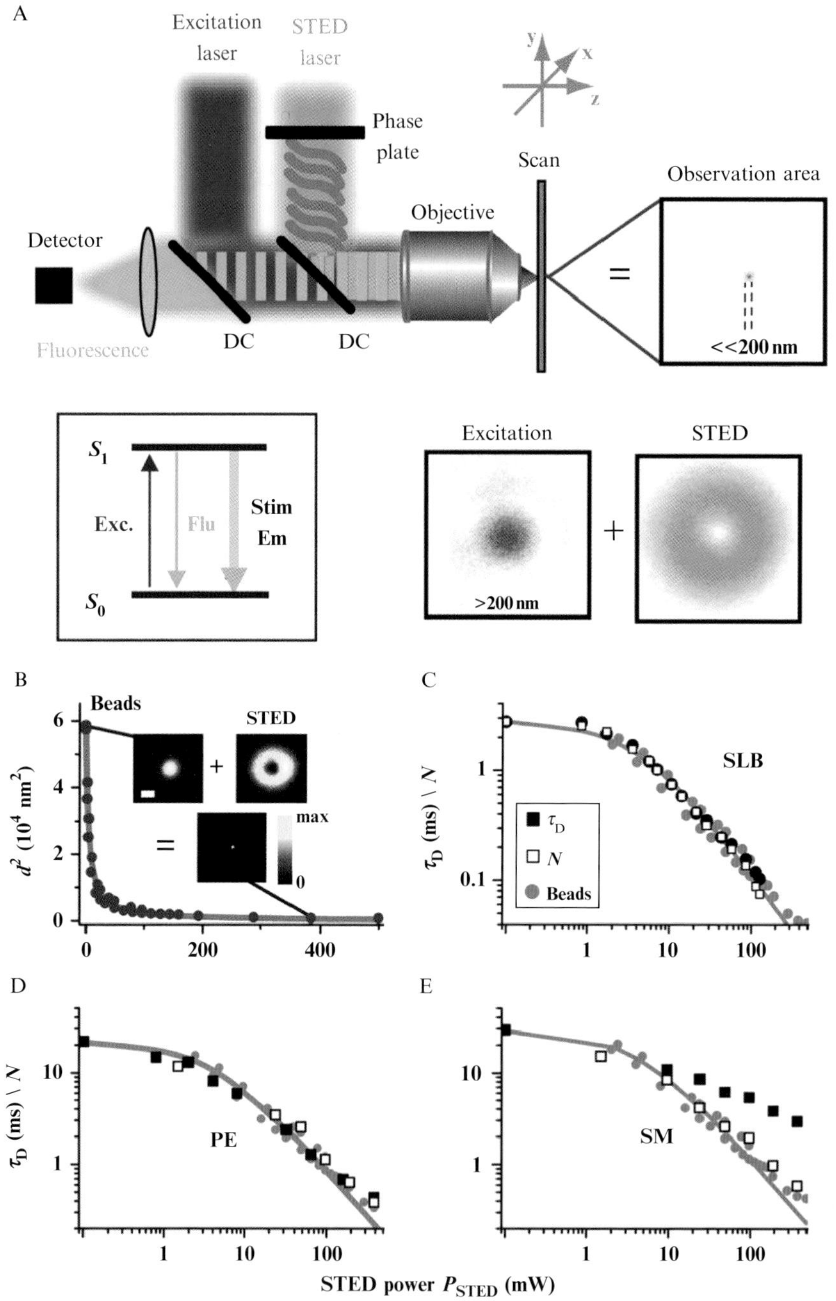


**Veronika Mueller *et al.*, Figure 1.3** Principle of STED nanoscopy and STED-FCS. (A) Besides the fluorescence excitation laser, a second STED laser is introduced into a conventional microscope (such as a confocal microscope), whose wavefront is (by introducing a phase plate) altered in such a way that, for example, a doughnut-like intensity distribution of the focused beam is realized. The overlay of the excitation and STED foci inhibits fluorescence emission everywhere but at the focal center leaving a subdiffraction sized

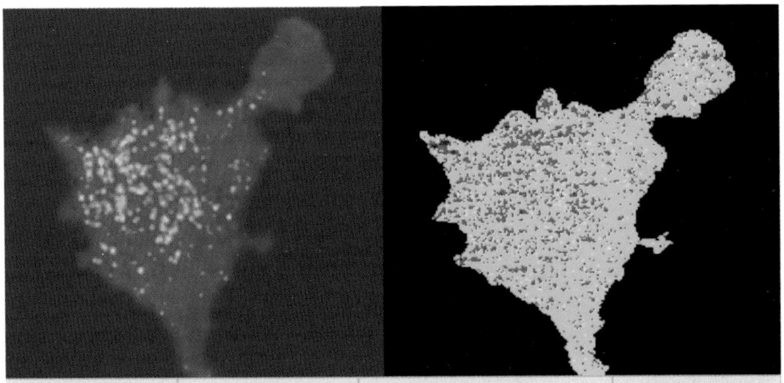

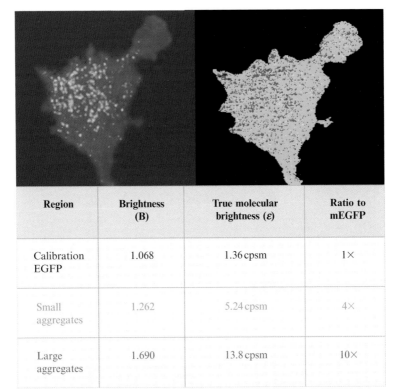

| Region | Brightness (B) | True molecular brightness ($\varepsilon$) | Ratio to mEGFP |
|---|---|---|---|
| Calibration EGFP | 1.068 | 1.36 cpsm | 1× |
| Small aggregates | 1.262 | 5.24 cpsm | 4× |
| Large aggregates | 1.690 | 13.8 cpsm | 10× |

**David M. Jameson *et al.*, Figure 3.4** Results of Ross et al. (2011) illustrating N&B analysis of TIRF data on dynamin2-EGFP oligomerization state in the plasma membrane of a mouse embryo fibroblast. Specifically, the analysis indicates that the majority of the dynamin2-EGFP is present as a tetramer. The image in the upper left corresponds to the TIRF intensity, while the image in the upper right corresponds to the N&B analysis, which shows pixels (green) with brightness levels four times those of the monomer standard.

---

area where emission is still allowed: the new observation area. Inhibition of fluorescence emission by the STED laser is realized by stimulated emission, where the fluorophore in its excited state $S_1$ is de-excited to its ground state $S_0$ more efficiently than the spontaneous fluorescence. (B) The overlay of the foci of the excitation (upper left inset) and of the STED light (upper right inset) create observation areas of subdiffraction size (lower inset); scale bar = 200 nm. The time-averaged STED power $P_{STED}$ tunes the observation area $d^2$. The diameter $d$ has been inferred from scanning fluorescent crimson beads (gray circles). Fitting of Eq. (1.1) (gray line) to the data results in $P_{SAT} = 4$ mW. (C–E) Relative decrease of the transit time $\tau_D$ (black squares) and the particle number $N$ (white squares) of the STED-FCS measurements on SLBs (C), Atto647N-labeled PE in living PtK2 cells (D), and Atto647N-labeled SM in living PtK2 cells (E) in comparison to the decrease of the observation areas $d^2$ of the bead measurements (gray circles). Most data can be described by the STED resolution scaling law (Eq. 1.1 with $P_{SAT} = 4$ mW, gray line), confirming the STED effect and in the case of $\tau_D$ indicating free Brownian diffusion. Deviation in the case of SM depicts strong anomalous subdiffusion.

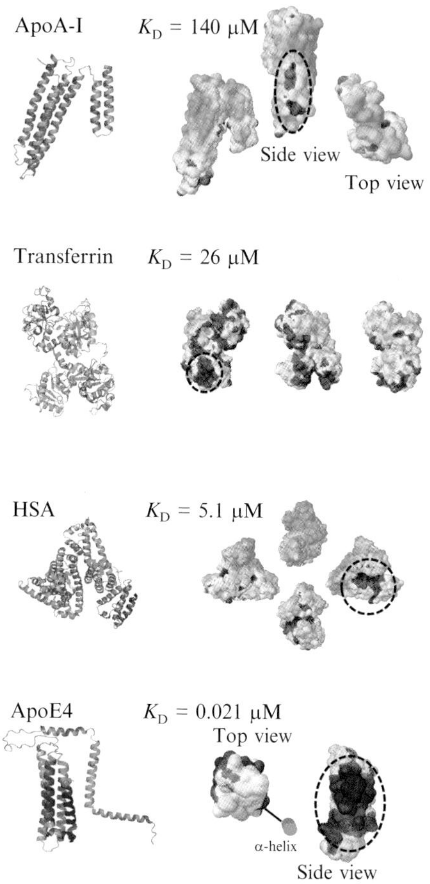


**G. Ulrich Nienhaus *et al.*, Figure 4.6** Electrostatic interactions govern protein adsorption. Left column: cartoon representations of apoA-I (protein data bank accession (pdb) code 2A01), apo transferrin (pdb code 2HAU), HSA (pdb code 1UOR), and apoE4 (pdb code 1GS9). For apoE4, only the structure of the 22-K domain (4-helix bundle) has been solved. Right column: space-filling models colored to indicate their surface electrostatics at pH 7.4 (blue/light gray: negative potential, red/dark gray: positive potential; range: −5 to +5 kT/e; calculated online at http://kryptonite.nbcr.net/pdb2pqr/ (Dolinsky, Nielsen, McCammon, & Baker, 2004)). The positively charged patches are marked by the dashed ellipses.

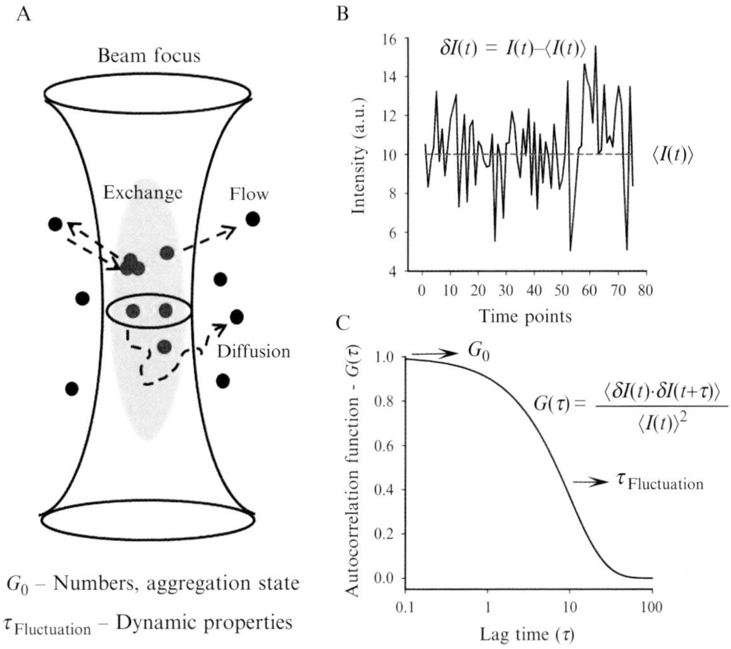


**Alexia I. Bachir *et al.*, Figure 6.1** Principles underlying fluorescence correlation spectroscopy (FCS). (A) Examples of dynamic processes (diffusion, flow, and exchange) that give rise to fluorescence intensity fluctuations in a focal volume. (B) Fluorescence intensity fluctuations over time. (C) Autocorrelation analysis of the intensity fluctuations quantifies the transport processes underlying the fluorescence fluctuations and the particle number densities and aggregation states.

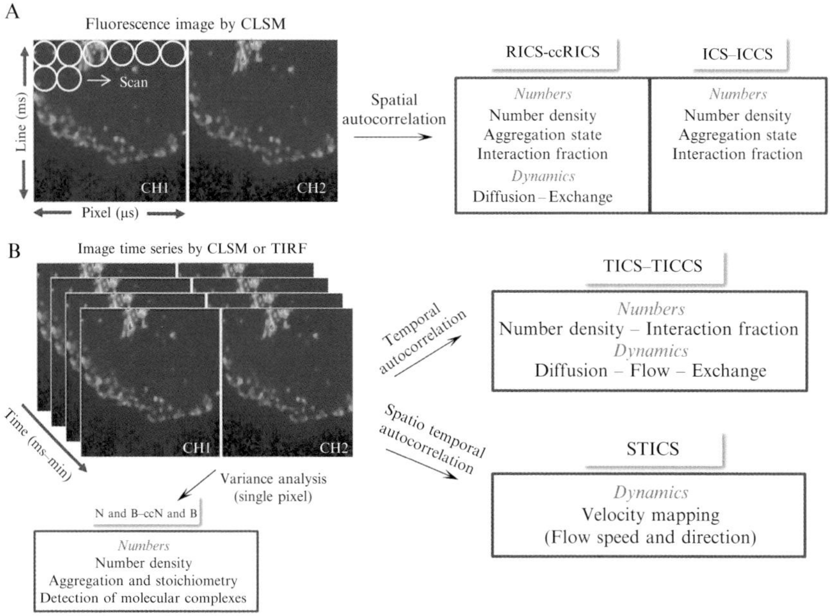


**Alexia I. Bachir *et al.*, Figure 6.2** Schematic illustration of the fluorescence fluctuation toolbox. (A) CLSM dual-color fluorescence images of a cell expressing two proteins, each tagged with a different color fluorescent probe visualized in channels 1 and 2 (CH1 and CH2). In a confocal setup, fluorescence excitation is achieved as a laser beam is scanned across a sample with microsecond pixel dwell time and millisecond line-scan time. Image correlation spectroscopy (ICS) and raster-image correlation spectroscopy (RICS) are implemented on CLSM images (or selected regions in an image) to measure molecular numbers and fast dynamics using spatial autocorrelation analysis. (B) Image time series acquired using either CLSM or TIRFM are used for temporal image correlation spectroscopy (TICS) and spatiotemporal image correlation analysis (STICS) analysis to study transport dynamics (diffusion, flow) and exchange kinetics. Number density information can also be recovered from TICS analysis. Single pixel variance analysis of fluorescence fluctuation in the time series provides high-resolution mapping of number densities and stoichiometry of molecular aggregates (N&B). All of these techniques can be applied to two fluorescence detection channels to detect and quantify molecular interactions in complexes.

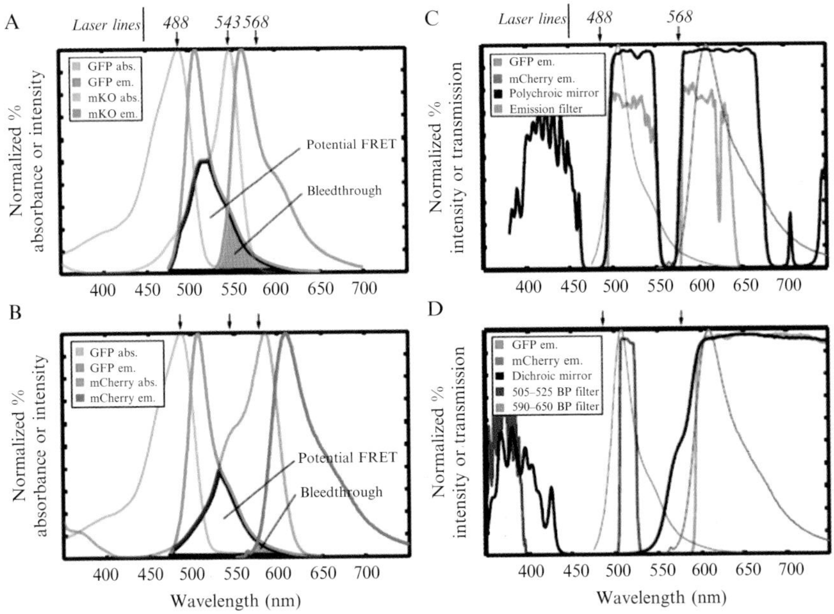


**Alexia I. Bachir *et al.*, Figure 6.3** (A) Absorbance and emission spectra of EGFP (green) and mKusabira-Orange (mKO; orange). Spectral overlaps potentially leading to FRET (black outlined area) or signal bleedthrough (gray area) are indicated. (B) Absorbance and emission spectra of EGFP (green) and mCherry (red) with spectral overlaps indicated as in (A). Note that the red-shifted mCherry spectra have less overlap with EGFP but also require a long-wavelength laser (568 nm) for optimal excitation. (C) Emission spectra of EGFP and mCherry, overlayed with the transmission spectra of the polychroic mirror (black) and emission filter (gray) used for TIRF microscopy. The two channels are further separated and projected onto different areas of a CCD chip using a Dual-View adapter (spectra of these mirrors not shown). (D) Emission spectra of EGFP and mCherry, overlaid with the transmission spectra of the dichroic mirror (black) and band-pass filters for the EGFP (dark gray) and mCherry (light gray) channels used in confocal microscopy. Notch filters at specific wavelengths remove reflected light from the lasers (spectra not shown).

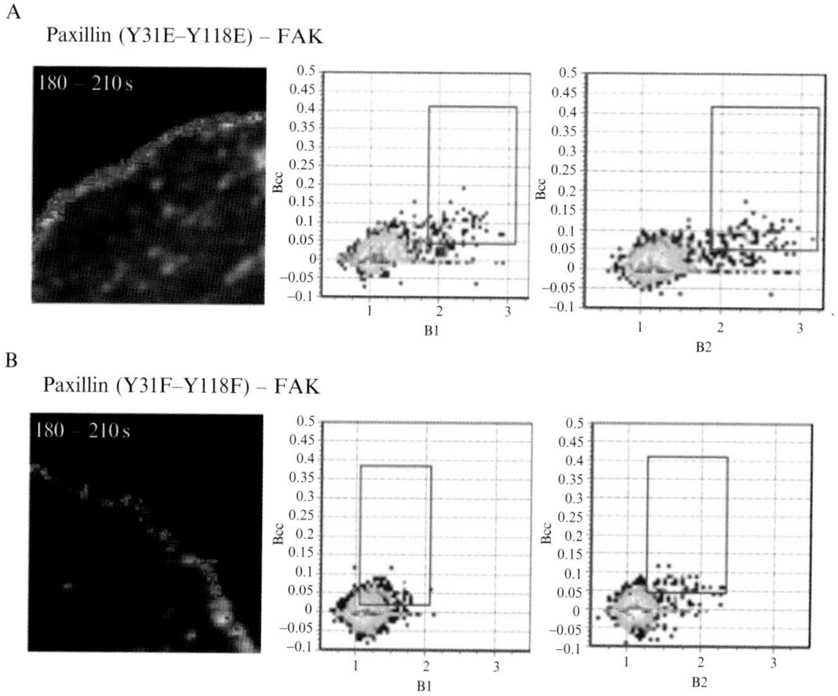


**Alexia I. Bachir *et al.*, Figure 6.6** Cross-number and -brightness (ccN&B) analysis of Paxillin–FAK interactions in nascent adhesions (Choi et al.). Highlighted pixels (red) indicate the presence of complexes and are selected for positive cross-brightness (Bcc) parameter values as indicated in the highlighted box in the histogram plots of pixels with a given Bcc value and corresponding B1 or B2 for each channel, respectively. For pixels that show complexes, the B1 and B2 values will reflect the size of stoichiometry of the individual species in the complex. More molecular complexes are detected for the phosphomimetic (Y31E, Y118E) versus the nonphosphorylatable (Y31F, Y118F) mutants of paxillin. TIRF image time series of CHO.K1 cells expressing paxillin–GFP and mCherry–FAK were acquired with 100 ms exposure time using stream acquisition for ~3.5 min. *Figure and caption reproduced from Choi et al. (2011).*

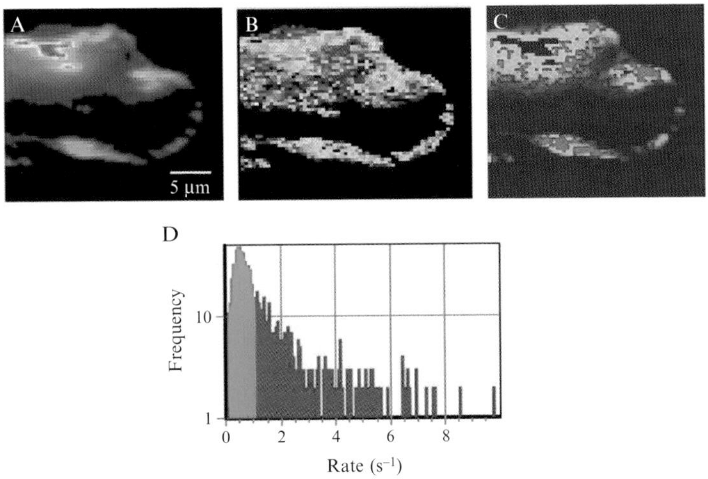


**Alexia I. Bachir *et al.*, Figure 6.7** Temporal pixel-autocorrelation analysis of CHO cells expression paxillin–EGFP. (A) Average fluorescence intensity of the image time series. Spatial mapping of the (B) amplitude ($G_0$) of the temporal autocorrelation function and (C) exchange rate constants ($s^{-1}$) obtained from the fits of the single pixel autocorrelation functions to an exponential decay function. Red and green pixels correspond to rates $>1$ and $<1\ s^{-1}$, respectively. (D) Histogram plot of the rate constants obtained from the single pixel temporal autocorrelation analysis. *Figure and caption reproduced from Digman, Brown, Horwitz, Mantulin, and Gratton (2008).*

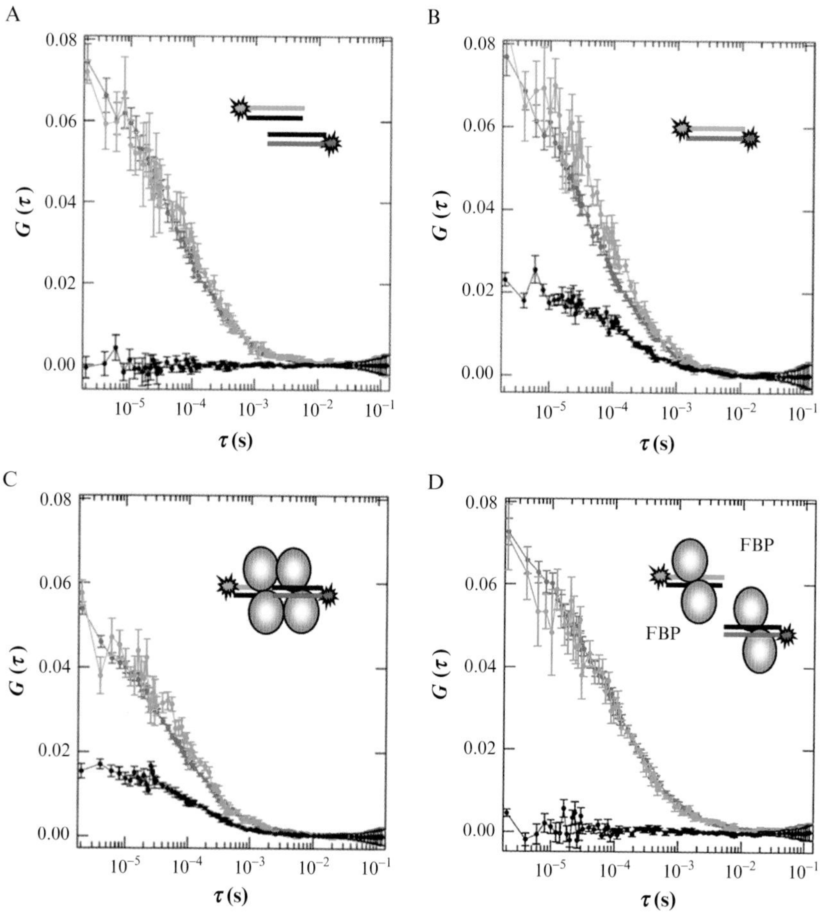


**Nathalie Declerck and Catherine A. Royer, Figure 7.5** Effect of FBP on CggR-operator interactions. Auto and cross-correlation profiles recovered from FCCS measurements with Atto-647N-labeled or fluorescein-labeled oligonucleotides corresponding to the CggR half-site operator in the presence/absence of the repressor and the inducer metabolite FBP. Schematics of the labeled DNAs and protein present in the sample chambers are shown. Red, green, and black curves correspond to the autocorrelation traces recorded in the red (675 nm) and green (525 nm) detection channels and the cross-correlation curve, respectively. (A) A mixture of the singly labeled dsDNA fragments showing the absence of cross-correlation signal. (B) A doubly labeled DNA hybrid serving as positive control of cross-correlation. The difference in the maximum amplitude of the auto and cross-correlation function (*G*(0)) denotes the partial labeling and hybridization of the two labeled DNA strands. (C) Cross-correlation upon addition of the repressor protein to the singly labeled DNA mixture of (A), demonstrating the CggR-mediated assembly of the DNA fragments. (D) Loss of cross-correlation signal upon addition of the inducer metabolite to the protein/DNA mixture of (C), demonstrating the FBP-induced disruption of the CggR/DNA ternary complex. Concentration was 60 n*M* for the labeled DNA fragments, 300 n*M* for CggR (monomer unit), and 0.5 m*M* for FBP. *Figure taken from Chaix et al. (2010).*

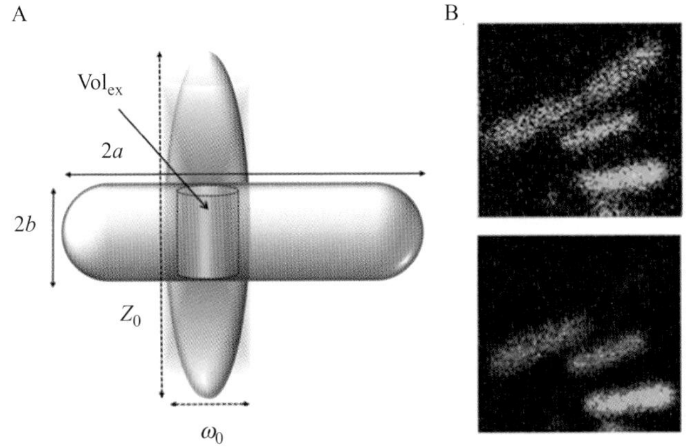


**Nathalie Declerck and Catherine A. Royer, Figure 7.7** Geometric constraints of FFM in bacteria. Size and geometry of the point spread function (PSF) relative to the bacterial cytoplasm. (A) A schematic showing the size and orientation of the infrared PSF relative to a 1-fL bacterial cell. The vol$_{ex}$ corresponds to the intersection between the two volumes, taking into account the quadratic dependence of excitation with laser power. (B) Fluorescent images of *B. subtilis* cells expressing *gfpmut3*, before (above) and after (below) a typical FCS experiment showing complete photobleaching of the targeted cell. *Figures adapted from Ferguson et al. (2011).*

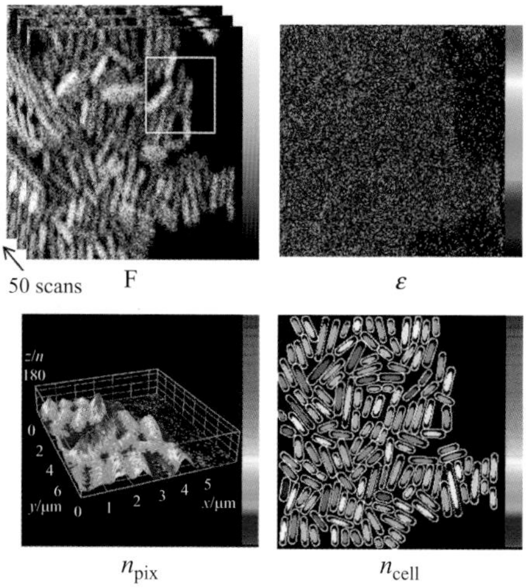


**Nathalie Declerck and Catherine A. Royer, Figure 7.8** Schematic of 2psN&B experiments in bacteria. A stack of 50 raster scans of agarose-immobilized live cells of *B. subtilis* expressing *gfpmut3* are recorded using infrared (930 nm) laser excitation and a dwell time of 50 μs at each pixel (faster than GFP diffusion); full scale of fluorescence intensity (*F*) is 10 photon counts/pixel/50 μs laser dwell time. The fluorescence fluctuations relative to the mean at each pixel are used to calculate the pixel-based maps of the true (shot noise corrected) molecular brightness ($\varepsilon$, full scale 1 photon/molecule/50 μs dwell time) and the number ($n_{pix}$) of the fluorescent particles detected in the two-photon excitation volume ($vol_{ex} = 0.07$ fL inside *B. subtilis*); a 3D surface plot of $n_{pix}$ is shown for the white-delineated area of the above intensity panel. Bottom right panel: cartoon representation of the individual cells auto-detected using PaTrack (Espenel et al., 2008) and showing the 50% central pixels used for averaging the particle number in each cell ($n_{cell}$); the full scale for the $n_{pix}$ and $n_{cell}$ maps is 180 molecules/$vol_{ex}$. *Figure taken from Ferguson et al. (2012).*

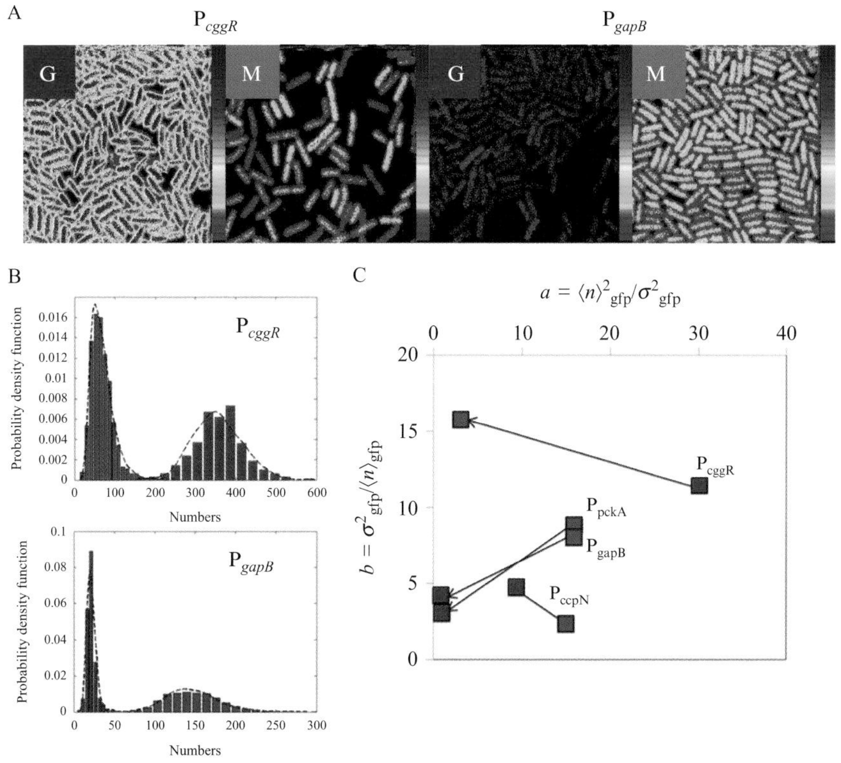


**Nathalie Declerck and Catherine A. Royer, Figure 7.10** Cell-by-cell quantification and noise in the activity of *B. subtilis* promoters implicated in the main switch between glycolysis and gluconeogenesis. (A) Pixel-based fluorescent particle number maps of *B. subtilis* cells expressing *gfpmut3* transcriptional fusion from $P_{ccgR}$ and $P_{gapB}$. Cells harvested from liquid cultures containing 0.5% glucose (G) or 0.5% malate (M) as the sole carbon source were immobilized on agarose pads and imaged by 2psN&B. The full scale is 360 molecules/$vol_{ex}$. (B) Cell-based particle number ($n_{cell}$) distributions for the indicated promoter–*gfpmut3* fusion strains grown on glucose (red - or grey) or malate (blue or black). Models of stochastic gene expression fitting the experimental data (dotted lines) were generated for the repressible promoters based on available knowledge about their control mechanism. (C) Noise patterns obtained by plotting two parameters of stochastic gene expression, computed from the first two moments of the protein number distribution: the apparent frequency of protein production burst per cell cycle ($a = \langle n \rangle^2/\sigma_n^2$) and the average number of protein molecules produced per burst, related to the Fano factor ($b = \sigma_n^2/\langle n \rangle$). Distinct changes in noise patterns were observed upon a switch in carbon source, reflecting distinct underlying molecular mechanisms of repression (see main text). *Figure adapted from Ferguson et al. (2012).*

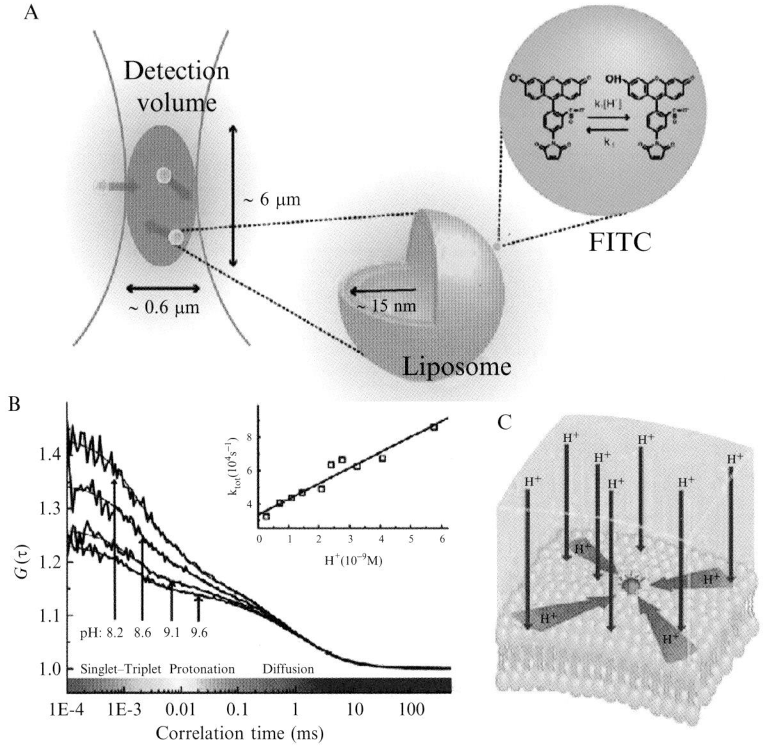


**Jerker Widengren, Figure 8.3** (A) Principal design of FCS experiments to study proton exchange kinetics at biological membranes. Liposomes were labeled with one pH-sensitive fluorophore undergoing fluorescence fluctuations due to protonation/deprotonation. (B) Set of FCS curves recorded from the vesicles at different pH. The FCS curves reflect singlet–triplet transitions in the μs time range, protonation kinetics in the 10–100 μs time range and translational diffusion in the ms time range. Inset: measured protonation relaxation rates versus proton concentration (C) Principle of the proton-collecting antenna effect for a proton acceptor in a biological membrane. *Modified from Brändén et al. (2006). Copyright (2006) National Academy of Sciences, USA.*

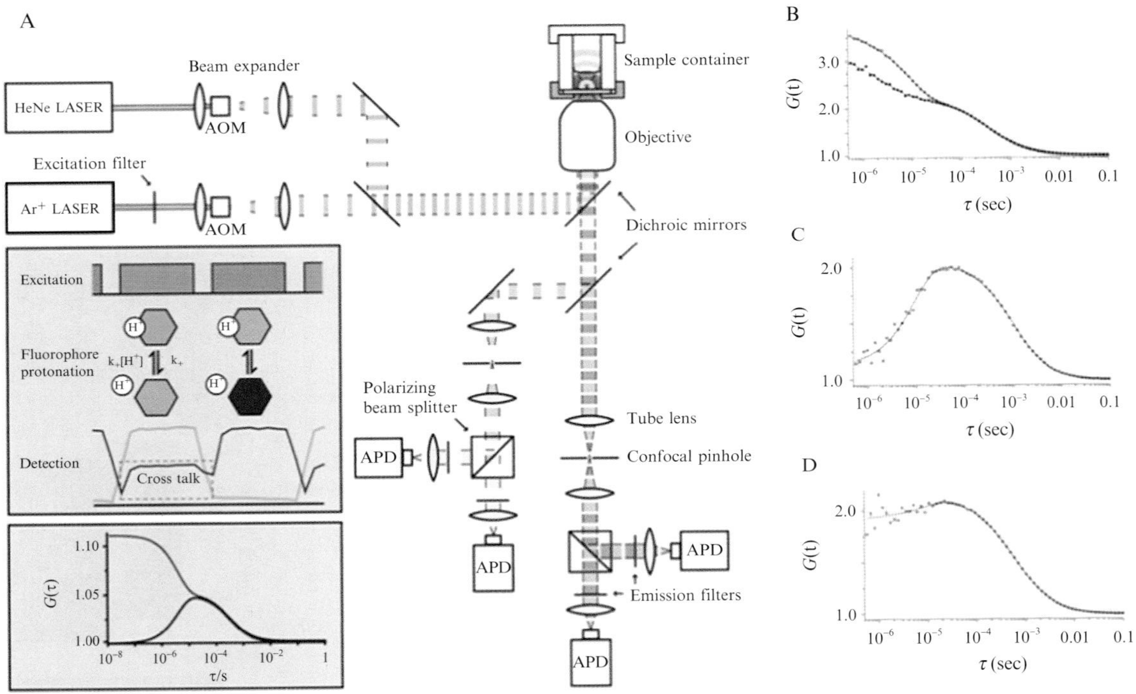


**Jerker Widengren, Figure 8.4** (A) Setup used for FCCS with alternating excitation. Two lasers, emitting at 488 and 594 nm, respectively, are modulated by acousto-optical modulators and serve as excitation sources. The setup has two separate detection pathways. The upper gray box shows how the blue laser preferably excites the protonated species of the fluorophore which is mainly detected in the green detection channels and the yellow laser almost exclusively excites the deprotonated species which is detected in the red detection channels. The lower gray box shows examples of how the switching between the protonated and deprotonated states contributes to the correlation in the case of autocorrelation (upper red curve) and gives rise to an anticorrelation when cross-correlating the signals form the two spectral detection ranges (lower black curve). (B–D) Measurements on the fluorophore NK 138 in 500 μ*M* HEPES buffer at a pH of 7.7. (B) Autocorrelation curves from measurements with CW excitation and detection of either the deprotonated (upper curve, 594 nm excitation) or the protonated species (lower curve, 488 nm excitation). (C) Cross-correlation of the signals from the two spectral ranges in a measurement with alternating excitation. Gating was used to remove cross talk. (D) Cross-correlation without gating. The solid lines in (C) and (D) represents model fits, using Eq. (8.3), with $R(\tau) = 1 - H\exp(-k_p\tau)$, where $H$ is a constant between 0 and 1. $H$ decreases with increasing cross talk. *Figure is modified from Person et al. (2009), reproduced by permission of the PCCP Owner societies.*

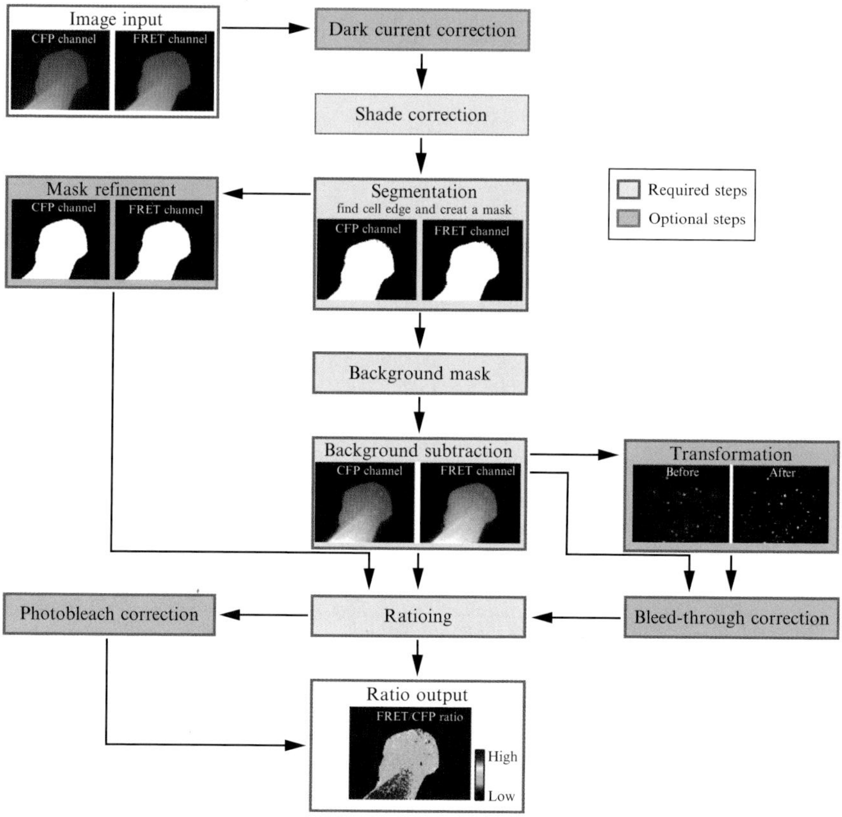


**Marco Vilela *et al.*, Figure 9.2** Image corrections and processing required for FRET-based biosensor readouts of signaling activities. The end product of the workflow is a ratio image that indicates the spatial biosensor activity at each frame of the movie.

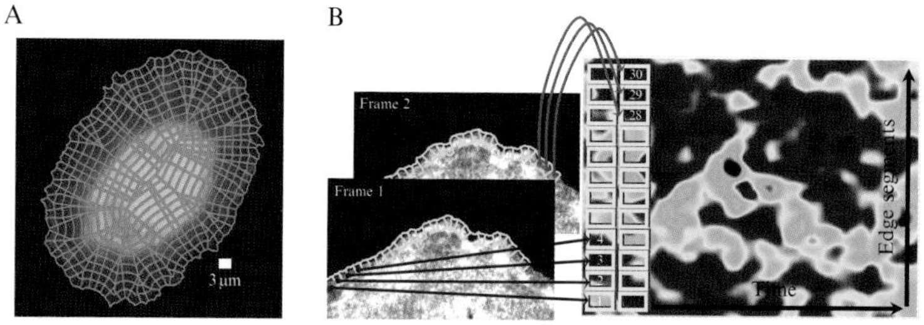


**Marco Vilela *et al.*, Figure 9.3** Windowing process. (A) Segmentation of a cell into sampling windows. (B) Sampling of the fluorescence signal and construction of the spatiotemporal activity map. *Figure is reproduced, with permission, from references Lim et al. (2010) and Welch, Elliott, Danuser, and Hahn (2011).*